Methods in Molecular Biology

Series Editor
John M. Walker
School of Life and Medical Sciences
University of Hertfordshire
Hatfield, Hertfordshire, AL10 9AB, UK

For further volumes:
http://www.springer.com/series/7651

Selenoproteins

Methods and Protocols

Edited by

Laurent Chavatte

Centre International de Recherche en Infectiologie, CIRI, Lyon, France
INSERM U1111, Lyon, France
CNRS/ENS/UCBL1 UMR5308, Lyon, France
Laboratoire de Chimie Analytique Bio-Inorganique et Environnement, LCABIE
CNRS/Univ Pau & Pays Adour, Institut des Sciences
Analytiques et de Physico-Chimie Pour l'Environnement
et Les Matériaux, UMR5254, Pau, France

Humana Press

Editor
Laurent Chavatte
Centre International de Recherche en Infectiologie, CIRI
Lyon, France

INSERM U1111
Lyon, France

CNRS/ENS/UCBL1 UMR5308
Lyon, France

Laboratoire de Chimie Analytique Bio-Inorganique et Environnement, LCABIE
CNRS/Univ Pau & Pays Adour, Institut des Sciences Analytiques et de Physico-Chimie Pour l'Environnement et Les Matériaux, UMR5254
Pau, France

ISSN 1064-3745 ISSN 1940-6029 (electronic)
Methods in Molecular Biology
ISBN 978-1-4939-7257-9 ISBN 978-1-4939-7258-6 (eBook)
DOI 10.1007/978-1-4939-7258-6

Library of Congress Control Number: 2017950666

Cover image: Front cover image is credited to Dr. Jérôme Frayret

Printed on acid-free paper

This Humana Press imprint is published by Springer Nature
The registered company is Springer Science+Business Media LLC
The registered company address is: 233 Spring Street, New York, NY 10013, U.S.A.

Preface

Selenium (Se) was discovered in 1817 by John Jacob Berzelius, a Swedish chemist who named it after the goddess of the moon *Selene*, since this element had many similarities with Tellurium ("tellus" meaning earth in Latin), found at the same period. For more than hundred years, selenium was viewed as a toxic element. It is only rather recently (in the 1950s) that selenium has been recognized as an essential nutrient for mammals. Selenium is implicated in many facets of human health and diseases, including cancer prevention, cardiovascular function, immunity, and brain function [1–3]. Selenium is incorporated in a small but vital group of proteins, the selenoproteins, in the form of a rare amino acid, the selenocysteine (Sec). Often located in the catalytic site of enzymes, selenocysteine is a key component of oxido-reduction reactions. Twenty-five selenoprotein genes have been discovered so far in human [4, 5], and a unified nomenclature has been recently proposed [6] (*see* Table 1). Selenoproteins are mostly involved in redox homeostasis and signaling, antioxidant defense, and selenoprotein biosynthesis, although about one third of the selenoproteome remains without precise function (for review *see* Ref. 7). Remarkably, in the 1990s selenium has amended the genetic code deciphered in the 1960s since selenocysteine is encoded by a UGA codon in selenoprotein mRNAs, which is otherwise read as a stop codon in other cellular mRNAs. In this regard, cells have evolved a unique and complex mechanism for UGA recoding as selenocysteine that is regulated at many levels, giving rise to a prioritized synthesis of selenoproteins. This noncanonical translational pathway for selenoprotein biosynthesis relies on two pivotal RNA molecules, namely the selenocysteine-$tRNA^{[Ser]Sec}$ (Fig. 1) and the SElenoCysteine Insertion Sequence (SECIS) within selenoprotein mRNA (Fig. 2), and their protein interacting partners (as reviewed in [2, 3, 8–12]). Due to the low levels of selenium in the body and complex biosynthesis, research on selenoproteins has been particularly challenging. Therefore, selenoproteins offer a fascinating playground in a broad range of fields in biology, including bioinformatics, genetics, RNA structure, translational control, biophysics, enzymology, animal models, and human health and diseases. This book aims at providing an update on state-of-the-art methodologies to study these various aspects of selenoprotein biology.

Table 1
List of the human selenoprotein genes identified to date, using the unified nomenclature

Gene name	Protein name	Synonyms
DIO1	Iodothyronine deiodinase 1	D1
DIO2	Iodothyronine deiodinase 2	D2
DIO3	Iodothyronine deiodinase 3	D3
GPX1	Glutathione peroxidase 1	Glutathione peroxidase 1
GPX2	Glutathione peroxidase 2	Glutathione peroxidase 2
GPX3	Glutathione peroxidase 3	Glutathione peroxidase 3
GPX4	Glutathione peroxidase 4	Glutathione peroxidase 4
GPX6	Glutathione peroxidase 6	Glutathione peroxidase 6
MSRB1	Methionine sulfoxide reductase B1	SelR, SelX, SEPX1
SELENOF	Selenoprotein F	Selenoprotein 15, Sel15, Sep15
SELENOH	Selenoprotein H	SelH
SELENOI	Selenoprotein I	SelI
SELENOK	Selenoprotein K	SelK
SELENOM	Selenoprotein M	SelM
SELENON	Selenoprotein N	SelN
SELENOO	Selenoprotein O	SelO
SELENOP	Selenoprotein N	SelP, SEPP1, SEPP, SeP
SELENOS	Selenoprotein S	SelS, SEPS1, VIMP
SELENOT	Selenoprotein T	SelT
SELENOV	Selenoprotein V	SelV
SELENOW	Selenoprotein W	SelW
SEPHS2	Selenophosphate 2	Selenide, water dikinase 2, SPS2
TXNRD1	Thioredoxin reductase 1	TR1, TRXR1
TXNRD2	Thioredoxin reductase 3	TR3, TRXR2, mitochondrial thioredoxin reductase
TXNRD3	Thioredoxin reductase 3	TR2, TRXR3, TGR

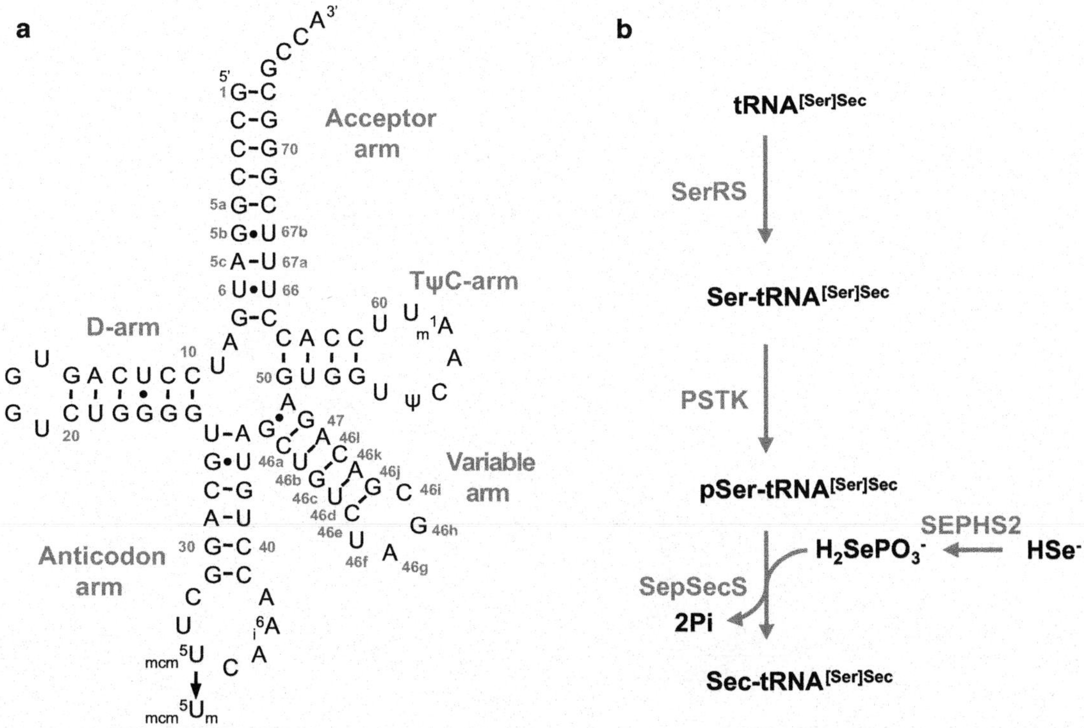

Fig. 1 Structure and aminoacylation of Sec-tRNA$^{[Ser]Sec}$. (**a**) Cloverleaf structure of the two isoforms of the tRNA$^{[Ser]Sec}$ differing from modification at position U34 between $_{mcm}{}^{5}U$ and $_{mcm}{}^{5}U_{m}$. (**b**) The aminoacylation of tRNA$^{[Ser]Sec}$ necessitates four enzymes instead of one for other tRNAs, including seryl-tRNA synthetase (SerRS), the phosphoseryl-tRNA kinase (PSKT), Sec synthase (SepSecS, also referred to as SLA/LP), and selenophosphate 2 synthetase (SEPHS2), which generates selenophosphate ($H_2SePO_3^-$) from selenide (HSe^-)

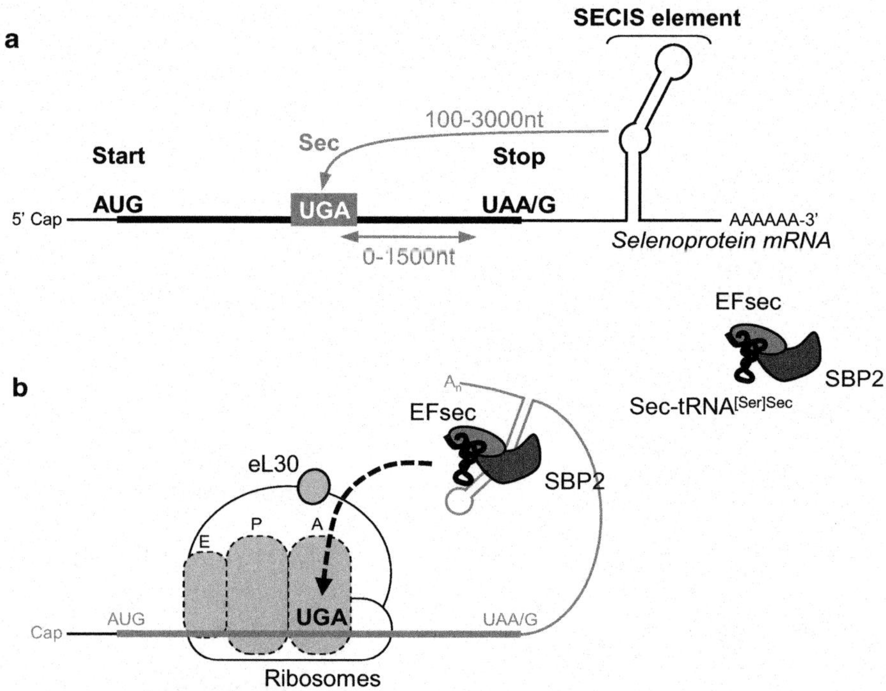

Fig. 2 Mechanism for selenoprotein biosynthesis in eukaryotes. (**a**) Schematic representation of a selenoprotein mRNA with its different *cis*-acting features: the SECIS element located in the 3′UTR and the UGA-selenocysteine codon present in frame in the Open Reading Frame (ORF). (**b**) Representation of the *trans*-acting factors needed for the efficient recoding of UGA as selenocysteine. Essential factors include the ribosomes, the Sec-tRNA$^{[Ser]Sec}$, a dedicated elongation factor EFsec, and a SECIS binding protein SBP2 (or SECIBP2). Accessory and/or regulatory proteins have also been described (for review *see* Refs. 8–12)

Pau, France *Laurent Chavatte*

References

1. Rayman MP (2012) Selenium and human health. Lancet 379(9822):1256–1268. doi:10.1016/S0140-6736(11)61452-9. S0140–6736 (11)61452–9 [pii]
2. Papp LV, Lu J, Holmgren A, Khanna KK (2007) From selenium to selenoproteins: synthesis, identity, and their role in human health. Antioxid Redox Signal 9(7):775–806
3. Hatfield DL, Gladyshev VN (2002) How selenium has altered our understanding of the genetic code. Mol Cell Biol 22(11):3565–3576
4. Lobanov AV, Hatfield DL, Gladyshev VN (2009) Eukaryotic selenoproteins and selenoproteomes. Biochim Biophys Acta 1790(11):1424–1428. doi:10.1016/j.bbagen.2009.05.014. S0304–4165(09)00148–2 [pii]
5. Kryukov GV, Castellano S, Novoselov SV, Lobanov AV, Zehtab O, Guigo R, Gladyshev VN (2003) Characterization of mammalian selenoproteomes. Science 300(5624):1439–1443
6. Gladyshev VN, Arner ES, Berry MJ, Brigelius-Flohe R, Bruford EA, Burk RF, Carlson BA, Castellano S, Chavatte L, Conrad M, Copeland PR, Diamond AM, Driscoll DM, Ferreiro A, Flohe L, Green FR, Guigo R, Handy DE, Hatfield DL, Hesketh J, Hoffmann PR, Holmgren A, Hondal RJ, Howard MT, Huang K, Kim H-Y, Kim IY, Kohrle J, Krol A, Kryukov GV, Lee BJ, Lee BC, Lei XG, Liu Q, Lescure A, Lobanov AV, Loscalzo J, Maiorino M, Mariotti M, Sandeep Prabhu K, Rayman MP, Rozovsky S, Salinas G, Schmidt EE, Schomburg L, Schweizer U, Simonovic M, Sunde RA, Tsuji PA, Tweedie S, Ursini F,

Whanger PD, Zhang Y (2016) Selenoprotein Gene Nomenclature. The Journal of biological chemistry 291(46):24,036–24,040

7. Labunskyy VM, Hatfield DL, Gladyshev VN (2014) Selenoproteins: molecular pathways and physiological roles. Physiol Rev. 94(3):739–777. doi:10.1152/physrev.00039.2013. 94/3/739 [pii]
8. Bulteau AL, Chavatte L (2015) Update on Selenoprotein Biosynthesis. Antioxid Redox Signal. doi:10.1089/ars.2015.6391
9. Donovan J, Copeland PR (2010) Threading the needle: getting selenocysteine into proteins. Antioxid Redox Signal 12(7):881–892. doi:10.1089/ars.2009.2878
10. Allmang C, Wurth L, Krol A (2009) The selenium to selenoprotein pathway in eukaryotes: more molecular partners than anticipated. Biochim Biophys Acta 1790(11):1415–1423. doi:10.1016/j.bbagen.2009.03.003. S0304-4165(09)00056-7 [pii]
11. Squires JE, Berry MJ (2008) Eukaryotic selenoprotein synthesis: mechanistic insight incorporating new factors and new functions for old factors. IUBMB Life 60(4):232–235
12. Driscoll DM, Copeland PR (2003) Mechanism and regulation of selenoprotein synthesis. Annu Rev. Nutr 23:17–40

Contents

Contributors

CHRISTINE ALLMANG • *Architecture et Réactivité de l'ARN, Centre National de la Recherche Scientifique, Institut de Biologie Moléculaire et Cellulaire, Université de Strasbourg, Strasbourg, France*

ELIAS S.J. ARNÉR • *Division of Biochemistry, Department of Medical Biochemistry and Biophysics, Karolinska Institutet, Stockholm, Sweden*

LISA BAIRD • *Department of Human Genetics, University of Utah, Salt Lake City, UT, USA*

J. SABINE BECKER • *Zentralinstitut für Engineering, Elektronik und Analytik, Analytik (ZEA-3), Forschungszentrum Jülich, Jülich, Germany*

KATARZYNA BIERLA • *IPREM-LCABIE, Hélioparc Pau-Pyrénées, Pau, France*

BELÉN CALLEJÓN-LEBLIC • *Department of Chemistry, Faculty of Chemistry, University of Huelva, Huelva, Spain; Campus of International Excellence A3, University of Huelva, Huelva, Spain; Health and Environment Research Center (CYSMA), University of Huelva, Huelva, Spain*

BRADLEY A. CARLSON • *Molecular Biology of Selenium Section, Mouse Cancer Genetics Program, National Cancer Institute, National Institutes of Health, Bethesda, MD, USA*

KRISHNA CHATTERJEE • *Wellcome Trust-MRC Institute of Metabolic Science, University of Cambridge, Cambridge, UK*

LAURENT CHAVATTE • *Centre International de Recherche en Infectiologie, CIRI, Lyon, France; INSERM U1111, Lyon, France; CNRS/ENS/UCBL1 UMR5308, Lyon, France; Laboratoire de Chimie Analytique Bio-Inorganique et Environnement, LCABIE, CNRS/Univ Pau & Pays Adour, Institut des Sciences Analytiques et de Physico-Chimie Pour l'Environnement et Les Matériaux, UMR5254, Pau, France*

QINGQING CHEN • *Department of Chemistry and Biochemistry, University of Delaware, Newark, DE, USA*

QING CHENG • *Division of Biochemistry, Department of Medical Biochemistry and Biophysics, Karolinska Institutet, Stockholm, Sweden*

ERIC M. COCKMAN • *Department of Cellular and Molecular Medicine, Lerner Research Institute, Cleveland Clinic, Cleveland, OH, USA; Department of Molecular Medicine, Cleveland Clinic Lerner College of Medicine of Case Western Reserve University, Cleveland, OH, USA*

PAUL R. COPELAND • *Department of Biochemistry and Molecular Biology, Rutgers-Robert Wood Johnson Medical School, Piscataway, NJ, USA*

ELISA CASTAÑEDA SANTA CRUZ • *Laboratory of Bioanalytics and Integrated Omics (LaBIOmics), Department of Analytical Chemistry, Institute of Chemistry, University of Campinas (UNICAMP), Campinas, SP, Brazil*

BRIAN K. DALLEY • *Huntsman Cancer Institute, University of Utah, Salt Lake City, UT, USA*

DONNA M. DRISCOLL • *Department of Cellular and Molecular Medicine, Lerner Research Institute, Cleveland Clinic, Cleveland, OH, USA; Department of Molecular Medicine, Cleveland Clinic Lerner College of Medicine of Case Western Reserve University, Cleveland, OH, USA*

JEAN DUMONT • *UT2A, Hélioparc Pau-Pyrénées, Pau, France*

GILBERT ERIANI • *Architecture et Réactivité de l'ARN, Centre National de la Recherche Scientifique, Institut de Biologie Moléculaire et Cellulaire, Université de Strasbourg, Strasbourg, France*
JOSÉ LUIS GÓMEZ-ARIZA • *Department of Chemistry, Faculty of Chemistry, University of Huelva, Huelva, Spain; Campus of International Excellence A3, University of Huelva, Huelva, Spain; Health and Environment Research Center (CYSMA), University of Huelva, Huelva, Spain*
TAMARA GARCÍA-BARRERA • *Department of Chemistry, Faculty of Chemistry, University of Huelva, Huelva, Spain; Campus of International Excellence A3, University of Huelva, Huelva, Spain; Health and Environment Research Center (CYSMA), University of Huelva, Huelva, Spain*
VADIM N. GLADYSHEV • *Division of Genetics, Department of Medicine, Brigham and Women's Hospital, Harvard Medical School, Boston, MA, USA*
ANNE-SOPHIE GRIBLING-BURRER • *Architecture et Réactivité de l'ARN, Centre National de la Recherche Scientifique, Institut de Biologie Moléculaire et Cellulaire, Université de Strasbourg, Strasbourg, France; Faculty of Medicine, ZBMZ, Institute of Biochemistry and Molecular Biology, University of Freiburg, Freiburg, Germany*
RODERIC GUIGÓ • *Centre for Genomic Regulation (CRG), The Barcelona Institute for Science and Technology, Barcelona, Spain; Universitat Pompeu Fabra (UPF), Barcelona, Spain; Institut Hospital del Mar d'Investigacions Mediques (IMIM), Barcelona, Spain*
DOLPH L. HATFIELD • *Molecular Biology of Selenium Section, Mouse Cancer Genetics Program, National Cancer Institute, National Institutes of Health, Bethesda, MD, USA*
MICHAEL T. HOWARD • *Department of Human Genetics, University of Utah, Salt Lake City, UT, USA*
LIANG JIANG • *College of Life Sciences and Oceanography, Shenzhen University, Shenzhen, People's Republic of China*
GALINA KARPOVA • *Institute of Chemical Biology and Fundamental Medicine, Siberian Branch of the Russian Academy of Sciences, Novosibirsk, Russia; Department of Molecular Biology, Novosibirsk State University, Novosibirsk, Russia*
OLGA KOSSINOVA • *Institute of Chemical Biology and Fundamental Medicine, Siberian Branch of the Russian Academy of Sciences, Novosibirsk, Russia*
ALAIN KROL • *Architecture et Réactivité de l'ARN, UPR9002, Université de Strasbourg, CNRS, Strasbourg, France*
BYEONG JAE LEE • *School of Biological Sciences, Seoul National University, Seoul, South Korea*
QIONG LIU • *College of Life Sciences and Oceanography, Shenzhen University, Shenzhen, People's Republic of China*
JUN LIU • *Department of Chemistry and Biochemistry, University of Delaware, Newark, DE, USA*
RYSZARD LOBINSKI • *IPREM-LCABIE, Hélioparc Pau-Pyrénées, Pau, France*
CATHERINE MÉPLAN • *School of Biomedical Sciences, Faculty of Medical Sciences, Newcastle University, Newcastle Upon Tyne, UK*
ALEXEY MALYGIN • *Institute of Chemical Biology and Fundamental Medicine, Siberian Branch of the Russian Academy of Sciences, Novosibirsk, Russia; Department of Molecular Biology, Novosibirsk State University, Novosibirsk, Russia*
MARCO MARIOTTI • *Brigham and Women's Hospital, Harvard Medical School, Boston, MA, USA; Centre for Genomic Regulation (CRG), The Barcelona Institute for Science and Technology, Barcelona, Spain; Universitat Pompeu Fabra (UPF), Barcelona, Spain; Institut Hospital del Mar d'Investigacions Mediques (IMIM), Barcelona, Spain*

SANDRA MOUNICOU • *Laboratoire de Chimie Analytique Bio-Inorganique et Environnement, LCABIE, CNRS/ Univ Pau & Pays Adour, Institut des Sciences Analytiques et de Physico-Chimie Pour l'Environnement et Les Matériaux, UMR5254, Pau, France*
EMMANUEL OHEIX • *Centrale Marseille, CNRS, iSm2 UMR 7313, Aix-Marseille Université, Marseille, France*
ZALÁN PÉTERFI • *Division of Genetics, Department of Medicine, Brigham and Women's Hospital, Harvard Medical School, Boston, MA, USA*
MARK H. PINKERTON • *Department of Biochemistry and Molecular Biology, Rutgers-Robert Wood Johnson Medical School, Piscataway, NJ, USA*
GEMA RODRÍGUEZ-MORO • *Department of Chemistry, Faculty of Chemistry, University of Huelva, Huelva, Spain; Campus of International Excellence A3, University of Huelva, Huelva, Spain; Health and Environment Research Center (CYSMA), University of Huelva, Huelva, Spain*
SHARON ROZOVSKY • *Department of Chemistry and Biochemistry, University of Delaware, Newark, DE, USA*
DIDAC SANTESMASSES • *Centre for Genomic Regulation (CRG), The Barcelona Institute for Science and Technology, Barcelona, Spain; Universitat Pompeu Fabra (UPF), Barcelona, Spain; Institut Hospital del Mar d'Investigacions Mediques (IMIM), Barcelona, Spain*
ERIK SCHOENMAKERS • *Wellcome Trust-MRC Institute of Metabolic Science, University of Cambridge, Cambridge, UK*
ULRICH SCHWEIZER • *Institut für Biochemie und Molekularbiologie, Rheinische Friedrich-Wilhelms-Universität Bonn, Bonn, Germany*
JORDAN SONET • *Laboratoire de Chimie Analytique Bio-Inorganique et Environnement, LCABIE, CNRS/ Univ Pau & Pays Adour, Institut des Sciences Analytiques et de Physico-Chimie Pour l'Environnement et Les Matériaux, UMR5254, Pau, France*
J. SUSANNE BECKER • *Konstanz, Germany*
ALESSANDRA SUSSULINI • *Laboratory of Bioanalytics and Integrated Omics (LaBIOmics), Department of Analytical Chemistry, Institute of Chemistry, University of Campinas (UNICAMP), Campinas, SP, Brazil*
JOANNA SZPUNAR • *IPREM-LCABIE, Hélioparc Pau-Pyrénées, Pau, France*
LIONEL TARRAGO • *Laboratoire de Bioénergétique Cellulaire, Institut de Biosciences et Biotechnologies Aix-Marseille (BIAM), Commissariat à l'Energie Atomique et aux Energies Alternatives (CEA), Saint-Paul-lès-Durance, France; UMR 7265, Centre National de Recherche Scientifique, Saint-Paul-lès-Durance, France; Aix Marseille Université, Marseille, France*
RYUTA TOBE • *College of Life Sciences, Ritsumeikan University, Kusatsu, Japan*
PETRA A. TSUJI • *Department of Biological Sciences, Towson University, Towson, MD, USA*
ANTON A. TURANOV • *Division of Genetics, Department of Medicine, Brigham and Women's Hospital, Harvard Medical School, Boston, MA, USA*
VÉRONIQUE VACCHINA • *UT2A, Hélioparc Pau-Pyrénées, Pau, France*
SUN HEE YIM • *Division of Genetics, Department of Medicine, Brigham and Women's Hospital, Harvard Medical School, Boston, MA, USA*
ZHENGQI ZHANG • *Department of Chemistry and Biochemistry, University of Delaware, Newark, DE, USA*

Part I

Bioinformatic Tools for Selenoprotein Identification and Evolution

Chapter 1

SECISearch3 and Seblastian: In-Silico Tools to Predict SECIS Elements and Selenoproteins

Marco Mariotti

Abstract

The computational identification of selenoprotein genes is complicated by the dual meaning of the UGA codon as stop and selenocysteine. SECIS elements are RNA structures essential for selenocysteine incorporation, which have been used as markers for selenoprotein genes in many bioinformatics studies. The most widely used tool for eukaryotic SECIS finding has been recently improved to its third generation, SECISearch3. This program is also a component of Seblastian, a pipeline for the identification of selenoprotein genes that employs SECIS finding as the first step. This chapter constitutes a practical guide to use SECISearch3 and Seblastian, which can be run via webservers at http://seblastian.crg.eu/ or http://gladyshevlab.org/SelenoproteinPredictionServer/.

Key words SECIS, Selenoprotein, Selenocysteine, Selenium, Gene finding, Bioinformatics, Genomics

1 Introduction

In selenoprotein synthesis, a recoding event occurs during protein translation so that a specific UGA codon supports insertion of selenocysteine (Sec), instead of causing termination [1]. The SECIS (SElenoCysteine Insertion Sequence) is a key element present in every selenoprotein transcript acting as required and sufficient signal for Sec recoding [2–4]. SECISes are RNA structures that show fundamental differences in bacteria, archaea, and eukaryotes [5]. Their structure bears no apparent similarity across these lineages, and their location is also different: in eukaryotes and archaea it resides in the 3′UTR, while in bacteria it is found within the coding sequence, next to the Sec UGA. The eukaryotic SECIS structure, already present in the quasi-eukaryotic archaeal lineage of *Lokiarchaeota*, appears to have evolved from an archaeal precursor, the SECIS of selenoprotein VhuD [6]. Bacterial SECISes are very diverse structures, and may be phylogenetically related to eukaryotic Sec insertion "helper" structures known as SRE

Laurent Chavatte (ed.), *Selenoproteins: Methods and Protocols*, Methods in Molecular Biology, vol. 1661,
DOI 10.1007/978-1-4939-7258-6_1, © Springer Science+Business Media LLC 2018

(Selenocysteine Recoding Elements) [7]. SRE are present in some, but not all, eukaryotic selenoprotein genes [8, 9].

The computational identification of selenoproteins is complicated by the dual meaning of the UGA codon (stop or Sec), coupled with their low frequency in proteomes [10]. For this reason, selenoprotein gene finding is typically carried out with computational tools developed on purpose. Most of the strategies utilized throughout the years took advantage of SECIS identification as evidence to support selenoprotein gene presence nearby (although exceptions exist: see [11] and Chapter 2 in this book). For this reason, the computational identification of SECIS elements is fundamental to selenoprotein research in the genomics era. Among the early strategies formulated [12–14], SECISearch emerged as the most widely used tool for predicting SECIS elements in genomes (e.g., [15–29]). The original SECISearch algorithm [12] was based on sequence patterns as descriptors of eukaryotic SECIS structure, and included the evaluation of the thermodynamic stability of candidates [30]. Similar pattern-based tools were developed for SECIS prediction in bacteria and archaea (bSECISearch [31], aSECISearch [32]). SECISearch was later enhanced by incorporating a procedure to assign a covariation-based score to SECIS candidates [33].

Lately, we further improved SECISearch by exploiting the increased availability of sequences and including state-of-the-art techniques for homology-based RNA motif finding. **SECISearch3** [34] is based on the Infernal suite (INFERence of RNA ALignment) [35], and has at its core a manually curated alignment of more than a thousand eukaryotic SECIS elements. SECISearch3 is part of a program called **Seblastian**, a pipeline to predict selenoprotein genes in nucleotide sequences which employs the identification of SECIS elements as a first step [34]. SECISearch3 and Seblastian are designed for the prediction of ***eukaryotic*** SECIS elements and selenoprotein genes. Analogous pipelines dedicated to bacteria and archaea are currently under development. This chapter is designed as a guide for understanding and running SECISearch3 and Seblastian. These programs can be executed online through webservers at http://seblastian.crg.eu/ or http://gladyshevlab.org/SelenoproteinPredictionServer/. The webpage let the user choose whether to run Seblastian (default), or SECISearch3 alone.

2 Input and Output Specifications

2.1 Input

Both SECISearch3 and Seblastian accept nucleotide sequences as input. Since the typical eukaryotic SECIS is ~65–75 nucleotide long, this is the minimal input length for SECISearch3. Seblastian requires more extended regions, comprising the SECIS elements as well as their upstream selenoprotein coding sequence. Input sequences can be either uploaded as a single file, or pasted in the dedicated input form in the main page of the webserver. The webserver accepts without distinction DNA or RNA characters ("U" symbols are converted to "T"), uppercase or lowercase. The preferred input format is as FASTA, optionally with multiple sequences per submission (i.e., multiple FASTA headers, each starting with a ">" symbol). Plain sequences are also accepted; in that case the default header "your_sequence" is assigned. Formatting characters, such as tab and line separators, are tolerated and ignored. By default, the search is performed not only on the input sequence, but also on its reverse complement. The user can deactivate this behavior by unchecking the relevant option in the input form (described later). SECISearch3 and Seblastian are designed to work on genomic sequences. Transcript data is also accepted and it is actually preferred, since it limits the search space and obviates the issue of intron prediction.

To avoid server overload, there is a limit for total sequence size when submitting Seblastian jobs. In practice, this prohibits to run whole-genome searches for selenoprotein genes on the webserver. If the limit is too restrictive for your intended use of Seblastian, do not hesitate in contacting us explaining your purpose. We may either offer to run the analysis for you, suggest alternative ways to predict selenoproteins, or provide the scripts necessary to run it yourself.

2.2 Output

Once the input data is accepted and computation is over, the webserver shows a page with the results of the analysis requested. At the top of the page, links are provided to download the output files in batches. Sequences and structures of all SECIS candidates are provided as FASTA files. If Seblastian was run, the nucleotide and protein sequences of the predicted selenoprotein genes can also be downloaded as FASTA. Their coordinates relative to the input sequences are available in a GFF format. Below the top section for output download, the results page contains a summary for each prediction whose composition is described later in this chapter, separately for SECISearch3 and Seblastian. The page of results for any given run is generally maintained for a week or more on the webserver.

3 SECISearch3: Identification of Eukaryotic SECIS Elements

3.1 SECISearch3 Overview

The purpose of SECISearch3 [34] is the identification of eukaryotic SECIS elements in nucleotide sequences. Its workflow is depicted in Fig. 1. SECISearch3 combines up to three distinct methods for SECIS detection. By default, the webserver employs only the best

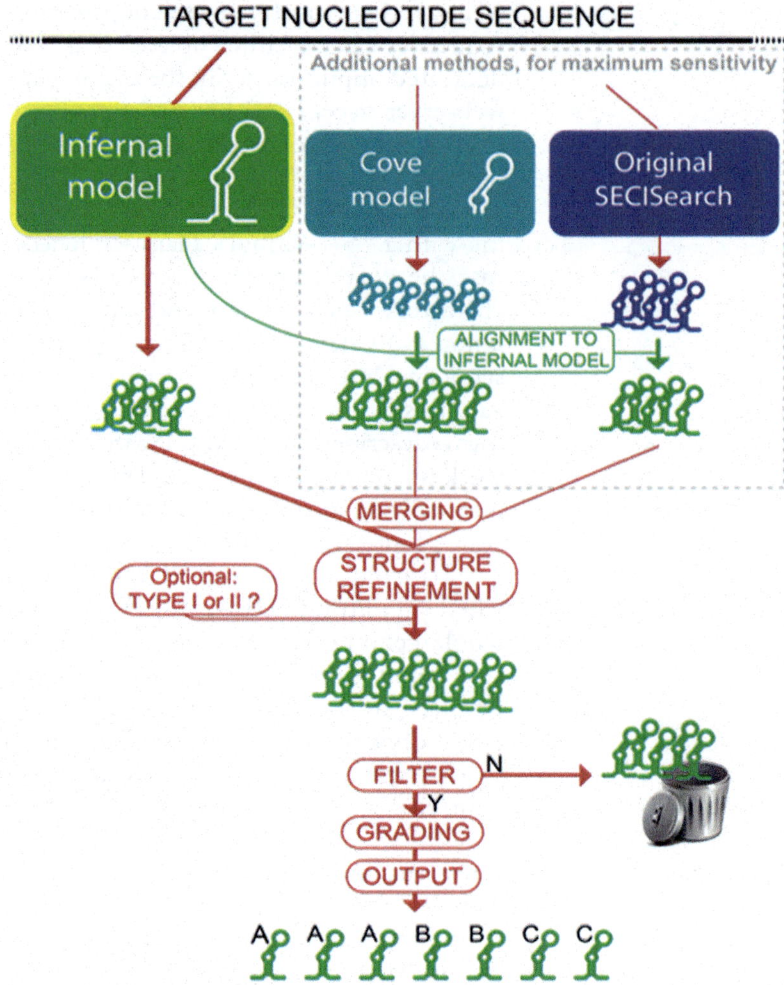

Fig. 1 Workflow of SECISearch3. Reproduced, with permission, from [34]

performing method, based on the Infernal suite [35]. The second available method is based on a program considered a predecessor of Infernal, named cove [36], and employs an earlier covariance model. This method allows us to achieve higher sensitivity than Infernal, but its speed is not suitable for large database searches, and its specificity is lower. The last available method is the "original" SECISearch, in its first pattern-based formulation presented in [12]. If multiple search methods are selected at once, their predictions are merged in a nonredundant set. SECIS candidates are then put through a procedure of structural refinement, in which base pairings in the stems may be further extended or shortened.

Next, each candidate is classified as either a type I or type II SECIS, with type II being characterized by an extra stem at the apex [37]. A specific routine must be run for this purpose, since the Infernal model underlying SECISearch3 is generic for all SECISes

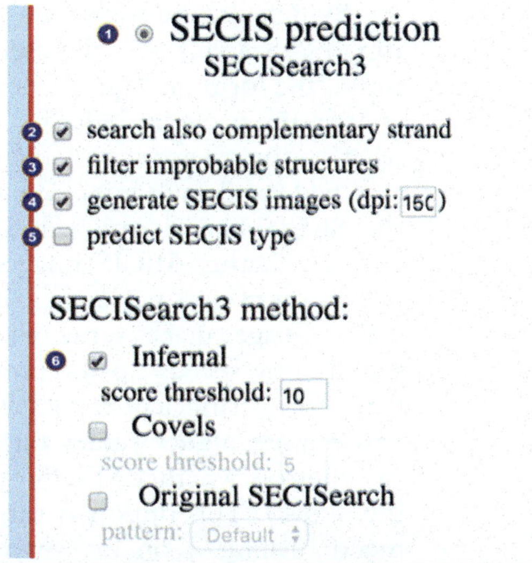

Fig. 2 Snapshot of the webserver's input form with the options of SECISearch3. The items indexed 1–6 are referred in the text below

and does not include the extra stem. The program attempts to fold the apical loop with RNAfold [38]. If successful, the extra stem is included in the structural prediction and the candidate is classified as type II. Otherwise, the SECIS is classified as type I. The **SECIS type classification** functionality was introduced after the method was published in 2013, and it **is not run by default** in the webserver (Fig. 2). When it is not active, unfortunately, all SECISes appear as type I, even when this is not the case [39, 40].

Next, the set of SECIS candidates is processed with a filter designed to optimize specificity. Optionally, this can be turned off (Fig. 2). Lastly, SECISearch3 executes a procedure to categorize candidates with a "grade," which can be A, B, or C. This labeling procedure was implemented through the manual inspection of hundreds of candidates, designing heuristic criteria to capture the degree of confidence in the SECIS prediction. The grade takes into account the covariance score, the presence of insertions or nonstandard pairs in stems, the nucleotide identity conserved positions in the apical loop, and other features (see [34]). The grade can be used as a practical measure of how well a candidate fits with the expectations for eukaryotic SECIS. Since these are mostly based on model animal organisms, it is not surprising to observe lowly graded SECISes in protozoans.

3.2 Practical Guide of SECISearch3 Webserver

The left portion of the webserver input page is dedicated to SECISearch3, and is shown in Fig. 2. In the figure, indices from 1 to 6 mark the position of the available options, described hereafter.

By default, the webserver is configured to run Seblastian. To run SECISearch3 only, the user must click the round button at the top of the form (1, Fig. 2). Below, various options are available to control the program behavior. The first checkbox (2) determines whether the reverse complement of the input sequence is also to be searched (by default it is). The second option (3) controls whether the SECIS filter is utilized. In the next line (4), the user can choose whether SECIS images are produced, and what is their resolution in dots per inches (dpi). Turning off the generation of SECIS images makes sense when scanning large databases, since it significantly speeds up the process. The default image resolution (150 dpi) is sufficient for visualization on screen, but many journals require higher values for publication. The fourth checkbox (5) controls whether SECIS candidates are classified by type (see the above). This option is turned off by default, but we recommend turning it on in every case in which the SECIS type is relevant to the analysis. The last section in the form (6) allows choosing which search method is employed, and specifying their parameters.

In practical terms, the default search mode is generally sufficient to detect SECIS elements in any eukaryotic species, keeping the number of false positives reasonably low [34]. Alternatively, the user may want to **maximize detection power**, trading off sensitivity for specificity. This is obtained by activating all methods for SECIS search (6), and by deactivating the filter (3). We recommend using this mode only if there is strong external evidence for a SECIS being present in the input sequence (i.e., a selenoprotein coding sequence was detected nearby), or if additional filtering criteria are applied afterward.

3.3 Guide to SECISearch3 Output

The output page of SECISearch3 contains one section for each SECIS candidate. Essential information about the prediction is provided: its sequence, secondary structure, position on the input sequence, grade, covariance-based score, and free energy predicted by RNAeval [38]. The sequence may be copied in the online tool SECISaln [41] to compare the candidate with a large collection of SECISes (*see* **Note 1**). If requested (option 5 in Fig. 2), the type of SECIS is also indicated here. On the right, a graphical representation of the SECIS structure is displayed (unless option 4 in Fig. 2 was deactivated). Examples of SECIS images produced by SECISearch3 are depicted in Fig. 3. Colors are used to highlight the various components: conserved SECIS features (kink-turn core and unpaired nucleotides in loop 2 [5, 41]) are shown in green; unpaired nucleotides within stems (insertions and incompatible pairs) are displayed in red; standard stem pairs are shown in cyan.

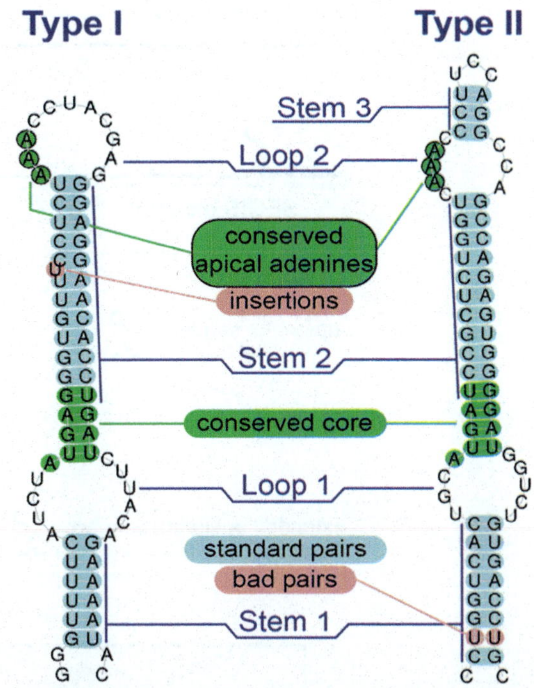

Fig. 3 Two examples of SECIS images by SECISearch3, annotated here to illustrate the meaning of colors. The type I SECIS displayed is that of human GPx1, while the type II SECIS belongs to human MsrB. Note that prediction of stem3 is attempted only if re-quested (option 5 in Fig. 2 activated)

4 Seblastian: Identification of Selenoprotein Genes

4.1 Seblastian Overview

Seblastian [34] is a program for the identification of eukaryotic selenoprotein genes in nucleotide sequences. Its workflow is depicted in Fig. 4. Seblastian runs SECISearch3 as the first step to predict SECIS elements. Then, it scans upstream of each potential SECIS for selenoprotein coding sequences. This is performed by homology, searching for similar proteins annotated in a given database. Blastx [42] is used to translate the input sequences to amino acids in the three possible frames, and find matches with significant scores in the protein database. The resulting hits are then analyzed, searching for UGA codons in the input aligned to selenocysteine (Sec or U) or cysteine (Cys or C) residues in any annotated protein. These constitute candidate Sec-UGA sites in putative selenoprotein genes. To improve the gene structure predictions, the program exonerate [43] is used next. Finally, a strict filter is applied to every candidate, requiring the proper positioning of the SECIS in respect to the coding sequence, and the presence of signatures of conservation at both the sides of the Sec-UGA between the selenoprotein candidate and the annotated homolog (see [34]).

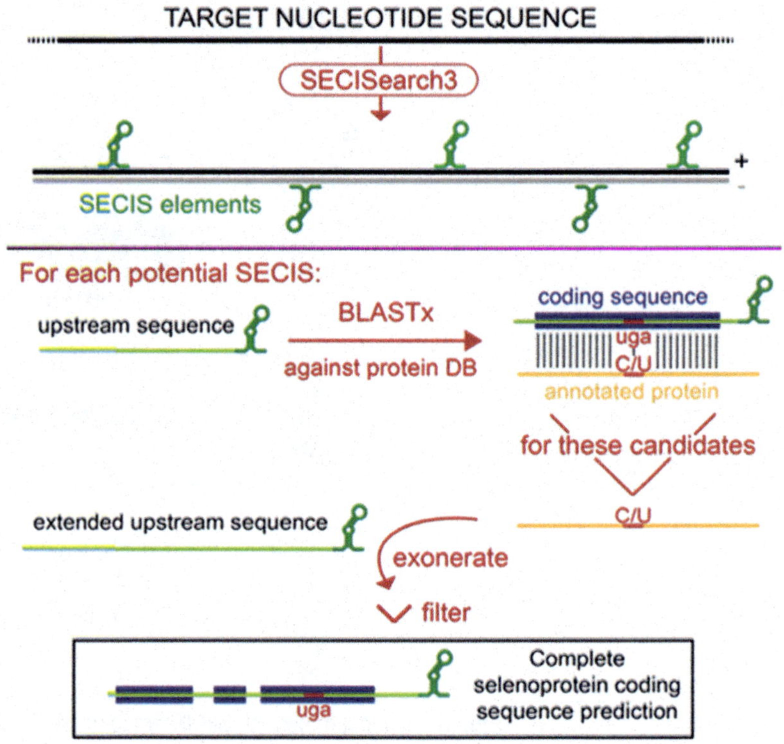

Fig. 4 Workflow of the Seblastian program. Reproduced, with permission, from [34]

Seblastian can be used for two distinct purposes:

1. Prediction of selenoproteins which have Sec-containing homologs in other organisms ("known selenoproteins").
2. Prediction of selenoproteins which have only Cys-homologs annotated in protein databases ("new selenoproteins").

In the first case, the blast matches accepted are those with a UGA aligned with a Sec residue in one annotated selenoprotein. In the second case, the accepted blast alignments are those with a UGA aligned to a Cys residue in any annotated protein. For the first purpose, only the selenoproteins present in the protein database are effectively considered, while for the second purpose a comprehensive database with all known proteins must be used. Since selenoproteins constitute a tiny fraction of all known proteins, the second purpose is considerably more difficult, and requires considerably longer computation time.

4.2 Practical Guide of Seblastian Webserver

The right portion of the webserver input page is dedicated to Seblastian, and is shown in Fig. 5. In the figure, indices from 1 to 6 mark the position of the available options, described

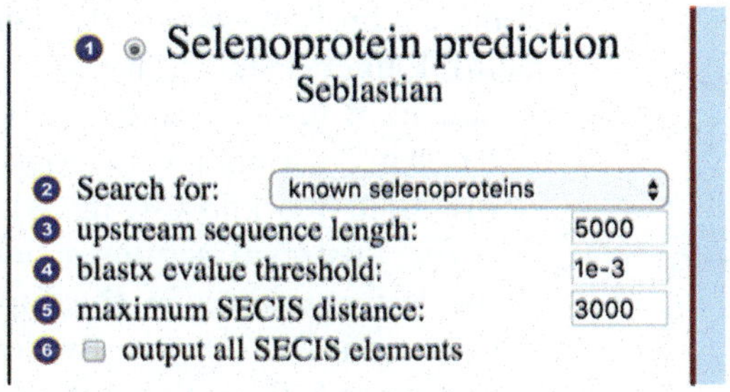

Fig. 5 Snapshot of the input form with the options of Seblastian. The items indexed 1–6 are referred in the text below

hereafter. Note that, since SECISearch3 is employed as the first step, its options (Fig. 2) must also be considered when running Seblastian.

At the top of the input form (1, Fig. 5), a round button controls whether Seblastian must be run. This button is mutually exclusive with its counterpart for SECISearch3 (*see* Fig. 2). Just below, a drop down menu (2) determines if Seblastian searches for "known selenoproteins," or for "known and new selenoproteins" (see the explanation above). This choice greatly affects the total time required for computation. When searching for "known selenoproteins," the protein database used by Seblastian consists of the collection of all known selenoproteins. Instead, the protein database used when searching for "known and new selenoproteins" is the full NCBI NR, which is more than 25,000 times larger in size (*see* **Note 2**). We recommend running searches uniquely for "known selenoproteins" (default option), unless justified by scientific criteria (e.g., for a newly sequenced lineage in which novel selenoproteins are expected). The confidence in the prediction of "new selenoproteins" is much lower than for "known selenoproteins" [34], and this should be taken into account in the interpretation of results (*see* **Note 3**). Next in the input form, the user can specify the length of sequence upstream of each SECIS considered when running blastx (3). For Seblastian to work properly, this value must be larger than the distance between the Sec-UGA and the SECIS in the input sequence. Excessively high values of this parameter increase computational time, and may produce confusing results if the sequence window becomes large enough to include the Sec-UGA of another selenoprotein gene, upstream of the actual gene to which the SECIS belongs. In practice, the default value works well to identify most vertebrate selenoprotein genes [34], and it can be further reduced to speed up the analysis in more compact genomes. Next, the user can decide the e-value threshold to filter blast matches (4). Although the default is rather permissive, it could be further lowered down for the detection of more remote homologs (trading off sensitivity for lower specificity and longer total runtime). After that, the user can input the maximum distance allowed between coding sequence and SECIS (5). This parameter is considered by the filter at the end of Seblastian workflow (Fig. 4), and should be adjusted according to the upstream sequence length previously chosen (option 3). The last checkbox in the input form (6) controls which SECIS predictions are output. By default, the Seblastian output includes only those SECIS elements for which a selenoprotein coding gene was detected upstream. Check this box to have instead all SECIS candidates in output.

4.3 Guide to Output

At the end of computation, Seblastian displays a page with the results of the analysis. Its content partially depends on the input options (Figs. 2 and 5). A prototypical example is shown in Fig. 6.

Each selenoprotein gene prediction is separated into two sections, one for the coding sequence and the other for its SECIS. The header of the coding sequence section (1, Fig. 6) contains basic information about the identity of the query and target sequences (*see* **Note 4**), and the coordinates of the gene prediction. The query is the annotated protein automatically selected from the protein database, and used for the prediction on the target, which is the input sequence. The query identifier is generally informative to recognize the gene family of the candidate. In the example in Fig. 6, the query is selenoprotein M, recently renamed SELENOM [44]. Below the header, the gene prediction alignment is depicted in green font (2) in the p2g (protein-to-genome) format of the program Selenoprofiles v2 [11]. The query protein is shown on the top. The target is shown below, with both the nucleotide sequence and its protein translation. When present, gaps are shown as "-" characters. Intron predictions are explicitly indicated if present. In the alignment, Sec positions are marked in bold font and highlighted with a star ("*" symbol) underneath (2). The "U" character in the alignment indicates a Sec residue. The corresponding position in the query protein may either carry a "U" or a "C" residue. The title in red font at the top reflects these two possibilities ("known selenoprotein" or "new selenoprotein," respectively). The SECIS section (3) is analogous to that displayed when SECISearch3 alone is run, enriched with the distance of the SECIS from the coding sequence and from the Sec-UGA site. An image representing the SECIS structure (*see* Fig. 3) is also normally included here (but missing from Fig. 6 for space constraints).

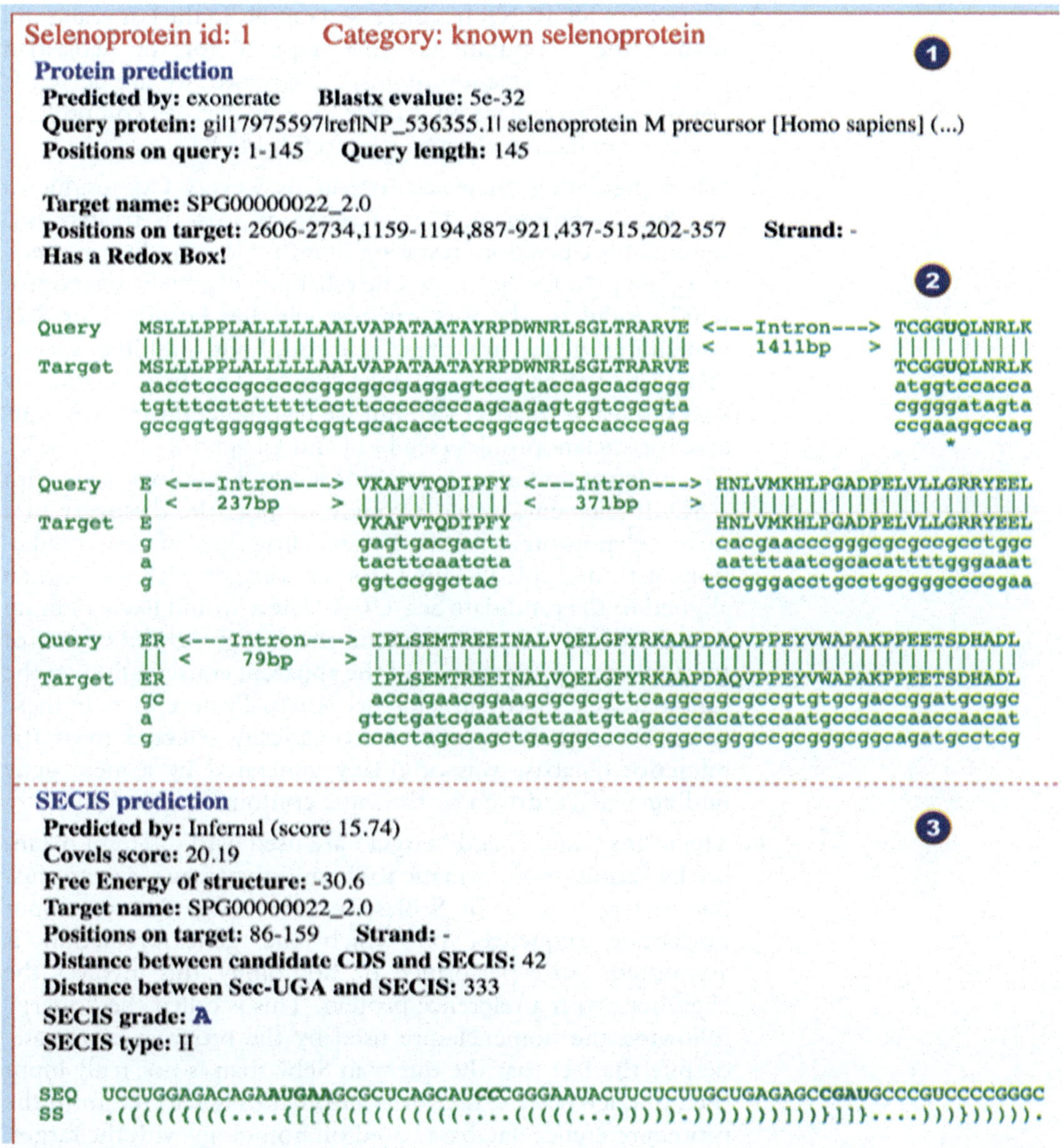

Fig. 6 Example of a selenoprotein gene prediction displayed in the output page of Seblastian. Items indexed 1–3 are referred in the text below

5 Notes

1. SECISaln [41] can be found at http://genome.crg.es/software/secisaln/. Note that this tool is based on the original version of SECISearch [12]. If a candidate SECIS cannot be predicted by this method, it cannot currently be visualized with SECISaln.

2. The size of NCBI NR has increased rapidly in the last years, and with it the computational time required for the Seblastian search for "new selenoproteins." Therefore, we are currently considering to replace NCBI NR for a smaller, yet comprehensive protein database, such as Uniref50 [45].

3. When predicting "new selenoproteins," every Cys residue of any known protein in the reference database is treated as a potential Sec position, resulting in rather low prediction specificity [34]. To further assess the reliability of a "new selenoprotein" candidate, the user can test whether homologous Sec-containing genes are present throughout multiple species related to the original sequence source. This may be roughly checked using tblastn [42], but we recommend the more suitable tool Selenoprofiles (see [11] and Chapter 2 in this book). The independent identification of several homologs, each with a SECIS element, would strongly support the discovery of a novel selenoprotein family. Beware, however, of gene predictions with multiple in-frame stop codons, or with tryptophan aligned to the candidate Sec-UGA. These would likely indicate that these loci contain a gene on the ***opposite strand*** of that of the Seblastian prediction, and the apparent conservation on the candidate selenoprotein sequence is actually deceptive. In these cases, the "query" protein automatically selected from the reference database was originally generated by a mere gene finding artifact, driven by the same confounding factor.

4. The terms "query" and "target" are used with different meaning by various bioinformatic tools, and they may result confusing to some users. In Seblastian, the "target" is the input nucleotide sequence, on which the gene prediction is attempted. Since performed by homology, this involves the alignment with a reference protein. This is called the "query" following the nomenclature used by the program exonerate, despite the fact that the query in Seblastian is not truly input (i.e., queried) by the user: it is automatically selected from the protein reference database based on homology with the target.

References

1. Labunskyy VM, Hatfield DL, Gladyshev VN (2014) Selenoproteins: molecular pathways and physiological roles. Physiol Rev 94:739–777. doi:10.1152/physrev.00039.2013
2. Berry MJ, Banu L, Chen Y et al (1991) Recognition of UGA as a selenocysteine codon in type I deiodinase requires sequences in the 3′ untranslated region. Nature 353:273–276
3. Berry MJ, Banu L, Harney JW, Larsen PR (1993) Functional characterization of the eukaryotic SECIS elements which direct selenocysteine insertion at UGA codons. EMBO J 12:3315–3322
4. Hubert N, Walczak R, Sturchler C et al (1996) RNAs mediating cotranslational insertion of selenocysteine in eukaryotic selenoproteins. Biochimie 78:590–596
5. Krol A (2002) Evolutionarily different RNA motifs and RNA-protein complexes to achieve selenoprotein synthesis. Biochimie 84:765–774

6. Mariotti M, Lobanov AV, Manta B et al (2016) Lokiarchaeota marks the transition between the Archaeal and eukaryotic Selenocysteine encoding systems. Mol Biol Evol 33:2441–2453. doi:10.1093/molbev/msw122
7. Mariotti M, Santesmasses D, Capella-Gutierrez S et al (2015) Evolution of selenophosphate synthetases: emergence and relocation of function through independent duplications and recurrent subfunctionalization. Genome Res 25:1256–1267. doi:10.1101/gr.190538.115
8. Howard MT, Moyle MW, Aggarwal G et al (2007) A recoding element that stimulates decoding of UGA codons by Sec tRNA [Ser] Sec. RNA 13:912
9. Howard MT, Aggarwal G, Anderson CB et al (2005) Recoding elements located adjacent to a subset of eukaryal selenocysteine-specifying UGA codons. EMBO J 24:1596–1607. doi:10.1038/sj.emboj.7600642
10. Driscoll DM, Chavatte L (2004) Finding needles in a haystack. In silico identification of eukaryotic selenoprotein genes. EMBO Rep 5:140–141. doi:10.1038/sj.embor.7400080
11. Mariotti M, Guigó R (2010) Selenoprofiles: profile-based scanning of eukaryotic genome sequences for selenoprotein genes. Bioinformatics 26:2656–2663. doi:10.1093/bioinformatics/btq516
12. Kryukov GV, Kryukov VM, Gladyshev VN (1999) New mammalian selenocysteine-containing proteins identified with an algorithm that searches for selenocysteine insertion sequence elements. J Biol Chem 274:33888
13. Lambert A, Lescure A, Gautheret D (2002) A survey of metazoan selenocysteine insertion sequences. Biochimie 84(9):953
14. Lescure A, Gautheret D, Carbon P, Krol A (1999) Novel Selenoproteins identified in Silico andin vivo by using a conserved RNA structural motif. J Biol Chem 274:38147
15. Novoselov SV, Rao M, Onoshko NV et al (2002) Selenoproteins and selenocysteine insertion system in the model plant cell system, Chlamydomonas reinhardtii. EMBO J 21:3681
16. Castellano S, Morozova N, Morey M et al (2001) In silico identification of novel selenoproteins in the Drosophila melanogaster genome. EMBO Rep 2:697
17. Martin-Romero FJ, Kryukov GV, Lobanov AV et al (2001) Selenium metabolism in Drosophila. J Biol Chem 276:29798
18. Castellano S, Novoselov S V, Kryukov G V et al (2004) Reconsidering the evolution of eukaryotic selenoproteins: a novel nonmammalian family with scattered phylogenetic distribution. EMBO Rep 5:71–77. doi:10.1038/sj.embor.7400036
19. Kryukov G, Gladyshev V (2000) Selenium metabolism in zebrafish: multiplicity of selenoprotein genes and expression of a protein containing 17 selenocysteine residues. Genes Cells 5:1049
20. Castellano S, Lobanov AV, Chapple C et al (2005) Diversity and functional plasticity of eukaryotic selenoproteins: identification and characterization of the SelJ family. Proc Natl Acad Sci U S A 102:16188
21. Shchedrina V, Novoselov S, Malinouski MY, Gladyshev VN (2007) Identification and characterization of a selenoprotein family containing a diselenide bond in a redox motif. Proc Natl Acad Sci 104:13919
22. Taskov K, Chapple C, Kryukov GV et al (2005) Nematode selenoproteome: the use of the selenocysteine insertion system to decode one codon in an animal genome? Nucleic Acids Res 33:2227
23. Lobanov AV, Gromer S, Salinas G, Gladyshev VN (2006) Selenium metabolism in Trypanosoma: characterization of selenoproteomes and identification of a Kinetoplastida-specific selenoprotein. Nucleic Acids Res 34:4012
24. Lobanov AV, Delgado C, Rahlfs S et al (2006) The plasmodium selenoproteome. Nucleic Acids Res 34:496. doi:10.1093/nar/gkj450
25. Novoselov S, Lobanov A, Hua D et al (2007) A highly efficient form of the selenocysteine insertion sequence element in protozoan parasites and its use in mammalian cells. Proc Natl Acad Sci U S A 104:7857–7862. doi:10.1073/pnas.0610683104
26. Novoselov SV, Hua D, Lobanov AV, Gladyshev VN (2006) Identification and characterization of Fep15, a new selenocysteine-containing member of the Sep15 protein family. Biochem J 394:575. doi:10.1042/BJ20051569
27. Jiang L, Ni J, Liu Q (2012) Evolution of selenoproteins in the metazoan. BMC Genomics 13:446. doi:10.1186/1471-2164-13-446
28. da Silva MTA, Caldas VEA, Costa FC et al (2013) Selenocysteine biosynthesis and insertion machinery in Naegleria gruberi. Mol Biochem Parasitol 188:87–90. doi:10.1016/j.molbiopara.2013.04.002
29. Chapple CE, Guigó R (2008) Relaxation of selective constraints causes independent selenoprotein extinction in insect genomes. PLoS One 3(8):e2968
30. Hofacker I, Fontana W, Stadler P et al (1994) Fast folding and comparison of RNA secondary structures. Monatsh Chem 125:167–188

31. Zhang Y, Gladyshev VN (2005) An algorithm for identification of bacterial selenocysteine insertion sequence elements and selenoprotein genes. Bioinformatics 21:2580–2589. doi:10.1093/bioinformatics/bti400
32. Kryukov GV, Gladyshev VN (2004) The prokaryotic selenoproteome. EMBO Rep 5:538
33. Kryukov GV, Castellano S, Novoselov SV et al (2003) Characterization of mammalian selenoproteomes. Science 300:1439
34. Mariotti M, Lobanov AV, Guigo R, Gladyshev VN (2013) SECISearch3 and Seblastian: new tools for prediction of SECIS elements and selenoproteins. Nucleic Acids Res 41:e149. doi:10.1093/nar/gkt550
35. Nawrocki EP, Kolbe DL, Eddy SR (2009) Infernal 1.0: inference of RNA alignments. Bioinformatics 25:1335–1337. doi:10.1093/bioinformatics/btp157
36. Eddy SR Cove. http://eddylab.org/software.html
37. Grundner-Culemann E, Martin GW, Harney JW, Berry MJ (1999) Two distinct SECIS structures capable of directing selenocysteine incorporation in eukaryotes. RNA 5:625–635
38. Lorenz R, Bernhart SH, Höner Z, Siederdissen C et al (2011) ViennaRNA Package 2.0. Algorithms Mol Biol 6:26. doi:10.1186/1748-7188-6-26
39. Otero L, Romanelli-Cedrez L, Turanov AA et al (2014) Adjustments, extinction, and remains of selenocysteine incorporation machinery in the nematode lineage. RNA 20:1023–1034. doi:10.1261/rna.043877.113
40. Otero L, Romanelli-Cedrez L, Turanov AA et al (2014) Erratum to: adjustments, extinction, and remains of selenocysteine incorporation machinery in the nematode lineage. RNA 20:1653
41. Chapple CE, Guigó R, Krol A (2009) SECISaln, a web-based tool for the creation of structure-based alignments of eukaryotic SECIS elements. Bioinformatics 25(5):674. doi:10.1093/bioinformatics/btp020
42. Altschul S, Madden T, Schaffer A et al (1997) Gapped BLAST and PSI-BLAST: a new generation of protein database search programs. Nucleic Acids Res 25:3389
43. Slater GSC, Birney E (2005) Automated generation of heuristics for biological sequence comparison. BMC Bioinformatics 6:31. doi:10.1186/1471-2105-6-31
44. Gladyshev VN, Arnér ES, Berry MJ et al (2016) Selenoprotein gene nomenclature. J Biol Chem. doi: 10.1074/jbc.M116.756155
45. Suzek BE, Wang Y, Huang H et al (2015) UniRef clusters: a comprehensive and scalable alternative for improving sequence similarity searches. Bioinformatics 31:926–932. doi:10.1093/bioinformatics/btu739

Chapter 2

Selenoprofiles: A Computational Pipeline for Annotation of Selenoproteins

Didac Santesmasses, Marco Mariotti, and Roderic Guigó

Abstract

Selenoproteins contain selenocysteine (Sec or U), the 21st amino acid, inserted in response to an in-frame UGA codon. UGA normally terminates translation, but in selenoprotein mRNAs it is recoded to specify Sec insertion. For this reason, standard gene prediction programs fail to predict Sec codons, and selenoproteins are usually misannotated in protein databases and genome projects. Selenoprofiles is a computational pipeline able to correctly annotate selenoprotein genes in genomic sequences. This program uses a SECIS-independent approach, based on homology searches, and employs curated built-in profile alignments for all known selenoprotein families. Selenoprofiles constitutes the most accurate method for predicting selenoprotein genes belonging to known families.

Key words Selenoprotein prediction, Gene annotation, Selenocysteine, Recoding, UGA codon

1 Introduction

Identification of selenoprotein genes in nucleotide sequences is challenged by the presence of an in-frame UGA codon, normally a stop codon [1]. Selenoprotein transcripts possess an RNA stem-loop structure called SECIS [2] that promotes Sec insertion in response to a UGA codon. For this reason, selenoproteins are generally missed or wrongly annotated by standard gene finding tools. Selenoprofiles [3] is aimed to correctly identify UGA-Sec codons using a homology-based strategy. The key concept is that Sec positions are known a priori in the input selenoprotein family, and thus ad-hoc scoring schemes are employed specifically for the identification of the homologous sites. Selenoprofiles includes a built-in set of manually curated profiles for known selenoprotein families and other proteins related to synthesis of selenoproteins (Table 1), allowing the prediction of these families by a SECIS-independent approach. Selenoprofiles can be used for gene finding in genomes or other nucleotide sequences from species across the whole tree of life. The package also includes additional programs to

Laurent Chavatte (ed.), *Selenoproteins: Methods and Protocols*, Methods in Molecular Biology, vol. 1661,
DOI 10.1007/978-1-4939-7258-6_2, © Springer Science+Business Media LLC 2018

Table 1
List of selenoprotein built-in profiles

Family ID	Description
Machinery	
SBP2	SECIS binding protein 2
SecS	Selenocysteine Synthase, eukaryotic
eEFsec	Sec specific elongation factor, eukaryotic
pstk	Phosphoseryl-tRNASec kinase
secp43	tRNA Selenocysteine 1 Associated Protein; Trnau1ap
SPS[a]	Selenophosphate synthetase, eukaryotic
Metazoa	
DI	Iodothyronine deiodinases
GPx	Glutathione peroxidases, eukaryotic
TR	Thioredoxin reductases
MsrA	Methionine sulfoxide reductase A, eukaryotic
SelR	Methionine sulfoxide reductase B
Sel15	Selenoprotein 15; SELENOF
Fep15	Fish selenoprotein 15; SELENOE
SelKi	Insect selenoprotein K; SELENOK
SelK	Non-insect selenoprotein K; SELENOK
FrnE	FrnE/DsbA oxidoreductase
SelW	Selenoproteins W and V; SELENOW and SELENOV
AhpC[b]	Alkyl hydroperoxide reductase C
Sel*	Selenoprotein *; SELENO* (* is any of HIJLMNOPSTU)
Protist	
EhSEP2	Emiliania huxleyi disulfide-isomerase like selenoprotein
Lmsel1	Leishmania major selenoprotein 1
SelQ	Toxoplasma gondii selenoprotein Q
SelTryp	Kinetoplastida SelTryp
MSP	Membrane selenoprotein
Ostsp*	Ostreococcus selenoprotein * (* is any of 123)
Sel*	Plasmodium selenoprotein * (* is any of 1234)
Prokarya	
seld	Selenophosphate synthetase, prokaryotic
gpx_b	Glutathione peroxidases, prokaryotic
msra_b	Methionine sulfoxide reductase A, prokaryotic
di_b	Deiodinase-like prokaryotic protein
fdha	FdhA formate dehydrogenase, alpha subunit
frha	FrhA/MvhA/VhuU hydrogenases
frhd	FrhD/MvhD/VhuD hydrogenases
hdra	Heterodisulfide reductase, subunit A
grx	Prokaryotic glutaredoxin 3
gst	glutathione S-transferase
prdb	D-proline reductase, PrdB subunit; selenoprotein B
ars_s	Arsenite S-adenosylmethyltransferase
arsc	Arsenate reductase
bbd	BFD/(2Fe-2S)-binding domain-containing protein
cytc	Cytochrome c-like selenoprotein
dsre	TusD/DsrE sulfurtransferase
duf1858	DUF1858-containing protein (unknown function domain)
fesor	Fe-S oxidoreductase

(continued)

Table 1 (continued)

Family ID	Description
fmdb	FmdB family regulatory protein
ftrb	FtrB ferredoxin thioredoxin reductase beta
frx	Ferredoxin-thioredoxin reductase like
hesb_like	Similar to HesB iron-sulfur cluster biosynthesis protein
imp	IMP dehydrogenase/GMP reductase
mucd	MucD putative serine proteinase
nadh_ox	FAD/NADH-dependent oxidoreductase
pp_sp1	Plesiocystis pacifica Trx-like selenoprotein
prx	Peroxiredoxin family
prx_like	Prx-like thiol:disulfide oxidoreductase
rhor	Rhodanese like sulfurtransferase
rsam	Radical SAM domain-containing oxidoreductase
soret	Split soret cytochrome c precursor
tdip	Thiol:disulfide interchange selenoprotein
ugc	UGC-containing hypothetical selenoprotein
ugsc	UGSC-containing hypothetical selenoprotein
uos_hp3	Hypotheical selenoprotein OS_HP3
usha	UshA UDP-sugar hydrolase
yeee	YeeE/YedE hypothetical selenoprotein
ahp*	Alkyl hydroperoxide reductase subunit * (* is any of df)
dsb*	Thiol-disulfide isomerase subunit * (* is any of ag)
grd*	Glycine reductase complex component * (* is any of ab)
mer*	Mercuric transport protein mer* (* is any of pt)
rnf*	NADH:ubiquinone oxidoreductase subunit * (* is any of bc)

Note that some protein families are split into two profiles for technical reasons (e.g., SelK and SelKi; MsrA and msra_b). The * character is used to indicate multiple families (see the description)

[a]SPS is also part of the metazoan set of profiles

[b]AhpC is also part of the prokaryotic set of profiles

collect and visualize the results in the context of the phylogenetic tree of the species analyzed. This chapter constitutes a practical guide for using Selenoprofiles for selenoprotein search. For more information, see the original publication [3] and the Selenoprofiles manual available in [4].

2 Selenoprofiles Overview

Selenoprofiles is a pipeline combining a number of "slave" programs for homology-based gene finding, whose predictions are analyzed and processed to produce complete multi-exonic gene predictions (Fig. 1). These programs are Blast (psitblastn) [5], Exonerate [6], and Genewise [7]. The three programs are based on the same principle: the target nucleotide sequence is translated in all possible frames and the protein query is aligned to such translated sequences, searching for high scoring matches. The

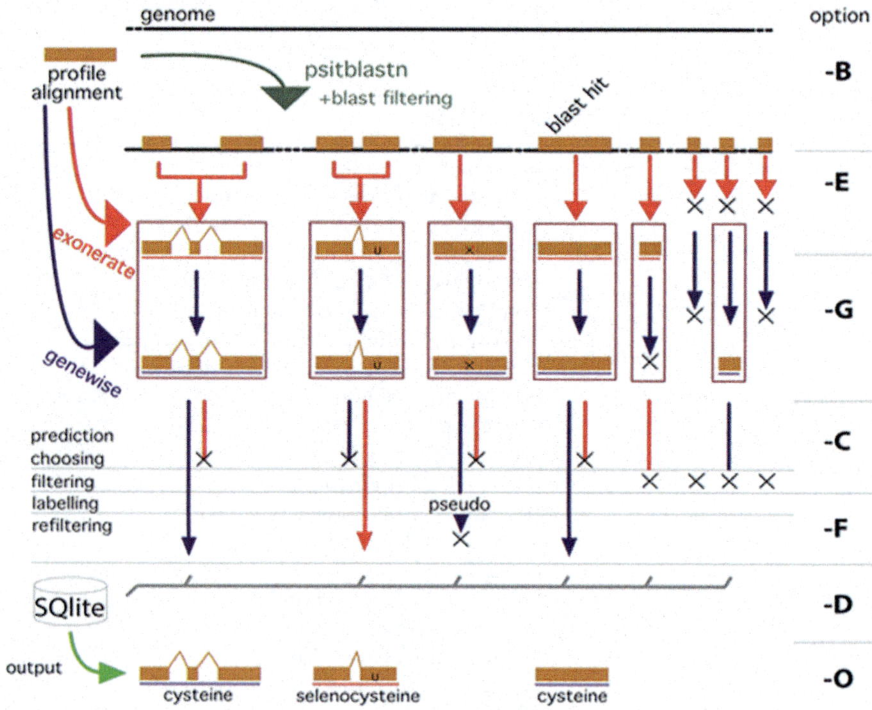

Fig. 1 Graphical summary of the Selenoprofiles pipeline. Reproduced and adapted from [3] by permission of Oxford University Press

program psitblastn is first used to scan the target sequence using a position-specific scoring matrix (PSSM) derived from the input profile alignment. The matches are then used through the two splice alignment programs, Genewise and Exonerate, to deduce the exonic structure of each candidate gene. The best prediction among the three programs is chosen and labeled by a dedicated procedure. Three layers of filtering are used to control the number of potential candidates and to exclude likely false positives. The filtering procedures are highly flexible and customizable, and can be adapted for each input profile independently.

The default filtering procedure is based on sequence similarity. A similarity score threshold is chosen based on the overall conservation present in the input profile sequence alignment; we call it Average Weighted Sequence Identity (AWSI) filter. In practice, this means that protein families with high sequence conservation have strict filtering procedures by default. When multiple profiles are searched, overlapping matches are assigned to one of the profiles, based on sequence similarity. The program produces nonoverlapping multi-exonic gene structures for all input profiles.

3 Installation

Selenoprofiles can be installed on any Unix/Linux system with python 2.6 or higher. All its "slave" programs must be previously installed by the user (*see* **Note 1**): Blast [5], Exonerate [6], Genewise [7], and Mafft [8]. The GNU AWK (GAWK) utility is also needed. The protein database UniRef50 is required for using the built-in profiles (*see* **Note 2**). The following commands should be executed in a terminal to install and test the Selenoprofiles package (*see* **Note 3**). Note that "/install_folder" is used to indicate the hypothetical desired location of installation. Administrator privileges are not required for installing or running Selenoprofiles.

```
cd /install_folder
git clone https://github.com/marco-mariotti/selenoprofiles
cd selenoprofiles
python install_selenoprofiles.py -full
python test_selenoprofiles.py
```

After installation, inside the folder "/install_folder/selenoprofiles," the "Selenoprofiles" executable is available (*see* **Note 4**), as well as a global configuration file named "selenoprofiles_3.config." This file contains all default options and parameters used by the program, each of which can be overridden via a command line option.

4 Input

The program takes two inputs: one or more profile alignments, representing the families to search for; and a "target" genome sequence (or any other nucleotide sequence database) which is to be searched.

The package includes a built-in set of manually curated profiles of known selenoprotein families and proteins related to selenoprotein synthesis. These profiles have been extensively tested and their filtering parameters optimized. The complete list of profiles can be found inside the "profiles/" folder in the Selenoprofiles installation directory. These profiles can be used out-of-the-box with the complete ("-full") installation of Selenoprofiles. Custom user-defined profiles can also be easily created from protein alignments and searched for. For a guide to create new profiles, refer to the Selenoprofiles manual available from [4].

5 Running Selenoprofiles

5.1 Basic Usage

Let us consider that we want to scan the genome of the species *Homo sapiens*, contained in the file "/genomes/Homo_sapiens/genome.fa," for the built-in profile GPx (glutathione peroxidases). Here is the basic command line:

```
Selenoprofiles results -t /genomes/Homo_sapiens/genome.fa -s
"Homo sapiens" -p GPx
```

The first argument (the "results" folder) is where all output and intermediate files are stored. The profile argument "-p" specifies the protein family to search for. Use a comma-separated list to specify more than one family (e.g., "-p GPx,SPS"). Specific option arguments are also accepted to specify convenient sets of profiles: "metazoa," "protist," "prokarya," "machinery" (these are defined in the Selenoprofiles configuration file). For example, this command will search for all selenoprotein profiles previously identified in metazoans genomes and for the Sec machinery (i.e., proteins involved in selenoprotein synthesis):

```
Selenoprofiles results -t /genomes/Homo_sapiens/genome.fa -s
"Homo sapiens" -p metazoa,machinery
```

For RNA sequences or prokaryotic genomes, which are not expected to contain introns, the option "-no_splice" is recommended. This option turns off the procedures for the identification of splicing events, and the pipeline runs faster.

At the end of computation, the Selenoprofiles output files can be found in the output folder located inside the results folder. For the example above, this is "results/Homo_sapiens.genome/output/." The results folder contains also additional intermediate files, which would rarely be inspected by the user. The contents of the output folder are described in the next section. Selenoprofiles attempts to minimize computation by not executing procedures for which the resulting files are detected. When the program is executed again with changed parameters, the user should take care of deleting the previous output files and add an option to specify which pipeline steps must be repeated (*see* Fig. 1). For example, if the final filtering parameters are changed, option "-F" must be provided.

5.2 Output

The output folder contains all output files. For each family with results, a ".ali" file is produced (e.g., "GPx.ali"). This fasta formatted file contains the alignment of all results for this family and the sequences from the profile. The fasta header of the results starts with the "output id" of the prediction, following the structure "family.index.label" (e.g., "GPx.30.selenocysteine"). The "output

Table 2
Output formats available in Selenoprofiles

Format	Description
p2g	Native output format (*see* Fig. 2)
fasta	Protein sequence in fasta
gff	Genomic coordinates in GFF[a]
gtf	Genomic coordinates in GTF[a]
cds	Coding sequence in fasta
dna	The full gene sequence, including introns, in fasta
three_prime	The sequence downstream of the prediction (default 6KB)
five_prime	The sequence upstream of the prediction ("-five_prime_length" must be specified)
introns	The sequence of all introns split in a multi-fasta file

[a]See http://www.ensembl.org/info/website/upload/gff.html for details

id" is a unique identifier for each result in the target, and it is used as a prefix for all output files, using the specific format of the file as a suffix (e.g., "GPx.30.selenocysteine.fasta") (*see* Table 2 for possible output formats). The index included in the "output id" has to be considered an arbitrary identifier; it does not indicate the protein subfamily (e.g., "GPx.4.selenocysteine" is not necessarily the protein known as GPx4). The "output id" also includes a label. The labeling procedure is the following: for selenoprotein families, the label is used to characterize the amino acid aligned to the Sec position. Possible labels are "selenocysteine," "cysteine" or any other amino acid. If the prediction does not span the Sec position in the profiles, the label "unaligned" is used. If it contains frameshifts or in-frame stop codons (apart from the Sec-TGA) the label "pseudo" is used. The label "uga_containing" is used if the only pseudogene feature is one or more TGA codons in any position other than the Sec position. For standard proteins (non-selenoprotein families), the possible labels are "homologue" or "pseudo."

By default, results are provided in the p2g format (Fig. 2), which contains extensive information about the candidate gene. Its header contains basic information about the gene prediction, most of which is self-explanatory. In addition, a few different metrics based on the comparison of the prediction and the sequences in the input profile are provided. The *ASI* corresponds to average sequence identity with the sequences in the profile. *AWSIc* and *AWSIw* are similar, but more sophisticated, scores (see the Selenoprofiles manual available from [4]). Next, the alignment section contains the query-target amino acid sequence alignment, showing the gene structure. Between the query and target

```
Output_id:  GPx.30.cysteine
----------  ---------------
-Target          /genomes/Homo_sapiens/genome.fa
-Chromosome (+)  chr6
-Program         genewise
-Query name      Chain A, Crystal Structure Of Human Glutathione Peroxidase 5
-Query range     23-215       length:215   coverage: 0.9
-Profile range   105-445      length:445   coverage: 0.77    sec_position: [192]
-ASI:            0.4008       (ignoring gaps: 0.4602)
-AWSIc:          0.7227       Z-score: 0.726
-AWSIw:          0.7302       Z-score: 0.855
-State           kept

------- alignment -------
Query    MKMDCHKDEKGTIYDYEAIALNKNEYVSFKQYVGKHILFVNVATYUGLTAQYPE <---Intron---> L
          /|||||||||||||||||||||||||||||||||||||||||||||| ||||||||| <   2173bp    > |
Target   FQMDCHKDEKGTIYDYEAIALNKNEYVSFKQYVGKHILFVNVATYCGLTAQYPE                L
         tcagtcaggagaatgtggagcaaagtgttactggacactgaggattgcagctcg                c
         tatagaaaaagctaaaactctaaaaatctaaatgaattttatccaggtccaac    gt      ag   at
         cggtccacgaccctctgccattgtattccggtgcgcccccgcccttgagatt                  aa
                                                              *

Query    NALQEELKPYGLVVLGFPCNQFGKQEPGDNKEILPGLK <---Intron---> YVRPGGGFVPSFQLFEK
         |||||||||||||||||||||||||||||||||||||| <   425bp    > |||||||||||||||||
Target   NALQEELKPYGLVVLGFPCNQFGKQEPGDNKEILPGLK                YVRPGGGFVPSFQLFEK
         agccggcactgcggtgtctactgacgcggaagaccgca                tgccgggtgcatcctga
         actaaatacagttttgtcgaatgaaacgaaaattcgta   gt     ag    atgcgggttcgtattaa
         tacgggggcttatggctcccatagaaaatcagtttgc                gtctagaatattcgttga

Query    GDVNGEKEQKVFSFLK <---Intron---> HSCPHPSEILGTFKSISWDPVKVHDIRWNFEKFLVGPDG
         |||||||||||||||| <   1540bp   > |||||||||||||||||||||||||||||||||||||||
Target   GDVNGEKEQKVFSFLK                HSCPHPSEILGTFKSISWDPVKVHDIRWNFEKFLVGPDG
         gggaggagcagtatta                cttccctgatgatatattgcgagcgactatgatcggcgg
         gatagaaaaattgtta   gt     ag    acgcaccattgctactcgactataatggataatttgcag
         gtgttaaagacctcgg                cctttctgtgcacatacgctagctcctgctagcgggtta

Query    IPVMRWSHRATVSSVKTDILAYLKQFKT-
         ||||||||||||||||||||||||||||
Target   IPVMRWSHRATVSSVKTDILAYLKQFKTK
         acgacttccgagatgaagacgttactaaa
         tcttggcagcctgctacattcataataca
         ctcgcgccgtgccacgaccggcggacaca

------- positions -------
Exon 1    28497222     28497381
Exon 2    28499555     28499672
Exon 3    28500098     28500197
Exon 4    28501738     28501941

-------- features -------
None
--------- 3' seq --------
Total sequence length available downstream >= 6000
Sequence until first stop codon:
TAG
 *
```

Fig. 2 The p2g format is the default output provided by Selenoprofiles sequences, bars are used to show identity "|" or similarity "/" of two aligned residues. The selenocysteine position, if present in the alignment, is marked by a "*," and corresponds to a U in the query sequence. If present, predicted in-frame stop codons, and frameshifts are marked in the alignment. Predicted introns are also included. Their length and the donor and acceptor splice sites are indicated. Next, the genomic positions for the exons are reported. The coordinates are 1-based (the first nucleotide in the chromosome is indexed as 1). Finally, the file reports the sequence at the 3′ of the prediction, up to the first stop codon encountered.

Several other output formats are available; they can be specified by using "-output_*format*" in the command line, where *format* is one of the formats listed in Table 2 (e.g., "-output_fasta").

5.3 SECIS Element

Eukaryotic selenoprotein genes possess a SECIS element in the 3′ UTR [2]. The 3′ UTR is not included in the predictions (which include the coding sequence only), but Selenoprofiles provides an option to easily retrieve those sequences from the genome. When specifying "-output_three_prime," Selenoprofiles fetches the sequence found downstream of each prediction, and generates a fasta file with the suffix ".three_prime" in the output folder. The sequence starts with the first nucleotide downstream of the prediction. The length in nucleotides of the retrieved sequence can be specified with the option "-three_prime_length" (6000 by default). The fasta file ".three_prime" can be uploaded to the Seblastian web server (see [9] and chapter 1 in this book) to search for eukaryotic SECIS elements. Since the SECIS must be in the same strand as the gene, the option "search also complementary strand" in the web server should be unchecked.

6 Searching Multiple Target Genomes

Selenoprofiles can be used to analyze multiple genomes and compare their results. The genomes have to be analyzed in different runs, but the results folder should be the same. For example:

```
Selenoprofiles results -t /genomes/Homo_sapiens/genome.fa -s
"Homo sapiens" -p metazoa,machinery
Selenoprofiles results -t /genomes/Drosophila_melanogaster/
genome.fa -s "Drosophila melanogaster" -p metazoa,machinery
```

After running the pipeline for all the targets, the program selenoprofiles_join_alignments.py is used to search for and to join all the ".ali" files inside each target subfolder.

```
selenoprofiles_join_alignments.py -d results -o joined_align-
ments
```

The command line above will join all the alignments inside the results directory ("-d" option), and will generate the new joined alignments in the output folder ("-o" option). Then, the program selenoprofiles_tree_drawer.py can be used to visualize the results for multiple target species, placed in the corresponding phylogenetic tree. This program requires the python library for tree exploration ETE (http://etetoolkit.org/). The program selenoprofiles_tree_-drawer.py takes as input the joined alignments, which are provided as a list of multiple arguments separated by a blank space (the order in which the alignments are provided is used to show them in the plot). The tree of the investigated species is also required (*see* **Note 5**). It is specified using the "-t" parameter, which must be placed as the last argument in the command line. For example:

```
cd joined_alignments
selenoprofiles_tree_drawer.py eEFsec.ali SelU.ali -t species_-
tree.nw
```

The output of selenoprofiles_tree_drawer.py is the graphical representation of the phylogenetic tree annotated with the results obtained by Selenoprofiles (Fig. 3). Running "selenoprofiles_tree_-drawer.py -h" will show additional options for the program.

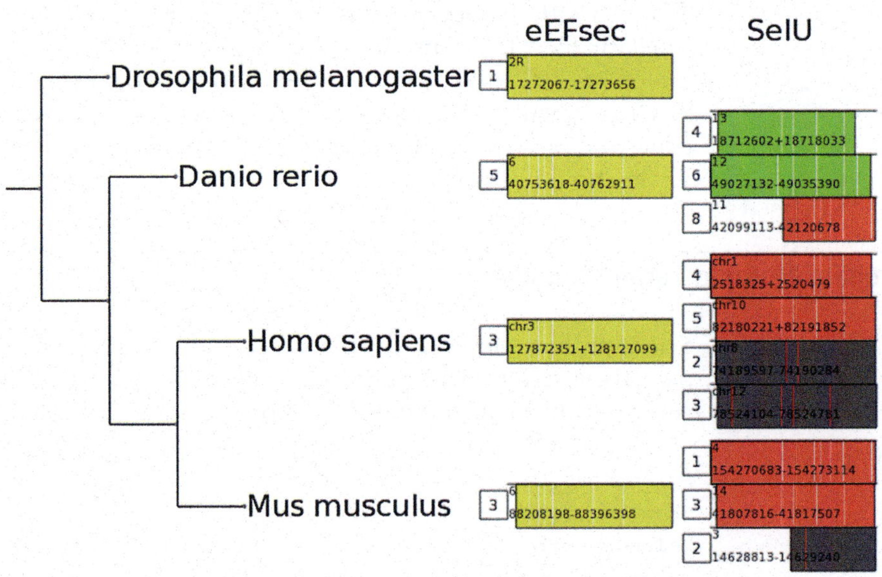

Fig. 3 Species tree annotated with Selenoprofiles results. Each column corresponds to a family, and each colored rectangle corresponds to a result. Multiple results for the same species and family are presented as piled up rectangles. The numeric tag on the left corresponds to the Selenoprofiles numeric index. The color depends on the label: *green* for selenocysteine; *red* for cysteine; *yellow* for homologue (non-selenoprotein families); *gray* for pseudo. Other colors are used for additional labels. The rectangle length and position indicate the coverage of the result aligned to the profile. The information inside the rectangle corresponds to the chromosome, genomic coordinates, and strand. The position of the introns, relative to the protein alignment, is indicated with *vertical white lines*. *Vertical red lines* correspond to insertions causing frameshifts

For further analyses using the alignments produced by selenoprofiles_join_alignments.py, it is strongly recommended to realign the sequences with a multiple sequence alignment program, such as T-Coffee [10] (http://tcoffee.crg.cat/).

7 Notes

1. The following web page contains information on how to install the "slave" programs and possible problems you may encounter during the installation: http://big.crg.cat/news/20110616/installing_programs_and_modules_needed_by_selenoprofiles.
2. The script install_selenoprofiles.py, if executed in "-full" mode, will attempt to download the protein database UniRef50 through the Internet. If the database is already in your system, you can skip this step by specifying its location with the option "-db." Running "install_selenoprofiles.py -h" will show more information.
3. If the program GIT is not installed in your system, you can point your web browser to https://github.com/marco-mariotti/selenoprofiles and download the package manually.
4. In order to run Selenoprofiles and the other programs provided with the package, the "selenoprofiles" folder needs to be present in your $PATH variable. You have to type "export PATH=$PATH:/install_folder/selenoprofiles" (where *install_folder* is the folder where you installed the package) every time you open a new terminal, or include that line in the ".bashrc" file in your home directory.
5. The most commonly used format for phylogenetic trees is the newick format, where the nodes of the tree are defined by parentheses. The tree used in Fig. 3 would be "(Drosophila melanogaster (Danio rerio (Mus musculus, Homo sapiens)));". ETE toolkit (http://etetoolkit.org) provides with automated tools to extract species trees from the NCBI taxonomy database and write them into a newick format file.

References

1. Driscoll DM, Chavatte L (2004) Finding needles in a haystack. In silico identification of eukaryotic selenoprotein genes. EMBO Rep 5:140–141. doi:10.1038/sj.embor.7400080
2. Berry MJ, Banu L, Chen YY et al (1991) Recognition of UGA as a selenocysteine codon in type I deiodinase requires sequences in the 3′ untranslated region. Nature 353:273–276. doi:10.1038/353273a0
3. Mariotti M, Guigó R (2010) Selenoprofiles: profile-based scanning of eukaryotic genome sequences for selenoprotein genes. Bioinformatics 26:2656–2663. doi:10.1093/bioinformatics/btq516
4. Mariotti M (2016) Selenoprofiles 3 | Bioinformatics and Genomics @ CRG http://big.crg.cat/services/selenoprofiles. Accessed 1 Nov 2016

5. Altschul SF, Madden TL, Schäffer AA et al (1997) Gapped BLAST and PSI-BLAST: a new generation of protein database search programs. Nucleic Acids Res 25:3389–3402
6. Slater GSC, Birney E (2005) Automated generation of heuristics for biological sequence comparison. BMC Bioinformatics 6:31. doi:10.1186/1471-2105-6-31
7. Birney E, Clamp M, Durbin R (2004) GeneWise and Genomewise. Genome Res 14:988–995. doi:10.1101/gr.1865504
8. Katoh K, Misawa K, Kuma K, Miyata T (2002) MAFFT: a novel method for rapid multiple sequence alignment based on fast Fourier transform. Nucleic Acids Res 30:3059–3066. doi:10.1093/nar/gkf436
9. Mariotti M, Lobanov AV, Guigo R, Gladyshev VN (2013) SECISearch3 and Seblastian: new tools for prediction of SECIS elements and selenoproteins. Nucleic Acids Res 41:e149. doi:10.1093/nar/gkt550
10. Notredame C, Higgins DG, Heringa J (2000) T-coffee: a novel method for fast and accurate multiple sequence alignment. J Mol Biol 302:205–217. doi:10.1006/jmbi.2000.4042

Chapter 3

SelGenAmic: An Algorithm for Selenoprotein Gene Assembly

Liang Jiang and Qiong Liu

Abstract

Computational methods for identifying selenoproteins have been developed rapidly in recent years. However, it is still difficult to identify the open reading frame (ORF) of eukaryotic selenoprotein gene, because the TGA codon for a selenocysteine (Sec) residue in the active center of selenoprotein is traditionally a terminal signal of protein translation. A gene assembly algorithm SelGenAmic has been constructed and presented in this chapter for identifying selenoprotein genes from eukaryotic genomes. A method based on this algorithm was developed to build an optimal TGA-containing-ORF for each TGA in a genome, followed by protein similarity analysis through conserved sequence alignments to screen out selenoprotein genes from these ORFs. This method improved the sensitivity of detecting selenoproteins from a genome due to the design that all TGAs in the genome were investigated for its possibility of decoding as a Sec residue. The method based on the SelGenAmic algorithm is capable of identifying eukaryotic selenoprotein genes from their genomes.

Key words Selenoprotein, Selenocysteine, Gene assembly algorithm

1 Introduction

The active site of selenoprotein is selenocysteine (Sec or U), the 21st amino acid encoded by a TGA codon in the open reading frame (ORF) of the gene [1]. Traditionally, TGA codon only signals the termination of protein synthesis; however, it can also be translated into a Sec residue when a specific stem-loop structure, designated as the Sec insertion sequence (SECIS) element, is located in the 3′-untranslated region (UTR) of a selenoprotein gene in eukaryotes and archaea, or located immediately downstream of the Sec-decoding TGA (designated as Sec-TGA) in bacteria [2–5]. The amino acid sequences flanking the active Sec residue are more conserved than other regions less functionally or structurally important in selenoproteins. These conserved regions play key roles in redox balance, metal combination, Sec-Sec/Sec-S bond formation, and protein folding in vivo. Additionally, the Sec residue is highly analogous to the Cys in biochemical properties,

Laurent Chavatte (ed.), *Selenoproteins: Methods and Protocols*, Methods in Molecular Biology, vol. 1661,
DOI 10.1007/978-1-4939-7258-6_3, © Springer Science+Business Media LLC 2018

which accounts for the fact that in most homologs of a selenoprotein the active Sec is replaced by Cys residues.

Several methods for the *in-silico* prediction of selenoprotein genes have been developed over the past decade, based on the structural characteristics of selenoprotein genes. These methods have been used separately or together for the identification of selenoproteins from the genomes or expressed sequence tag (EST) libraries in many species, such as human, fish, insects, green algae, nematodes, and protozoa [6–11]. Recent application on the analyses of environmental metagenomic sequences has also been succeeded in finding prokaryotic selenoprotein genes [12, 13]. To identify selenoprotein genes from the genome, several computational methods have to be combined for use, including the methods of RNA secondary-structure prediction, eukaryotic or prokaryotic ORF prediction, sequence alignment analysis, and phylogenetic analysis. With the development and application of those methods, the size of the selenoprotein family is growing, for example, the members of human selenoproteome increased from 14 to 25. In addition, up to 58 selenoprotein families have been identified recently in the Global Ocean Sampling (GOS) Project, which shed light on the evolution of selenoproteins according to their distribution in different species and environments.

For the prediction of selenoprotein genes, it is indispensable to construct complete or partial ORFs containing the Sec-TGA codons. In a non-intron DNA sequence like prokaryotic genome, it is relatively easy to build a theoretical ORF containing a TGA codon termed "interrupted ORF" (i-ORF) as shown in Fig. 1. However, in eukaryotic genome, the intron-exon structure of gene makes it difficult to build an i-ORF. Most of the earlier studies on eukaryotic selenoprotein identification were performed by the following scheme. First, RNA prediction algorithms were used to predict SECIS elements. Second, the SECIS elements were used to inform gene prediction algorithms to predict i-ORFs, of which a suitable SECIS element must be downstream [14]. The disadvantage of this scheme is that if any special-structure SECIS elements, which have not been discovered so far and included into the known SECIS models, existed in a newly sequenced genome, then no SECIS information can be used to inform the gene prediction algorithms to find the upstream ORFs. Naturally thinking, to

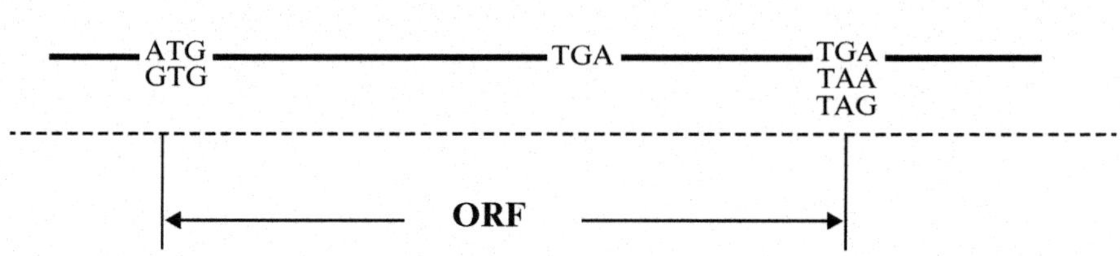

Fig. 1 A buildup of theoretical i-ORF containing a TGA codon in prokaryotic genomes

identify the selenoproteins with special-structure SECIS, it must be able to predict i-ORFs without the help of SECIS information.

For eukaryotic genome, a SECIS-independent gene prediction approach was previously introduced in 2004 [11]. A modified gene prediction program named geneid was developed to identify 20 human selenoprotein genes. The gene assembly algorithm GenAmic used by geneid only builds the optimal gene structures with the TGA-containing exons having higher coding potential scores. Selenoproteins with lower score TGA-containing exons, such as human selenoproteins K, S, and T, were not identified by the GenAmic-based method. Although the sensitivity of this SECIS-independent gene prediction approach is high (20/25), it is not so good as the SECIS-dependent approach that discovered all 25 human selenoproteins [14]. In this chapter, a new gene assembly algorithm named SelGenAmic was constructed to develop a similar SECIS-independent gene prediction method for the identification of selenoproteins. Compared with the GenAmic algorithm used by geneid, the SelGenAmic is more sensitive because its target is to build an optimal gene structure for each TGA. Thus, no TGA codon is neglected for building i-ORFs. Finally, amino acid conservation assessment is used to find the real selenoproteins from these i-ORFs.

2 Methods

2.1 Basic Idea

TGA coding for Sec is one of the key characters of selenoprotein genes. If we scan all TGA codons from a genome, all Sec codons will be included. It is relatively easy to build a theoretical i-ORF from a prokaryotic genome as shown in Fig. 1. This task can be carried out by finding a start codon and a stop codon in the nucleotide sequences flanking any TGA codon. All of i-ORFs can be translated into amino acid sequences and compared with known proteins to find potential selenoprotein genes. This method has been reported to be used for the discovery of rare amino acids, selenocysteine, and pyrrolysine, in prokaryotes [15].

However, building eukaryotic i-ORFs is much more difficult than that of prokaryotes. Many potential gene prediction signals like start codon ATG, stop codons TGA/TAA/TAG, and splice sites AG/GT can be found in the sequences flanking any TGA in a eukaryotic genome. It is rather complicated to choose suitable signals to build exons, and consequently suitable exons to build i-ORFs during selenoprotein identification.

A. c-exons in which no in-frame TGA codon is allowed. B. i-exons in which an in-frame TGA codon is present and codes for a Sec residue. The signals such as splice sites AG and GT, start codon, and stop codons are indicated in the figure.

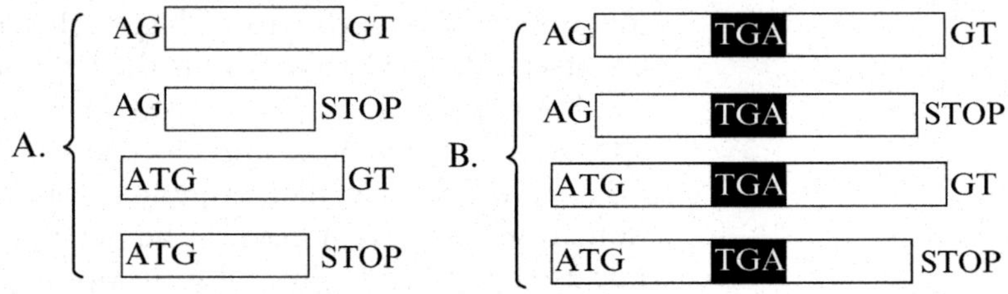

Fig. 2 Different types of exons used to assemble selenoprotein genes. (**a**) c-exons in which no in-frame TGA codon is allowed. (**b**) i-exons in which an in-frame TGA codon is present and codes for a Sec residue. The signals such as splice sites AG and GT, start codon, and stop codons are indicated in the figure

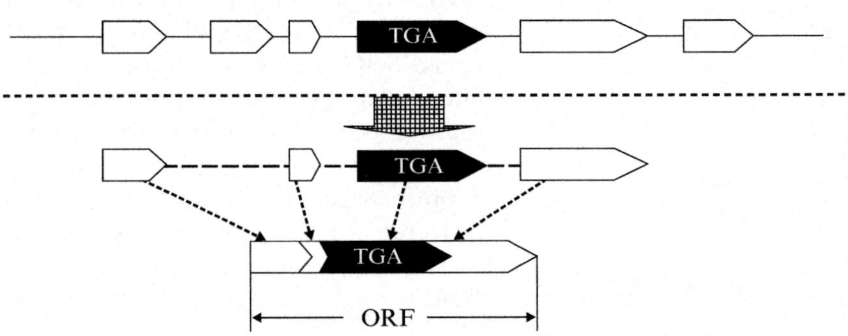

Fig. 3 Building an ORF containing a TGA codon. The i-exon is indicated by a *black arrow*, and the c-exons by *white arrows*. An ORF containing an i-exon is formed by concatenating suitable c-exons and the i-exon

To address this issue, a method was presented in this chapter. First, all the TGA codons were found from a genome, and supposed to be signals of Sec. All other signals such as start codon, stop codon, and splice sites are also predicted. Second, common exons (c-exons) were built with common signals as shown in Fig. 2a and interrupt exons (i-exons) containing TGA were built by concatenating common signals and TGA as shown in Fig. 2b. Third, the gene assembly algorithm SelGenAmic was used to build the best ORF for each i-exon. Figure 3 shows the process of building a best ORF for an i-exon. The best ORF that has the maximal coding potential is composed of this i-exon and other c-exons.

If all i-exons and i-ORFs were enumerated from a genome to build a set, theoretically it should include all potential selenoprotein genes with Sec-TGA codons. However, the vast majority of these i-genes (genes containing Sec-TGA) will be biologically meaningless. To filter out such meaningless i-genes, the conservation of amino acid sequence in the local regions flanking the Sec residue (shown in Fig. 4) was used to screen out i-genes that are more likely to be selenoprotein genes.

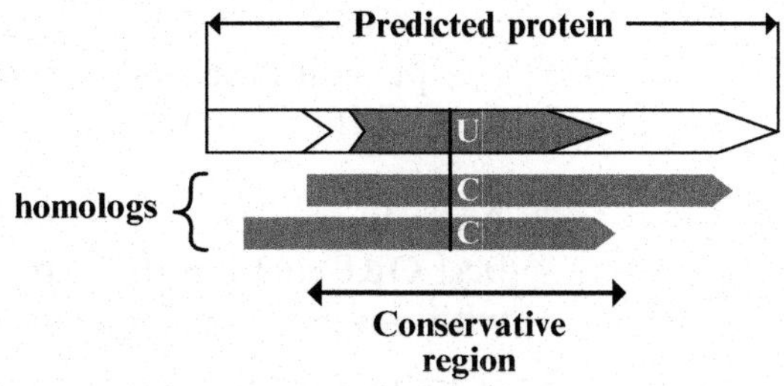

Fig. 4 A predicted protein that has conservative amino acid sequences flanking the Sec residue is more likely to be a selenoprotein. In its homologs, the Sec (U) residue is expected to be replaced by Cys (C) residues

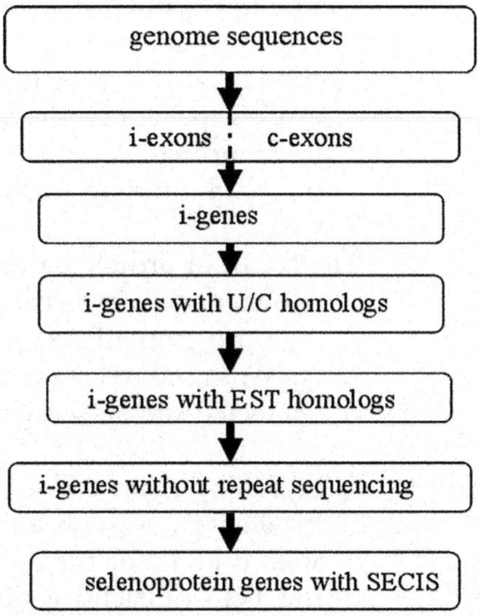

Fig. 5 Building an ORF containing a TGA codon. The i-exon is indicated by a *black arrow*, and the c-exons by *white arrows*. An ORF containing an i-exon is formed by concatenating suitable c-exons and the i-exon

2.2 General Identification Procedure

General procedures of our method are shown in Fig. 5 and described as follows.

1. Obtaining i-exons and c-exons from the genome. The whole genome sequence was scanned to find all TGA codons and other signals including ATG, TAA\TAG, and AG\GT. All i-exons and c-exons were built from these signals. The coding potential of any exon was calculated as the sum of the scores of the signals, plus the log-likelihood ratio of a Markov model for coding DNA.

The program Geneid (version 1.2a) [16] is used to obtain all common gene signals, such as start codon, splice sites, stop codons, and common potential exons, from genomic sequences, and calculate the coding potential of i-exons. Perl scripts edit based on our basic idea are used to obtain all TGA codons from genome sequences and build i-exons.

2. Assembling i-genes from i-exons and c-exons. For each i-exon, a best ORF that has the maximal coding potential score was built with our gene assembly algorithm SelGenAmic.
3. Searching for Sec/Cys pairing and the conservation of its flanking regions. All i-genes were translated into amino acid sequences. Local sequences flanking the Sec residue were extracted for detecting similarity in the NCBI nonredundant (nr) protein database by the BLASTp program in order to obtain multiple sequence alignments. Those sequences were screened out with conservation in the local regions flanking the Sec residue, and alignments containing Sec/Cys pairing (simplified as U/C pair), i.e., the Sec-containing local sequence must have its homologous sequences contain Cys residues in the position of Sec in multiple alignments.

The NCBI nr protein database and BLAST [17] programs could be downloaded from the NCBI ftp server. Search All i-genes by the program BLASTp with an E-value cutoff at 0.001. ClustalW is used to create multiple sequence alignments with all the detected similar sequences [18].

4. Searching against EST databases and splicing the ESTs. Similarity analysis was performed against EST databases to obtain spliced ESTs for the i-genes. The local DNA sequences flanking the TGA of each i-gene were searched by BLASTn against the EST database. Discard the i-genes in which TGAs were analyzed to be the products of sequencing errors. And those i-genes that were caused by repeated sequencing were also discarded. Finally, best i-genes were found to have complete ORFs and UTRs.

EST sequences could be downloaded and compared with all the predicted selenoprotein genes using the program BLASTn. Highly similar EST sequences are spliced by the SeqMan program from the DNASTAR package (http://www.dnastar.com/) and analyzed for the selenoprotein gene structure. The constructed genes are homologously compared to genomic sequences with the program Sim4 [19] to find the locations of exons and introns in the genome.

5. Checking for SECIS elements. All the i-genes were found to have downstream SECIS elements in their UTRs, which further confirmed them as selenoprotein genes.

RNAfold [20] and PatScan [21] are automatically used by online program SECISearch to detect SECIS-like structures from genome sequences. The SECIS patterns used are the same as that in the search of human SECIS [14].

2.3 Assembly Algorithm SelGenAmic

The algorithm SelGenAmic is developed from GenAmic to solve the problem of finding an optimal ORF for each i-exon. The word optimal here means that the coding potential score of such ORF is bigger than any other ORFs composed of this i-exon and other suitable c-exons.

The input data of SelGenAmic are all i-exons and c-exons along with their information such as coding potential, position, and coding frame. Let $E = \{e_1, e_2, ..., e_k\}$, ($k \geq 0$, k is an integer) be the set of exons. The coding potential of these exons is shown as $P(e)$ for each $e \in E$ in this chapter.

Let $C = \{c_1, \ldots, c_m\}$, ($m \geq 0$, m is an integer) and $T = \{t_1, \ldots, t_n\}$, ($n \geq 0$, n is an integer) be the sets of c-exons and i-exons. Obviously, $E = C \cap T$.

The principle to constrain the algorithm to choose suitable exons for concatenation is described as a function M.

$$M(e_a, e_b) = \begin{cases} 1, (a < b, e_a \text{ and } e_b \text{ can be concatenated legally}) \\ 0, (e_a \text{ and } e_b \text{ cannot be concatenated legally}) \end{cases}$$

The function M describes the relation of two exons e_a and e_b ($e_a \in E$, $e_b \in E$), which can be concatenated legally in a numerical order. The word legal here means e_a and e_b are frames compatible, nonoverlapping, and adjoining splice sites matched.

First, we recall the concepts of gene assembly. A gene assembly g is a sequence consisting of exons $e_{1'}, \ldots, e_{q'}$ from $E(e_{i'} \in E)$. Thus, a legal gene assembly can be described as

$$g = < e_{1'}, \ldots e_{q'} > ,$$

where for all $e_{i'}$ in g, $M(e_{i'}, e_{(i+1)'}) = 1$.

The coding potential of a gene assembly is the sum of scores of assembled exons:

$$P(g) = P(e_{1'}) + \ldots + P(e_{q'}).$$

The problem to find an optimal g could thus be interpreted so as to search for the gene assembly g with maximum $P(g)$, i.e., for all other genes g' constructed from E, $P(g) > = P(g')$.

Thus, the target of SelGenAmic is that, for each $t_i \in T$, finding an optimal gene assembly $g(t_i) = <c_{1'}, \ldots, c_{k'}, t_i, c_{1''}, \ldots, c_{k''}>$, where $c_{i'} \in C$, $c_{i''} \in C$, $M(c_{k'}, t_i) = 1$, $M(t_i, c_{1''}) = 1$, $M(c_{i'}, c_{(i+1)'}) = 1$, $M(c_{i''}, c_{(i+1)''}) = 1$, i.e., for the set $T = \{t_1, \ldots, t_n\}$, build a set of optimal gene assemblies $\{g(t_1), \ldots, g(t_n)\}$.

This problem is equivalent to finding the best upstream assembly $g_u = <c_{1'}, \ldots c_{k'}, t_i>$ and the best downstream assembly $g_d = <t_i, c_{1''}, \ldots c_{k''}>$ for each t_i.

To solve such a problem, we introduced six concepts. (1) The best upstream assembly(BUA); (2) the coding potential of BUA (CpBUA); (3) the best upstream adjoining exon (BUE); (4) the best downstream assembly (BDA); (5) the coding potential of BDA (CpBDA); and (6) the best downstream adjoining exon (BDE).

The concepts (1), (2), and (3) were used in the GenAmic algorithm of Geneid program to calculate the optimal assembly from common exons [22]. The concept (1) BUA was described as that for each exon c_i from C it is possible to find a best assembly ended with c_i, i.e., $g_u(c_i) = <c_{1'}, \ldots, c_i>$, where for all other assembly g' ended with c_i, $P(g_u(c_i)) >= P(g')$. The coding potential $P(g_u(c_i))$ is the CpBUA (concept (3)) of c_i.

The concept (2) BUE was described as that for each c_i from C, it is possible to find a best upstream adjoining exon c_j, if and only if

1. $M(c_j, c_i) = 1$, and
2. for all other $c' \in C$, if $M(c', c_i) = 1$, $P(g_u(c_j)) >= P(g_u(c'))$.

Here, we use function G to describe the relationship between a c-exon c_i and its BUE c_j: $G(c_i) = c_j$.

Obviously, the BUA $g_u(c_i)$ of c_i can be obtained by concatenating its BUE c_i and the BUA $g(c_j)$ of c_i:

$$g_u(c_i) = < c_{1'}, \ldots, c_j, c_i > = < g_u(c_j), c_i > = < g_u(G(c_i)), c_i >.$$

So that, it can be easily concluded that if all the BUE for each $c \in C$ were known, the BUA can be easily calculated as follows:

$$g_u(c_i) = < c_0, \ldots, G(G(c_i)), G(c_i), c_i >,$$

(c_0 is used to describe the first exon of this assembly).

In Geneid, the author of GenAmic algorithm used dynamic programming to obtain all the BUE $G(c)$ for every $c \in C$ [17]. Using the GenAmic, we can easily build three sets for $C = \{c_1, c_2, \ldots, c_k\}$:

the set of BUE $\{G(c_1), G(c_2), \ldots, G(c_k)\}$,
the set of BUA $\{g_u(c_1), g_u(c_2), \ldots, g_u(c_k)\}$, and
the set of CpBUA $\{P(g_u(c_1)), P(g_u(c_2)), \ldots, P(g_u(c_k))\}$.

Knowing how to calculate the BUA $g_u(c_i)$ for each c-exon $c_i \in C$, we can use the same method to build the BUA $g_u(t_i)$ for each i-exon $t_i \in T$:

$$g_u(t_i) = < c_{1'}, \ldots c_{k'}, t_i > = < g_u(G(t_i)), t_i > = < g_u(c_u), t_i >,$$

where c_u is the BUE of t_i.

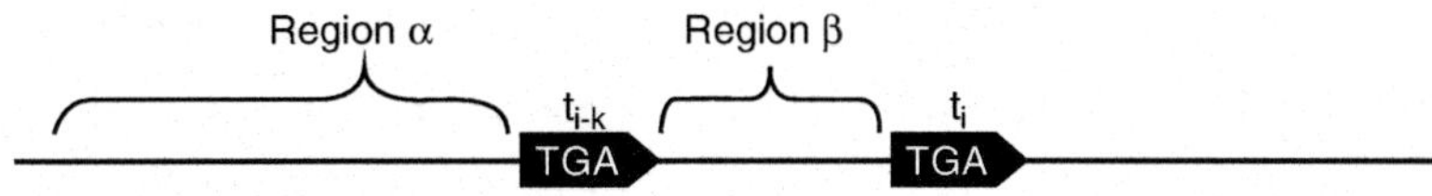

Fig. 6 The regions in which c-exons are used to search for the best upstream adjoining exon of t_i

The BUE $c_u \in C$ is a c-exon, and the set of BUA $\{g_u(c_1), g_u(c_2), \ldots, g_u(c_k)\}$ for each $c \in C$ can be obtained with GenAmic algorithm in linear time. Thus, if all the BUE c_u for each i-exon $t_i \in T$ can be obtained, its BUA can also be produced.

The BUE $G(t_i) = c_u$ can be found in the following way. We find all exons c' satisfying $M(c', t_i) = 1$, and their CpBUA from $\{P(g_u(c_1)), P(g_u(c_2)), \ldots, P(g_u(c_k))\}$. By comparing their coding potentials, the c_u with maximal $P(g_u(c))$ can be found.

As shown in Fig. 6, all c-exons $c'' \in C$ located upstream of t_i and satisfying $M(c'', t_i)$ will be searched for the c_u of t_i. An i-exon t_{i-k} can be found to divide the genome into two regions. The t_{i-k} is a exon for which all c-exons c'' satisfying $M(c'', t_{i-k}) = 1$ are also satisfying $M(c'', t_i) = 1$. Obviously, the c-exon with maximal CpBUA in region α is the BUE of t_{i-k}. Then only c' in region β satisfying $M(c', t_i) = 1$ will be searched for the c-exon with maximal CpBUA. Let ς and B be the sets of c-exons in the regions α and β, respectively, then to find the c_u for t_i, can be described as follows:

$$P(g_u(G(t_i))) = \max\left\{ \begin{array}{l} P(g_u(G(t_i - k))) \\ \max\{P(g_u(c)) : c \in B\} \end{array} \right\}$$

With this equation, all BUE(s) of $t_i \in T$ can be calculated recursively with one scan of the genome in linear time, due to the reason that only the region β for each t_i is needed to be searched.

Then the set of BUE for T is obtained as $\{G(t_1), G(t_2), \ldots, G(t_k)\}$ and

the set of BUA for T is obtained as $\{g_u(G(t_1)), g_u(G(t_2)), \ldots, g_u(G(t_k))\}$.

The BDA, BDE, and CpBDA for all $c \in C$ can be obtained in a similar way. Let $g_d(c_i)$ be the BDA of c_i, and let function G^r describe the relation of c_i and its BDE c_j:

$G^r(c_i) = c_j$, if and only if

1. $M(c_i, c_j) = 1$ and
2. for all other $c' \in C$, if $M(c_i, c') = 1$, then $P(g_d(c_j)) >= P(g_d(c'))$.

Thus the set of BDA, BDE, and CpBDA for all $c \in C$ can be produced similarly as follows:

$$\{G^r(c_1), G^r(c_2), \ldots, G^r(c_k)\},$$
$$\{g_d(c_1), g_d(c_2), \ldots, g_d(c_k)\}, \text{and}$$
$$\{P(g_d(c_1)), P(g_d(c_2)), \ldots, P(g_d(c_k))\}.$$

Then the set of BDA for T is obtained as $\{g_d(G^r(t_1)), g_d(G^r(t_2)), \ldots, g_d(G^r(t_k))\}$.

By concatenating BUA and BDA, the best assembly for t_i is constructed as

$$g = < g_u(G(t_i)), t_i, g_d(G^r(t_i)) >.$$

Thus, for each i-exon an optimal assembly can be constructed in linear time.

2.4 Application of SelGenAmic-Based Methods

The method based on the selenoprotein gene assembly algorithm SelGenAmic was used for the identification of more than 200 selenoprotein genes from various eukaryotic species such as *Ciona intestinalis, Amphimedon queenslandica, Trichoplax adhaerens, Nematostella vectensis, Lottia gigantea, Capitella teleta, Branchiostoma floridae, Schmidtea mediterranea, Schistosoma japonicum, Taenia solium*, and so on [23–26]. Research of those work released the evolutionary pattern of metazon animal selenoproteins, and found several special selenoprotein such as Sponge AphC.like protein, amphioxus N-terminal region selenoprotein P [23, 26]. The mechanism and application of SelGenAmic shows us, it is a suitable bioinformatics tool for assembling and identifying eukaryotic selenoprotein genes from various genomes.

Acknowledgments

This work was supported by the National Natural Science Foundation of China (Nos. 31401129), the Shenzhen Bureau of Science, Technological innovation projects of Education Bureau of Guangdong Province (No. 2014KTSCX109) and Natural Science Foundation of Shenzhen University (201426).

References

1. Hatfield DL, Berry MJ, Gladyshev VN (2001) Selenium: its molecular biology and role in human health. Springer, New York, pp 1–5
2. Kryukov GV, Kryukov VM, Gladyshev VN (1999) New mammalian selenocysteine-containing proteins identified with an algorithm that searches for selenocysteine insertion sequence elements. J Biol Chem 274 (48):33888–33897
3. Atkins JF, Gesteland RF (2000) The twenty-first amino acid. Nature 407(6803):463, 465
4. Bock A (2000) Biosynthesis of selenoproteins—an overview. Biofactors 11 (1–2):77–78
5. Hatfield DL, Gladyshev VN (2002) How selenium has altered our understanding of the genetic code. Mol Cell Biol 22 (11):3565–3576
6. Kryukov GV, Gladyshev VN (2000) Selenium metabolism in zebrafish: multiplicity of selenoprotein genes and expression of a protein containing 17 selenocysteine residues. Genes Cells 5(12):1049–1060

7. Castellano S, Morozova N, Morey M, Berry MJ, Serras F, Corominas M, Guigo R (2001) In silico identification of novel selenoproteins in the Drosophila melanogaster genome. EMBO Rep 2(8):697–702
8. Novoselov SV, Rao M, Onoshko NV, Zhi H, Kryukov GV, Xiang Y, Weeks DP, Hatfield DL, Gladyshev VN (2002) Selenoproteins and selenocysteine insertion system in the model plant cell system, Chlamydomonas reinhardtii. EMBO J 21(14):3681–3693
9. Taskov K, Chapple C, Kryukov GV, Castellano S, Lobanov AV, Korotkov KV, Guigo R, Gladyshev VN (2005) Nematode selenoproteome: the use of the selenocysteine insertion system to decode one codon in an animal genome? Nucleic Acids Res 33(7):2227–2238
10. Novoselov SV, Hua D, Lobanov AV, Gladyshev VN (2006) Identification and characterization of Fep15, a new selenocysteine-containing member of the Sep15 protein family. Biochem J 394(Pt 3):575–579
11. Castellano S, Novoselov SV, Kryukov GV, Lescure A, Blanco E, Krol A, Gladyshev VN, Guigo R (2004) Reconsidering the evolution of eukaryotic selenoproteins: a novel nonmammalian family with scattered phylogenetic distribution. EMBO Rep 5(1):71–77
12. Zhang Y, Fomenko DE, Gladyshev VN (2005) The microbial selenoproteome of the Sargasso Sea. Genome Biol 6(4):R37
13. Zhang Y, Gladyshev VN (2008) Trends in selenium utilization in marine microbial world revealed through the analysis of the global ocean sampling (GOS) project. PLoS Genet 4 (6):e1000095
14. Kryukov GV, Castellano S, Novoselov SV, Lobanov AV, Zehtab O, Guigo R, Gladyshev VN (2003) Characterization of mammalian selenoproteomes. Science 300 (5624):1439–1443
15. Fujita M, Mihara H, Goto S, Esaki N, Kanehisa M (2007) Mining prokaryotic genomes for unknown amino acids: a stop-codon-based approach. BMC Bioinformatics 8:225
16. Parra G, Blanco E, Guigo R (2000) GeneID in Drosophila. Genome Res 10(4):511–515
17. Altschul SF, Madden TL, Schaffer AA, Zhang J, Zhang Z, Miller W, Lipman DJ (1997) Gapped BLAST and PSI-BLAST: a new generation of protein database search programs. Nucleic Acids Res 25(17):3389–3402
18. Thompson JD, Higgins DG, Gibson TJ (1994) CLUSTAL W improving the sensitivity of progressive multiple sequence alignment through sequence weighting, position-specific gap penalties and weight matrix choice. Nucleic Acids Res 22(22):4673–4680
19. Florea L, Hartzell G, Zhang Z, Rubin GM, Miller W (1998) A computer program for aligning a cDNA sequence with a genomic DNA sequence. Genome Res 8(9):967–974
20. Hofacker IL, Stadler PF (2006) Memory efficient folding algorithms for circular RNA secondary structures. Bioinformatics 22 (10):1172–1176
21. Dsouza M, Larsen N, Overbeek R (1997) Searching for patterns in genomic data. Trends Genet 13(12):497–498
22. Guigo R (1998) Assembling genes from predicted exons in linear time with dynamic programming. J Comput Biol 5(4):681–670
23. Jiang L, Liu Q, Ni JZ (2010) In silico identification of the sea squirt selenoproteome. BMC Genomics 11:289
24. Jiang L, Ni JZ, Liu Q (2012) Evolution of selenoproteins in the metazoan. BMC Genomics 3:446
25. Chen H, Jiang L, Liu Q, Ni JZ (2012) Bioinformatic prediction of selenoprotein genes in the dolphin genome. Chin Sci Bull 57 (13):1533–1541
26. Jiang L, Zhu HZ, YZ X, Ni JZ, Zhang Y (2013) Comparative selenoproteome analysis reveals a reduced utilization of selenium in parasitic platyhelminthes. PeerJ 1:e202

Part II

Selenoprotein Biosynthesis

Chapter 4

Selenocysteine tRNA$^{[Ser]Sec}$, the Central Component of Selenoprotein Biosynthesis: Isolation, Identification, Modification, and Sequencing

Bradley A. Carlson, Byeong Jae Lee, Petra A. Tsuji, Paul R. Copeland, Ulrich Schweizer, Vadim N. Gladyshev, and Dolph L. Hatfield

Abstract

The selenocysteine (Sec) tRNA$^{[Ser]Sec}$ population consists of two isoforms that differ from each other by a single 2′-*O*-methylribosyl moiety at position 34 (Um34). These two isoforms, which are encoded in a single gene, *Trsp*, and modified posttranscriptionally, are involved individually in the synthesis of two subclasses of selenoproteins, designated housekeeping and stress-related selenoproteins. Techniques used in obtaining these isoforms for their characterization include extraction of RNA from mammalian cells and tissues, purifying the tRNA$^{[Ser]Sec}$ population by one or more procedures, and finally resolving the two isoforms from each other. Since some of the older techniques for isolating tRNA$^{[Ser]Sec}$ and resolving the isoforms are used in only a few laboratories, these procedures will be discussed briefly and references provided for more detailed information, while the more recently developed procedures are discussed in detail. In addition, a novel technique that was developed in sequencing tRNA$^{[Ser]Sec}$ for identifying their occurrence in other organisms is also presented.

Key words Chromatography, Gel electrophoresis, Selenium, Selenocysteine tRNA, Selenocysteine tRNA detection, Selenocysteine tRNA gene modification

1 Introduction

Selenocysteine (Sec) tRNA$^{[Ser]Sec}$ has been described as the central component [1], the key factor [2] and the quintessential constituent in selenoprotein biosynthesis [3]. Indeed, the targeted removal of the gene encoding this tRNA (*Trsp*) from cells or tissues results in the total loss of selenoprotein expression [4, 5]. Sec tRNA$^{[Ser]Sec}$ appears to be among the most essential factors of those that selenoprotein biosynthesis is dependent upon, as removal of *Trsp* completely eliminates expression of this protein class [4, 5], whereas deletion of another dependent factor, *SECIS-binding protein 2* (*Secisbp2*), reduces selenoprotein expression substantially, albeit not to the extent as *Trsp* [6].

Laurent Chavatte (ed.), *Selenoproteins: Methods and Protocols*, Methods in Molecular Biology, vol. 1661,
DOI 10.1007/978-1-4939-7258-6_4, © Springer Science+Business Media LLC 2018

Selenoproteins are the only known class of proteins that depend on a single tRNA for their expression. Thus, altering the expression of $tRNA^{[Ser]Sec}$ levels results in modulating the expression of selenoproteins. There are two species of Sec tRNAs that differ from each other by a single methylated base occurring on the 2′-*O*-hydroxyribosyl moiety at position 34, designated Um34 ([7] and *see* Fig. 1). The hypermodified base at position 34 is 5-methoxycarbonylmethyluridine and the Um34-methylated form is 5-methoxycarbonylmethyl-2′-*O*-methyluridine. These two isoforms have only four modified bases, mcm^5U, N6-isopentenyladenosine (i^6A), pseudouridine (ψ), and 1-methyladenosine (m^1A) at positions 34, 37, 55, and 58, respectively, and one nucleoside, mcm^5Um [7].

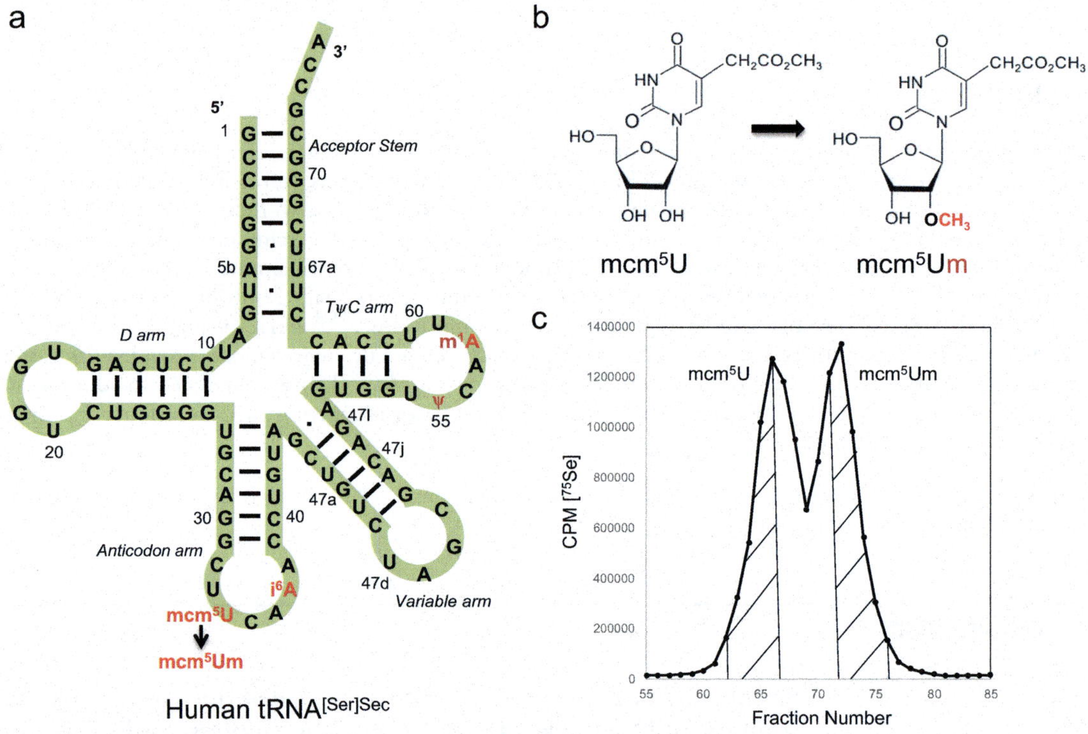

Fig. 1 Selenocysteine $tRNA^{[Ser]Sec}$. (**a**) The primary structure of human $tRNA^{[Ser]Sec}$ is shown in a cloverleaf model. Mammalian $tRNA^{[Ser]Sec}$ is 90 bases in length. The acceptor stem constitutes the paired 5′ and 3′ terminal bases, the D stem and loop constitute the six paired and four unpaired bases of the left portion of the tRNA, the anticodon stem and loop, the six paired and seven unpaired bases of the lower portion of the tRNA, the variable stem and loop, the five paired and four unpaired bases, and the TψC stem and loop, the four paired and seven unpaired bases of the right portion of the tRNA. Mammalian $tRNA^{[Ser]Sec}$ contains modifications at positions 34 (mcm^5U or mcm^5Um), 37 (i^6A), 55 (ψ), and 58 (m^1A). (**b**) The two $tRNA^{[Ser]Sec}$ isoforms (containing either mcm^5U or mcm^5Um at position 34) differ from each other by a single methyl group on the 2′-*O*-ribosyl moiety. (**c**) Representative fractionation of [75Se]selenocysteyl-$tRNA^{[Ser]Sec}$ isoforms by RPC5 chromatography. [75Se]selenocysteyl-$tRNA^{[Ser]Sec}$ was extracted from HL-60 cells and run on an RPC5 column as described in Subheading 3.6. The initial eluting peak from the column is the mcm^5U-containing isoform and the later eluting peak is the mcm^5Um-containing isoform. The isoforms are pooled as shown by the hatched areas and collected as described in Subheading 3.6 for further study

The two tRNA[Ser]Sec isoforms are involved in the expression of separate subclasses of selenoproteins, housekeeping and stress-related selenoproteins [8, 9], which provides an additional handle to control the expression of these two subclasses of proteins and regulate the expression of the two tRNA[Ser]Sec isoforms by specific base modifications in the *Trsp* gene transcript [8, 10]. This chapter describes the isolation of tRNA[Ser]Sec from mammalian cells and tissues, resolution of the two isoforms, identification and sequencing of tRNA[Ser]Sec to further elucidate the roles of these two unique tRNAs in selenoprotein synthesis and regulation. Traditional techniques used in isolating tRNA[Ser]Sec, such as isolating total tRNA by diethylaminoethyl (DEAE)-cellulose chromatography [11], resolving the two tRNA[Ser]Sec isoforms by RPC5 chromatography [12, 13], and aminoacylation of tRNA[Ser]Sec [14] are described herein, since they are highly significant in the characterization of the tRNA[Ser]Sec isoforms. However, the details regarding techniques such as enriching tRNA[Ser]Sec by benzoylated DEAE (BD)-cellulose chromatography [15–17], sequencing the isoforms by the [32P]-labeling techniques of Silberklang et al. [18] as modified [7], and identification of the modified bases and nucleoside by [32P]-labeling [18] can be found in the given references.

The present chapter emphasizes the more recently developed procedures. One of these protocols involves the isolation, partial purification, and sequencing of tRNA[Ser]Sec from lower eukaryotic cells by a novel procedure [19]. This procedure was previously used to determine the primary sequences of tRNA[Ser]Sec from *Chlamydomonas reinhardtii* [19, 20], *Dictyostelium discoideum*, and *Tetrahymena thermophile* [21] .

2 Materials

All the solutions were prepared with molecular biology grade H_2O and all the reagents were of the highest grade commercially available. Routine reagents with no listed supplier should be purchased from your supplier of choice. Buffers are stored at room temperature unless otherwise noted. Purchased reagents are stored as recommended by the supplier. Freshly grown cells or freshly harvested tissues were either frozen immediately by placing into liquid nitrogen and storing at −80 °C until ready for use, or by immediately beginning RNA extraction followed by total tRNA and subsequent tRNA[Ser]Sec purification. For large-scale tRNA preparations, calf liver was obtained from freshly euthanized animals, and the liver (~10 kg) maintained on ice until the isolation procedure.

2.1 Extraction and Isolation of Total RNA from Cells and Tissues

1. Extraction buffer A: 0.14 M sodium acetate, pH 4.5.
2. H_2O-saturated phenol, pH 4.5.
3. Chloroform ($CHCl_3$).

4. DE52 preswollen microgranular DEAE cellulose (GE Healthcare Life Sciences, Pittsburgh, PA).
5. Disposable chromatography columns (e.g., 20 mL, 1.5 × 12 cm columns from Bio-Rad, Hercules, CA).
6. DE52 column buffer A: 0.1 M Tris–HCl, 0.1 M NaCl, pH 7.4.
7. DE52 column buffer E: 0.1 M Tris–HCl, 1.0 M NaCl, pH 7.4.
8. Pure ethanol (EtOH).

2.2 RPC5 Chromatography

1. Reverse phase chromatography 5 (RPC5) column [13] (*see* Subheading 3.3 for preparation).
2. RPC5 starting buffer (+ $Mg2^{+}$): 10 mM NaOAc, 0.45 M NaCl, 10 mM $MgOAc_2$, 1 mM EDTA, pH 4.5.
3. RPC5 starting buffer (− $Mg2^{+}$): 10 mM NaOAc, 0.5 M NaCl, 1 mM EDTA, pH 4.5.
4. RPC5 buffer A: 10 mM NaOAc, 0.525 M NaCl, 10 mM $MgOAc_2$, 1 mM EDTA, pH 4.5.
5. RPC5 buffer B: 10 mM NaOAc, 0.675 M NaCl, 10 mM $MgOAc_2$, 1 mM EDTA, pH 4.5.
6. RPC5 buffer C: 10 mM NaOAc, 1.5 M NaCl, 10 mM $MgOAc_2$, 1 mM EDTA, pH 4.5.
7. RPC5 buffer D: 10 mM NaOAc, 0.6 M NaCl, 1 mM EDTA, pH 4.5.
8. RPC5 buffer E: 10 mM NaOAc, 0.825 M NaCl, 1 mM EDTA, pH 4.5.
9. 0.45 μm HA nitrocellulose filters (Millipore, Billerica, MA).
10. Column fraction collector.
11. Ecoscint A scintillation cocktail (National Diagnostics, Atlanta, GA) and scintillation vials.

2.3 Northern Blotting and Aminoacylation

1. Isotopes: [3H]serine (specific activity ~20 Ci/mmol; Moravek Biochemicals, Brea, CA); γ-[32P]-ATP (specific activity ~3000 Ci/mmol; Perkin Elmer, Waltham, MA).
2. Positively charged nylon blotting membrane.
3. 15% TBE/urea gels (handcast or precast).
4. Quikhyb hybridization solution (Stratagene, San Diego, CA).
5. Trizol® reagent (Thermo Fisher Scientific, Waltham, MA).
6. T4 polynucleotide kinase.
7. TAM (10×): 0.5 M Tris–HCl, 0.06 M ATP, 0.2 M $MgCl_2$, pH 7.4.
8. 19 unlabeled amino acid mix (minus serine): 10 mM final concentration of each amino acid.

9. Purified seryl-tRNA synthetase (MyBioSource, Inc., San Diego, CA).
10. Trichloroacetic acid (TCA).
11. Wash buffer 1: 2× SSC, 0.1% SDS.
12. Wash buffer 2: 0.1× SSC, 0.1% SDS.
13. 1× TBE buffer: 89 mM Tris, 89 mM boric acid, 2 mM EDTA, pH 7.6.
14. Betafluor scintillation cocktail (National Diagnostics, Atlanta, GA) and scintillation vials.

2.4 Extraction and Isolation of [75Se] Selenocysteyl-$tRNA^{[Ser]Sec}$ from Cells

1. [75Se]selenious acid (specific activity ~1000 Ci/mmol) in HNO_3 was obtained from the University of Missouri Research Reactor Center (MURR), Columbia, MO.
2. Phosphate-buffered saline (PBS): 9 mM Na_2HPO_4, 1.5 mM KH_2PO_4, 137 mM NaCl, 2.7 mM KCl, pH 7.4.
3. Sodium selenite (Sigma-Aldrich, St. Louis, MO).
4. Cycloheximide.
5. Extraction buffer B: 5 mM NaOAc, 225 mM NaCl, 5 mM $MgOAc_2$, 1 mM EDTA, 0.2% SDS, 6 mM 2-mercaptoethanol, pH 4.5.
6. H_2O-saturated phenol, pH 4.5.
7. $CHCl_3$.
8. DE52 column equilibration buffer: 5 mM NaOAc, 112.5 mM NaCl, 5 mM $MgOAc_2$, 1 mM EDTA, pH 4.5.
9. DE52 elution buffer: 10 mM NaOAc, 1.5 M NaCl, 10 mM $MgOAc_2$, 1 mM EDTA, pH 4.5.
10. Pure 200 proof EtOH.

2.5 Acid-Urea gel Electrophoresis

1. Molecular biology grade urea.
2. 1 M NaOAc, pH 4.5.
3. 40% acrylamide stock solution: acrylamide:bis-acrylamide (29:1).
4. Ammonium persulfate: 10% solution in molecular biology grade H_2O.
5. Tetramethylethylenediamine (TEMED).
6. Acid gel running buffer: 100 mM NaOAc, pH 4.5.
7. 2× acid-urea sample buffer: 8 M Urea, 100 mM NaOAc, pH 4.5.
8. Xylene cyanol and bromophenol blue dyes.
9. Gel fixative solution: 25% ethanol, 10% acetic acid.

2.6 Affinity Purification of $tRNA^{[Ser]Sec}$

1. Purified total tRNA.
2. Streptavidin-conjugated magnetic M-280 Dynabeads (Thermo Fisher Scientific, Waltham, MA).
3. Buffer A: 10 mM Tris–HCl, 2 M NaCl, 2 mM EDTA, pH 7.5.
4. 6× Saline Sodium Citrate (SSC) buffer: 90 mM sodium citrate, 0.9 M NaCl, pH 7.0.
5. RQ1 RNase-Free DNase (Promega, Madison, WI).

2.7 tRNA Sequencing

1. *Escherichia coli* (*E. coli*) poly(A) polymerase (New England Biolabs, Ipswich, MA).
2. RNase A (10 μg/μL) and RNase H (2 U/μL).
3. T4 polynucleotide kinase (New England Biolabs, Ipswich, MA).
4. Adenosine 5′-triphosphate (ATP) disodium salt hydrate.
5. Terminal Deoxynucleotidyl Transferase (TdT) and ddATP (Thermo Fisher Scientific, Waltham, MA).
6. T4 RNA ligase (New England Biolabs, Ipswich, MA).
7. DH5α competent *E. coli* cells (Thermo Fisher Scientific, Waltham, MA).
8. Luria broth (LB) and LB agar plates.
9. Taq DNA polymerase.
10. pCR2.1-TOPO vector (Thermo Fisher Scientific, Waltham, MA).
11. QIAprep Spin Miniprep Kit and QIAquick spin columns (Qiagen, Valencia, CA).

3 Methods

Carry out all the procedures at room temperature unless otherwise specified.

3.1 Extraction of Total RNA from Cells and Tissues

1. Combine equal parts of H_2O-saturated phenol and RNA extraction buffer A in a 50 mL polypropylene tube. Typically, ~20 mL of extraction buffer A mix is used per gram of frozen tissue or cells (*see* **Note 1**).
2. To frozen cells or tissues, add extraction buffer A mix and immediately homogenize for 3–4 min using an Omni Macro Homogenizer with a 50 mL stainless-steel chamber for tissues weighing less than 2 g (*see* **Note 2**).
3. Centrifuge homogenized sample at 4000 × *g* for 10 min.
4. Transfer the upper aqueous layer to a clean polypropylene tube.
5. Extract upper layer with an equal volume of $CHCl_3$. Vortex vigorously for 1–2 min.

6. Centrifuge at 4000 × *g* for 10 min.
7. Take top layer and transfer to clean polypropylene tube and load sample onto a prepared DE52 column (*see* Subheading 3.2).

3.2 Isolation of Total tRNA from Cells and Tissues

1. DE52 is prepared by adding DE52 to a glass beaker or flask. Gently add 4–5 volumes of molecular biology grade H_2O, let DE52 settle, pour off H_2O, and repeat wash 5–6 times to remove fine particles (*see* **Note 3**).
2. Gently pipette DE52 into empty chromatography column (*see* **Note 4**).
3. Equilibrate column with several column volumes of column buffer A.
4. Add the supernatant obtained following $CHCl_3$ extraction to the column.
5. Wash ~5–6 times with 25 mL of buffer A or until absorbance (A_{260}) of eluate is virtually undetectable.
6. Elute tRNA from column with ~12–15 mL of column buffer E.
7. Add $MgCl_2$ to the eluate to a final concentration of 10 mM.
8. Precipitate tRNA with 3 volumes of EtOH. Leave at −20 °C for at least 3 h.
9. Collect precipitated tRNA on Millipore 0.45 μm HA filter. Alternatively, precipitate in 3 volumes of EtOH and collect precipitant by centrifugation at 20,000 × *g* for 30 min at 4 °C.
10. Wash collected tRNA two to three times with 75% EtOH.
11. Gently dry, but do not over-dry the filter or pellet.
12. Elute tRNA from filter with molecular biology grade H_2O by placing filter in microcentrifuge tube, adding 250 μL of H_2O, vortexing and centrifuging.
13. Transfer eluted tRNA to a clean tube. Repeat elution with another 250 μL of H_2O. (Typically use a final volume of 500 μL for tRNA from 1 g of tissue or cells).
14. Deacylate purified Sec-tRNA$^{[Ser]Sec}$ in 1.0 M Tris–HCl, pH 8.0, at 37 °C for 1 h. Deacylated tRNA$^{[Ser]Sec}$ is precipitated with 3 volumes of EtOH, collected by centrifugation at 20,000 × *g* for 30 min, washed with 75% EtOH, dried and dissolved in nuclease-free H_2O (*see* **Note 5**).

3.3 Purification of tRNA$^{[Ser]Sec}$ Isoforms by RPC5 Column Chromatography and Northern Blot Analysis

RPC5 columns can be prepared using the technique outlined by Kelmers and Heatherly [13]. Briefly, 100 g of polychlorotrifluoroethylene (Plaskon CTFE) powder is coated in 4 mL of Adogen 464 (methyltrialkyl (C_8-C_{10}) ammonium chloride; Sigma-Aldrich, St. Louis, MO) that has been dissolved in 200 mL of chloroform. The solution is blended in a fume hood and ground vigorously with a mortar and pestle to break down the Plaskon powder to ~10 μ size particles and until the chloroform has evaporated. The coated Plaskon powder is then suspended in RPC5 starting buffer (+ $Mg2^+$) and used to pack a 30 cm × 1.5 cm glass chromatographic column (*see* **Note 6**).

1. Total tRNA (5–8 mg maximum) following the DE52 column is dissolved in RPC5 starting buffer (+ $Mg2^+$) and loaded onto an RPC5 column that has been equilibrated in RPC5 starting buffer (+ $Mg2^+$) (*see* **Note 7**).
2. A 160 mL linear gradient (80 mL of RPC5 buffer A as the starting buffer and 80 mL of RPC5 buffer B as the terminal buffer) is run at a flow rate of 2 mL/min and 2 mL fractions are collected in a fraction collector.
3. Wash column with 60 mL RPC5 buffer C and collect eluted fractions (*see* **Note 8**).
4. The absorbance at A_{260} in the eluted fractions is measured and $tRNA^{[Ser]Sec}$ is detected by dot blot hybridization.
5. Perform dot blot analysis by spotting 5 μL of each column fraction on a Hybond-N+ membrane followed by UV cross-linking of membrane.
6. Hybridize membrane using Quikhyb or similar hybridization solution for 4 h at 58 °C in a rotating hybridization oven with a [32P]-end labeled oligo probe complementary to the 3′-end of $tRNA^{[Ser]Sec}$ (5′-CGCCCGAAAGGTGGAATTGA-3′) prepared using T4 polynucleotide kinase and [γ-32P]-ATP according to the manufacturer's instructions.
7. Following hybridization, wash membrane three times with 2× SSC, 0.1% SDS, wash twice with 0.1×SSC, 0.1% SDS, expose to a PhosphorImager screen and quantitate spots using ImageQuant (GE Healthcare Life Sciences, Pittsburgh, PA) or similar software.
8. The two eluting peaks represent mcm^5U- (initial eluting peak) and mcm^5Um- (later eluting peak) containing isoforms of $tRNA^{[Ser]Sec}$. Pool column fractions containing mcm^5U or mcm^5Um-containing isoforms, precipitate the tRNA with 3 volumes of EtOH.
9. Collect the pooled samples either on Millipore 0.45 μm HA nitrocellulose filters or by centrifugation at 20,000 × *g* for 30 min at 4 °C.
10. Dissolve tRNA samples containing either mcm^5U or mcm^5Um isoforms (or both if purified isoforms are not needed) in RPC5 starting buffer (− $Mg2^+$) and individually load onto the RPC5 column. A 160 mL linear gradient (80 mL of RPC5 buffer D as the initial buffer and 80 mL of RPC5 buffer E) is run at a flow rate of 2 mL/min and 2 mL fractions collected, as described above.

11. The absorbance at A_{260} in the eluted fractions is measured and plotted, and tRNA$^{[Ser]Sec}$ is detected by dot blot hybridization as above.
12. Samples are pooled and collected as given above in **steps 8** and **9** (*see* **Note 9**).
13. tRNA$^{[Ser]Sec}$ isoforms can be further purified by running on a 15% TBE/urea gel in 1× TBE buffer according to standard techniques (*see* **Note 10**) or by binding to immobilized oligonucleotides complementary to tRNA$^{[Ser]Sec}$ (*see* Subheading 3.8), if more highly purified tRNA$^{[Ser]Sec}$ is needed for further examination (e.g., modified base analysis).

3.4 Aminoacylation of tRNA

Add in the following order to a final volume of 25 μL: 10× TAM (2.5 μL); H_2O (to 25 μL final volume); 19 unlabeled amino acid mix (2.5 μL); [3H]serine (2.5 μL); tRNA (0–5 μL); purified SARS (2.0 μL or ~0.5 μg).

1. tRNA is typically added in 0, 0.5, 1, 2, 3, and 5 μL volumes to generate a curve and determine linearity (*see* **Note 11**).
2. Incubate at 37 °C for 15 min.
3. Stop reaction by adding 0.5 mL of cold H_2O and immediately precipitate tRNA with ~5 mL of 10% TCA.
4. Collect each reaction on a Millipore 0.45 μm HA nitrocellulose filter.
5. Dry filter, place in liquid scintillation vial, add scintillation cocktail and count in liquid scintillation counter.

3.5 Quantification of tRNA$^{[Ser]Sec}$ from Cells or Tissues Using Total RNA

1. Isolate total RNA from tissue or cells using Trizol® according to the manufacturer's instructions.
2. Load total RNA (1–2 μg) on 15% TBE/urea gels and electrophorese in 1× TBE buffer according to the manufacturer's instructions.
3. Stain gel with ethidium bromide in 1× TBE buffer, photograph, and transfer to Hybond-N+ membrane in 0.5× TBE buffer at 25 V for 60 min.
4. UV-crosslink the membrane.
5. Hybridize membrane using Quikhyb solution for 3–4 h at 58 °C in a rotating hybridization oven using a [32P]-end labeled oligo probe complementary to the 3′-end of tRNA$^{[Ser]Sec}$ (5′-CGCCCGAAAGGTGGAATTGA-3′) prepared using T4 polynucleotide kinase and [γ-32P]-ATP according to the manufacturer's instructions.
6. Following hybridization, wash membrane three times with wash buffer 1, wash twice with wash buffer 2. Wrap membrane and expose to PhosphorImager. After exposure, strip

membrane by gentle rocking with ~100 mL wash buffer 2 that has been heated to 95–100 °C until radioactivity is no longer detected on the membrane (~5–10 min).

7. Remove the blot from the solution, shake off excess liquid, and wrap the membrane with plastic wrap or reuse immediately. Do not let the membrane dry.
8. Rehybridize stripped membrane with [32P]5′-end labeled $tRNA^{Ser1}$ (5′-CGTAGTCGGCAGGATTCGAA-3′) oligo probe as a loading control.
9. Expose to a PhosphorImager and quantitate spots using ImageQuant (GE Healthcare Life Sciences, Pittsburgh, PA) or similar software.

3.6 Isolation of [75Se]-Labeled Selenocysteyl-$tRNA^{[Ser]Sec}$

1. Grow several flasks of HL-60 cells (~5 g wet weight) in RPMI-1640 media in log phase and in the presence of 300 nM sodium selenite (*see* **Note 12**).
2. Collect cells by centrifugation at 300 × *g* for 5 min at 4 °C, wash with PBS and resuspend in 75 mL of RPMI-1640 media containing 1% fetal bovine serum without sodium selenite.
3. Add 5 mCi of neutralized 75Se and gently shake the cells for 3 h at 37 °C.
4. Add cycloheximide (100 μM final concentration) and incubate cells for an additional 45 min.
5. Collect cells by centrifugation at 300 × *g* for 5 min at 4 °C, wash with ice-cold PBS, and store at −80 °C until ready to use.
6. To frozen cell pellet, add H_2O-saturated phenol and extraction buffer B and immediately vortex vigorously for 3–4 min.
7. Centrifuge the homogenized sample at 4000 × *g* for 10 min.
8. Transfer the upper aqueous layer to a clean polypropylene tube.
9. Extract upper layer with an equal volume of $CHCl_3$. Vortex vigorously for 1–2 min.
10. Centrifuge at 4000 × *g* for 10 min.
11. Take the top layer and transfer to a clean polypropylene tube and load the sample onto a prepared DE52 column (*see* Subheading 3.2) equilibrated with several column volumes of DE52 column equilibration buffer.
12. Add the supernatant obtained following $CHCl_3$ extraction to the column.
13. Wash five to six times with 25 mL of DE52 column equilibration buffer or until absorbance (A_{260}) of eluate is virtually undetectable.
14. Elute tRNA from column with ~12–15 mL of DE52 elution buffer.

15. Dilute eluate to a final concentration of 0.45 M NaCl with ice-cold H_2O containing 0.02 M 2-mercaptoethanol, and load onto an RPC5 column that has been equilibrated with RPC5 starting buffer (+ $Mg2^+$).
16. Purification of [75Se]-labeled Sec-$tRNA^{[Ser]Sec}$ by RPC5 chromatography is carried out using a 160 mL linear gradient (80 mL of RPC5 buffer A as the starting buffer and 80 mL of RPC5 buffer B as the terminal buffer), run at a flow rate of 2 mL/min and 2 mL fractions collected.
17. Wash column with 60 mL RPC5 buffer C and collect eluted fractions.
18. Analyze all fractions for [75Se]-containing tRNA using a gamma counter (*see* Fig. 1c for a representative RPC5 profile).
19. Pool the mcm^5U and mcm^5Um-containing isoforms, precipitate with 2.5 volumes of EtOH, store at −20 °C for 2–3 h, and collect the resulting precipitate by centrifugation at 20,000 × *g* for 30 min at 4 °C.
20. Wash pellet in 75% EtOH, gently dry and dissolve in ~100 μL of H_2O. Measure A_{260} and cpm/μL and store the [75Se] selenocysteyl-$tRNA^{[Ser]Sec}$ at −80 °C after gently blowing nitrogen gas over the sample before sealing the tube.

3.7 Analysis of [75Se]-Labeled Selenocysteyl-$tRNA^{[Ser]Sec}$ by Acid-Urea Gel Electrophoresis

The following method is adapted from [22].

1. Prepare 7% acid-urea gel mix by dissolving 48 g of urea into a solution containing 35 mL of H_2O, 10 mL of 1 M NaOAc, pH 4.5 and 17.5 mL of a 40% acrylamide stock solution. Protected from light, this solution is stable for ~2 weeks.
2. Polymerize gel by adding 1/10 volume of 10% ammonium persulfate and 1/100 volume of TEMED. Pour 7.5 × 10 cm mini-gel with 0.75 mm or 0.5 mm comb and spacers. Allow polymerization to occur for at least 1 h.
3. Set up the gel with acid gel running buffer at 4 °C. Rinse wells thoroughly with running buffer using a 10 mL syringe and 21-gauge needle. Prerun the gel at 12 V/cm (~100 V) for at least 30 min.
4. Concentrate purified [75Se]selenocysteyl-$tRNA^{[Ser]Sec}$ to at least 500 cpm/μL and add 2 μL of tRNA to 2 μL of 2× acid-urea sample buffer and 0.05% (w/v) each of xylene cyanol and bromophenol blue dyes.
5. After the prerun, thoroughly rinse the wells again and load the entire 8 μL sample. Run gel at 100 V constant voltage for about 1 h until the bromophenol blue reaches the bottom of the gel.
6. Disassemble the rig and transfer the gel to gel fixative solution and incubate with gentle rocking for 30 min.

7. Dry the gel and expose to a PhosphorImager screen overnight. The [75Se]selenocysteyl-tRNA$^{[Ser]Sec}$ will appear as a single band migrating approximately two-thirds of the way down the gel, migrating slightly slower than a control [14C]phenylalanyl-tRNA^{Phe} [23].

3.8 Isolation of tRNA$^{[Ser]Sec}$ from Cells and Tissues by Affinity Capture

The following method is adapted from [24].

1. Total tRNA is purified from tissues or cells as described in Subheadings 3.1 and 3.2.
2. Immobilize 3′-biotinylated DNA oligonucleotide (5′-CGCCCGAAAGGTGGAATTGAACCACTCTGTCGCTA-biotin-3′) corresponding to the 3′-end of tRNA$^{[Ser]Sec}$ on streptavidin-conjugated magnetic M-280 Dynabeads as detailed below.
3. For nucleic acid purification, beads must be made RNase-free according to the manufacturer's instructions.
4. Wash desired amount of beads with an equal volume of buffer A using a magnet to pellet the beads.
5. Suspend beads in buffer A to a final concentration of 5 μg/μL.
6. Add 2 μM 3′-biotinylated oligonucleotide in H_2O to an equal volume of Dynabeads in buffer A and incubate at room temperature for 30 min with gentle mixing.
7. Wash the oligonucleotide-coated Dynabeads four to five times using a magnet in 0.5× buffer A.
8. Equilibrate in 6× Saline Sodium Citrate (SSC) buffer.
9. Incubate the oligonucleotide-coated Dynabeads and total tRNA suspended in 6× SSC (binding capacity using single-stranded biotinylated oligonucleotides is ~200 pmol/mg of Dynabeads) for 5 min at 75 °C.
10. Pool oligonucleotide-coated Dynabeads and tRNA and incubate for 5 min at 75 °C.
11. Bind tRNA by incubating at room temperature for 1–2 h with gentle mixing (*see* **Note 13**).
12. Wash the beads three times with 3× SSC, twice with 1× SSC, and three to four times with 0.1× SSC or until the absorbance (A_{260}) of the wash buffer is undetectable.
13. Elute tRNA from beads with 0.1× SSC at 65 °C for 5 min. Perform elution three times in succession and pool eluate.
14. Treat eluate with RNase-Free DNase to remove any bound oligonucleotide.
15. Concentrate eluted tRNA by adding $MgCl_2$ (10 mM final concentration) and precipitate with 3 volumes of EtOH.

16. Centrifuge at 20,000 × *g* for 15 min at 4 °C, wash with 75% EtOH, and centrifuge at 14,000 × *g* for 10 min at 4 °C. Dry pellet and dissolve in molecular biology grade H_2O or buffer of choice needed for further analysis of tRNA$^{[Ser]Sec}$, such as modified base analysis by mass spectrophotometry.

3.9 Identification and Sequencing of tRNA$^{[Ser]Sec}$ from Any Organism

This method can be used to identify and sequence tRNA$^{[Ser]Sec}$ from organisms whose tRNA$^{[Ser]Sec}$ has not been characterized [19–21] and can also be used to verify mutations in tRNA$^{[Ser]Sec}$, such as those that have been shown to occur in human tRNA$^{[Ser]Sec}$ [25]. *See* Fig. 2 for a schematic illustration of this procedure.

1. To isolate tRNA$^{[Ser]Sec}$ for sequencing from organisms whose tRNA$^{[Ser]Sec}$ has not been characterized, grow ~5 g of cells and isolate total tRNA as described in Subheadings 3.1 and 3.2.
2. Label ~0.5 g cells with 0.5 mCi of 75Se, isolate tRNA as described in Subheading 3.6, and combine 75Se-labeled tRNA with the unlabeled tRNA obtained from **step 1**.
3. Chromatograph the pooled tRNAs by RPC5 column chromatography, and collect [75Se]selenocysteyl-tRNA$^{[Ser]Sec}$-containing fractions as described in Subheading 3.6.
4. Polyadenylate the 3′-end of the partially purified tRNA$^{[Ser]Sec}$ using *E. coli* poly(A) polymerase according to the manufacturer's instructions.
5. Denature the polyadenylated tRNAs at 65 °C for 10 min along with primer 5′-TTGAATTCGCATTGAGCACCTGCTTTT TTTTTTTTTTTTTTGG-3′ (100 nM final concentration) and a mixture of dNTPs (0.5 mM final concentration/each), cool on ice, briefly centrifuge, and then transcribe with reverse transcriptase according to the manufacturer's protocol.
6. Add RNase A (50 μg/mL final concentration) and RNase H (1 U/nmol tRNA) and incubate at 37 °C for 20 min to remove any remaining tRNA.
7. Purify cDNA on QIAquick spin columns to separate fragments greater than 100 bp in length from the other components in the reaction mix.
8. Phosphorylate the anchor oligonucleotide, 3′-CCGTTAATTGGGAGTGATTTCT-5′, on its 5′-end using T4 polynucleotide kinase and ATP, and then block its 3′-end by incubating with ddATP and terminal deoxynucleotidyl transferase according to the manufacturer's protocols (*see* **Note 14**).
9. Ligate the phosphorylated, blocked anchor-oligonucleotide, 3′-ddACCGTTAATTGGGAGTGATTTCTp-5′, to the 3′-terminus of cDNA using T4 RNA ligase according to the manufacturer's instructions.

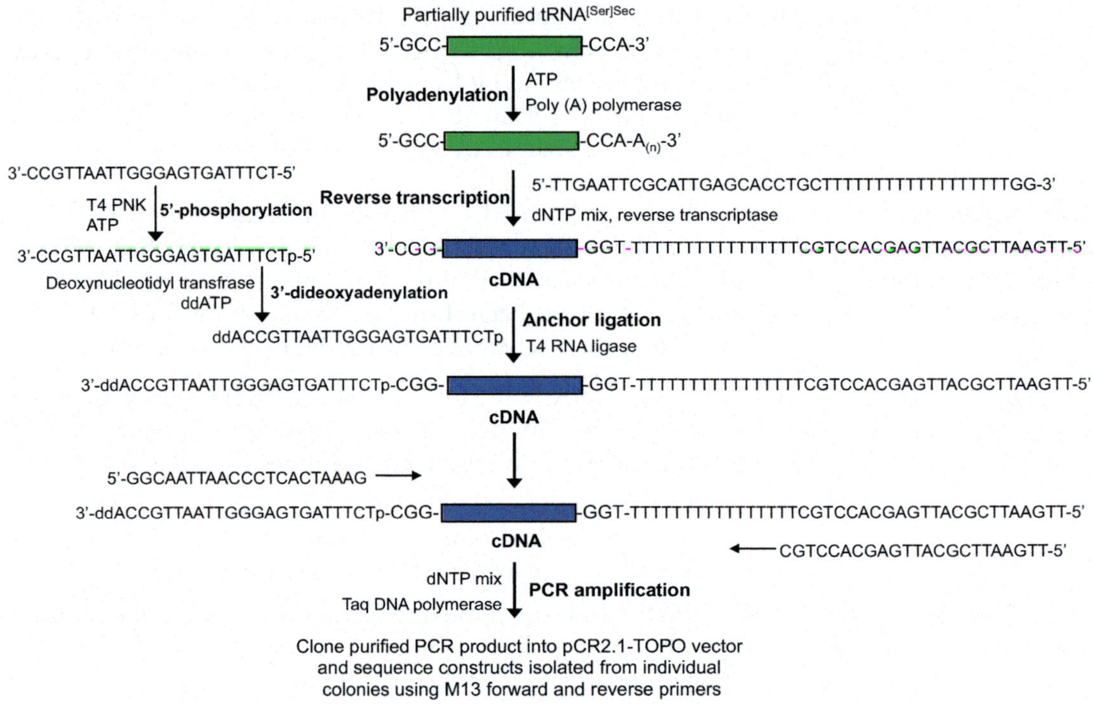

Fig. 2 Schematic diagram demonstrating the procedure for sequencing uncharacterized tRNA$^{[Ser]Sec}$ from any organism as described in the text

10. Purify the resulting cDNA-anchor-oligonucleotide product by loading on a QIAquick spin column to remove protein and unreacted primers.
11. Amplify the purified cDNA-anchor-oligonucleotide by PCR using Taq DNA polymerase according to the manufacturer's instructions with the forward primer 5′-TTGAATTCGCATT-GAGCACCTGC-3′ and reverse primer 5′-GGCAATTAACC CTCACTAAAG-3′ under the following conditions: 4 min at 94 °C; 5 cycles of 94 °C for 30 s, 40 °C for 1 min, 55 °C for 1 min; 30 cycles of 94 °C for 30 s, 50 °C for 1 min, 70 °C for 1 min; 10 min at 72 °C.
12. Purify the PCR products (100–200 bp) on 1% agarose gels.
13. Clone PCR products into the pCR2.1-TOPO vector using the TOPO TA cloning kit according to the manufacturer's instructions.
14. Transform competent DH5α *E. coli* cells with the construct by standard techniques and plate onto LB agar.
15. Isolate individual colonies and purify plasmids from each colony using a plasmid miniprep kit according to the manufacturer's instructions.

16. Sequence the plasmid DNA constructs encoding the cloned tRNAs from both ends using M13 forward (5′-GTAAAACGACGGCCAG-3′) and reverse (5′-CAGGAAACAGCTATGAC-3′) primers by standard techniques.

4 Notes

1. Procedure can be scaled up for isolating RNA from large amounts of tissues or cells and will require larger amounts of materials and equipment.
2. It is very important to keep the tissues/cells frozen during RNA extraction by adding the phenol/buffer mix directly onto the frozen sample and begin homogenizing IMMEDIATELY.
3. DE52 can be stored in molecular biology grade H_2O at 4 °C with a few drops of $CHCl_3$.
4. Typically, use 2 mL of bed volume for 1 g of tissue/cells used.
5. Transfer RNA can be stored at −20 °C, but should be stored at −80 °C, if storing for an extended period of time. Never store tRNA samples in liquid nitrogen as the extremely low temperature will shear off the CCA 3′-terminus.
6. RPC5 chromatography [13] is the only procedure described thus far for separating the two Sec tRNA isoforms. This procedure is currently actively being used in the laboratory of Dr. Byeong J. Lee. Dr. Lee can be contacted regarding any needs for RPC5 chromatography.
7. The amount of starting material will depend upon the investigator's objective. For example, if the objective is to (a) resolve small amounts of labeled $tRNA^{[Ser]Sec}$ isoforms, apply >100,000 cpm of [3H]seryl-tRNA; (b) resolve double-labeled samples, apply ~250,000 cpm and ~25,000 cpm of [3H]- and [14C]seryl-tRNA, respectively; and (c) resolve larger amounts of unlabeled $tRNA^{[Ser]Sec}$, add up to 5–8 mg of sample.
8. It is extremely important to regenerate the RPC5 column with 1.5 M NaCl in the wash buffer to remove any remaining $tRNA^{[Ser]Sec}$ from the column. Washes at concentrations of 1.0 M NaCl [13] will leave residual, late eluting $tRNA^{[Ser]Sec}$, which may affect ensuing runs; especially when carrying out multiple runs, wherein the first run was carried out with labeled material and the following run also involved material labeled with the same isotope yielding an eluted, contaminated product with false results.

9. It is recommended to precipitate the more highly purified mcm^{5}U- and mcm^{5}Um- containing samples in 10 mM $MgCl_2$ with 3 volumes of EtOH and collect by high speed centrifugation as substantial losses of highly purified samples can occur by collection on filters.
10. Although gel electrophoresis was employed numerous times attempting to separate the two isoforms without success, this procedure would still seem to be a fruitful area to explore as a potential technique to achieve their separation.
11. The aminoacylation reaction can be scaled up using total tRNA under limiting conditions to produce large amounts of [3H] seryl-tRNA$^{[Ser]Sec}$ for analysis by RPC5 chromatography.
12. HL-60 cells contain a relatively high level of tRNA$^{[Ser]Sec}$ compared to the total seryl-tRNA population [26], whereas other mammalian cell types contain relatively smaller amounts, but many other cell types and media may be used and can be adapted to this procedure.
13. This method can also be used to prepare a tRNA$^{[Ser]Sec}$-depleted tRNA population for use in in vitro translation systems or other assays requiring such a tRNA$^{[Ser]Sec}$-deficient population. It may take three to four passes over the column to completely remove tRNA$^{[Ser]Sec}$ from the total tRNA population depending on the amount of starting material and beads used (e.g., 200 μg total tRNA and 2 mL of coated beads yields a tRNA population depleted of tRNA$^{[Ser]Sec}$).
14. To achieve the complete sequence of tRNA, the 3′-end of the cDNA is extended by ligation using an anchor-oligonucleotide primer. This technique has also been used to examine total calf liver tRNA fractionated over a RPC5 column for minor tRNA$_{[Ser]Sec}$s that may have a different primary sequence than that encoded in Trsp. However, all tRNA$_{[Ser]Sec}$s detected had the same primary sequence as dictated by Trsp.

Acknowledgments

This work was supported by the Intramural Research Program of the National Institutes of Health, NCI, Center for Cancer Research to DLH, NIH grants CA080946, GM061603 and GM065204 to VNG, DFG Priority Program 1784 to US (Schw914/5-1) to US, Towson University's Jess and Mildred Fisher College of Science and Mathematics to PAT, who is a Jess and Mildred Fisher Endowed Chair of Biological Sciences, and NIH grants GM077073 and HD083616 to PRC.

References

1. Hatfield DL, Choi IS, Ohama T, Jung JE, Diamond AM (1994) Selenocysteine tRNA(Ser) sec isoacceptors as central components in selenoprotein biosynthesis in eukaryotes. In: Burk RF (ed) Selenium in biology and human health. Springer-Verlag, New York, NY, pp 25–44
2. Baron C, Sturchler C, XQ W, Gross HJ, Krol A, Böck A (1994) Eukaryotic selenocysteine inserting tRNA species support selenoprotein synthesis in Escherichia coli. Nucleic Acids Res 22(12):2228–2233
3. Carlson BA, Lee BJ, Tsuji PA, Tobe R, Park JM, Schweizer U, Gladyshev VN, Hatfield DL (2016) Selenocysteine tRNA[Ser]Sec: From nonsense suppressor tRNA to the quintessential constituent in selenoprotein biosynthesis. In: Hatfield DL, Schweizer U, Tsuji PA, Gladyshev VN (eds) Selenium: its molecular biology and role in human health, 4th edn. Springer, New York, NY, pp 3–12
4. Kumaraswamy E, Carlson BA, Morgan F, Miyoshi K, Robinson GW, Su D, Wang S, Southon E, Tessarollo L, Lee BJ, Gladyshev VN, Hennighausen L, Hatfield DL (2003) Selective removal of the selenocysteine tRNA [Ser]Sec gene (Trsp) in mouse mammary epithelium. Mol Cell Biol 23(5):1477–1488
5. Carlson BA, Novoselov SV, Kumaraswamy E, Lee BJ, Anver MR, Gladyshev VN, Hatfield DL (2004) Specific excision of the selenocysteine tRNA[Ser]Sec (Trsp) gene in mouse liver demonstrates an essential role of selenoproteins in liver function. J Biol Chem 279 (9):8011–8017
6. Seeher S, Carlson BA, Miniard AC, Wirth EK, Mahdi Y, Hatfield DL, Driscoll DM, Schweizer U (2014) Impaired selenoprotein expression in brain triggers striatal neuronal loss leading to co-ordination defects in mice. Biochem J 462 (1):67–75
7. Diamond AM, Choi IS, Crain PF, Hashizume T, Pomerantz SC, Cruz R, Steer CJ, Hill KE, Burk RF, McCloskey JA, Hatfield DL (1993) Dietary selenium affects methylation of the wobble nucleoside in the anticodon of selenocysteine tRNA([Ser]Sec). J Biol Chem 268 (19):14215–14223
8. Carlson BA, XM X, Gladyshev VN, Hatfield DL (2005) Selective rescue of selenoprotein expression in mice lacking a highly specialized methyl group in selenocysteine tRNA. J Biol Chem 280(7):5542–5548
9. Hatfield DL, Carlson BA, XM X, Mix H, Gladyshev VN (2006) Selenocysteine incorporation machinery and the role of selenoproteins in development and health. Prog Nucleic Acid Res Mol Biol 81:97–142
10. Carlson BA, Moustafa ME, Sengupta A, Schweizer U, Shrimali R, Rao M, Zhong N, Wang S, Feigenbaum L, Lee BJ, Gladyshev VN, Hatfield DL (2007) Selective restoration of the selenoprotein population in a mouse hepatocyte selenoproteinless background with different mutant selenocysteine tRNAs lacking Um34. J Biol Chem 282(45):32591–32602
11. Roe BA (1975) Studies on human tRNA. I. The rapid, large scale isolation and partial fractionation of placenta and liver tRNA. Nucleic Acids Res 2(1):21–42
12. Hatfield D, Diamond A, Dudock B (1982) Opal suppressor serine tRNAs from bovine liver form phosphoseryl-tRNA. Proc Natl Acad Sci U S A 79(20):6215–6219
13. Kelmers AD, Heatherly DE (1971) Columns for rapid chromatographic separation of small amounts of tracer-labeled transfer ribonucleic acids. Anal Biochem 44(2):486–495
14. Hatfield D, Matthews CR, Rice M (1979) Aminoacyl-transfer RNA populations in mammalian cells chromatographic profiles and patterns of codon recognition. Biochim Biophys Acta 564(3):414–423
15. Carlson BA, Hatfield DL (2002) Transfer RNAs that insert selenocysteine. Methods Enzymol 347:24–39
16. Diamond A, Dudock B, Hatfield D (1981) Structure and properties of a bovine liver UGA suppressor serine tRNA with a tryptophan anticodon. Cell 25(2):497–506
17. Sharp SJ, Stewart TS (1977) The characterization of phosphoseryl tRNA from lactating bovine mammary gland. Nucleic Acids Res 4 (7):2123–2136
18. Silberklang M, Gillum AM, RajBhandary UL (1979) Use of in vitro 32P labeling in the sequence analysis of nonradioactive tRNAs. Methods Enzymol 59:58–109
19. Rao M, Carlson BA, Novoselov SV, Weeks DP, Gladyshev VN, Hatfield DL (2003) Chlamydomonas reinhardtii selenocysteine tRNA[Ser] Sec. RNA 9(8):923–930
20. Novoselov SV, Rao M, Onoshko NV, Zhi H, Kryukov GV, Xiang Y, Weeks DP, Hatfield DL, Gladyshev VN (2002) Selenoproteins and selenocysteine insertion system in the model plant cell system, Chlamydomonas reinhardtii. EMBO J 21(14):3681–3693
21. Shrimali RK, Lobanov AV, XM X, Rao M, Carlson BA, Mahadeo DC, Parent CA, Gladyshev VN, Hatfield DL (2005) Selenocysteine

tRNA identification in the model organisms Dictyostelium discoideum and Tetrahymena Thermophila. Biochem Biophys Res Commun 329(1):147–151

22. Köhrer C, Rajbhandary UL (2008) The many applications of acid urea polyacrylamide gel electrophoresis to studies of tRNAs and aminoacyl-tRNA synthetases. Methods 44 (2):129–138
23. Merrick WC (1979) Assays for eukaryotic protein synthesis. Methods Enzymol 60:108–123
24. Songe-Møller L, van den Born E, Leihne V, Vagbo CB, Kristoffersen T, Krokan HE, Kirpekar F, Falnes PO, Klungland A (2010) Mammalian ALKBH8 possesses tRNA methyltransferase activity required for the biogenesis of multiple wobble uridine modifications implicated in translational decoding. Mol Cell Biol 30(7):1814–1827
25. Schoenmakers E, Carlson B, Agostini M, Moran C, Rajanayagam O, Bochukova E, Tobe R, Peat R, Gevers E, Muntoni F, Guicheney P, Schoenmakers N, Farooqi S, Lyons G, Hatfield D, Chatterjee K (2016) Mutation in human selenocysteine transfer RNA selectively disrupts selenoprotein synthesis. J Clin Invest 126(3):992–996
26. Hatfield D, Lee BJ, Hampton L, Diamond AM (1991) Selenium induces changes in the selenocysteine tRNA[Ser]Sec population in mammalian cells. Nucleic Acids Res 19(4):939–943

Chapter 5

Identification and Characterization of Proteins that Bind to Selenoprotein 3′ UTRs

Eric M. Cockman and Donna M. Driscoll

Abstract

This chapter explains the use of RNase-assisted RNA chromatography. RNA affinity chromatography is a powerful technique that is used to isolate and identify proteins that bind to a specific RNA ligand. The RNA of interest is attached to beads before protein lysates are passed over the column. In traditional RNA chromatography, bound proteins are eluted with high salt or harsh detergent, which can also release proteins that are nonspecifically bound to the beads. To avoid this, a new method was developed in which RNases are used to cleave RNA from the beads, releasing only RNA binding proteins (RBPs) and leaving behind proteins that are bound to the beads (Michlewski and Caceres, RNA 16(8):1673–1678, 2010). This chapter will describe the isolation of proteins that bind specifically to the distal region of the Selenoprotein S 3′ untranslated region (3′ UTR).

Key words RNA-binding protein, Affinity chromatography, Selenoprotein S, Selenoproteins, 3′ UTR, Selenium

1 Introduction

The 3′ UTRs of mRNA transcripts are known to bind and interact with RBPs. Many of these RBPs recognize and specifically bind to sequences and/or structures within 3′ UTRs. Once bound, the RBPs can mediate regulation of the mRNA transcript at multiple levels including RNA processing, transcript stability, and translation. To identify 3′ UTR binding proteins, RNA affinity chromatography approaches can be performed that use a 3′ UTR sequence of interest as a ligand for protein binding. This RNA ligand is fixed to agarose or sepharose beads by chemical modification with cyanogen bromide or biotinylation. The RNA/bead complex is then incubated with protein lysates from cells or tissues. Common techniques to elute bound proteins include boiling the beads in sodium dodecyl sulfate polyacrylamide gel electrophoresis (SDS-PAGE) loading dye or washing the beads with high salt buffers. Our lab has had success using many of these RNA affinity chromatography

Laurent Chavatte (ed.), *Selenoproteins: Methods and Protocols*, Methods in Molecular Biology, vol. 1661,
DOI 10.1007/978-1-4939-7258-6_5, © Springer Science+Business Media LLC 2018

approaches [1–4]. However, these elution methods share one common drawback: nonspecific proteins that bind to the beads often appear in the eluted fraction. In fact, the nonspecific proteins that are detected can vary depending on the type of beads, washing conditions, and elution method [5].

To avoid elution of nonspecific proteins, a new method called RNase-assisted RNA chromatography was developed [6]. This technique uses RNases to cleave the RNA from the beads, releasing only RBPs and leaving behind nonspecific proteins that are bound to the beads. The original method used small RNA oligomers as ligands. Our lab has modified the protocol to identify proteins that bind to longer 3′ UTRs. We have shown that the distal part of the Selenoprotein S 3′ UTR contains sequences that inhibit insertion of selenocysteine [7]. To identify proteins that bind the distal 3′ UTR, two RNA ligands were used. The region spanning nucleotides 335–574 of the Selenoprotein S 3′UTR contains the Selenocysteine Insertion Sequence (SECIS), which is required for selenocysteine insertion, as well as the inhibitory sequences. This region is referred to as SECIS-End and is the RNA ligand that will bind the proteins of interest. As a negative control, an RNA consisting of nucleotides 335–443, which contain only the SECIS, was used as a deletion of the inhibitory sequences.

This approach takes multiple steps. The first step involves preparation of protein lysates and the RNA ligands, followed by the treatment and binding of RNA to the beads, the binding of proteins to the RNA ligand, and, finally, elution of RBPs using RNases.

2 Materials

1. Nuclease-free water.
2. Jurkat E6-1 cells (ATCC).
3. 1× Phosphate-Buffered Saline, pH 7.4 (PBS).
4. Cytoplasmic extraction buffer: 10 mM HEPES, 60 mM KCl, 1 mM EDTA, 0.075% (v/v) NP40, 1 M dithiothreitol, 1 mM PMSF; pH 7.3.
5. pGEM3Zf (+) vector (Promega).
6. Restriction enzymes.
7. Agarose.
8. 1× TAE buffer: 40 mM Tris-acetate, 1 mM EDTA, pH 7.2.
9. Ethidium bromide (10 mg/mL).
10. RNA or DNA molecular weight markers.
11. 6× loading dye: 50% (w/v) glycerol, 1 mM EDTA, 0.04% (w/v) bromophenol blue, 0.04% (w/v) xylene cyanol.
12. Phenol:chloroform:isoamyl alcohol (25:24:1).

13. Micro Bio-Spin Columns with Bio-Gel 30 in Tris Buffer (BioRad).
14. RiboMAX Large Scale RNA Production System-T7 (Promega).
15. Adipic acid dihydrazide agarose beads, 50% slurry (Sigma).
16. 0.1 M NaOAc, pH 5.5.
17. 1× coupling buffer: 0.1 M NaOAc, 5 mM Sodium periodate ($NaIO_4$), pH 5.5 (store in black tube).
18. 2× coupling buffer: 0.2 M NaOAc, 10 mM Sodium periodate ($NaIO_4$), pH 5.5 (store in black tube).
19. 100% Ethanol.
20. 3 M NaOAc, pH 5.5.
21. Glycogen (5 mg/mL).
22. 75% Ethanol.
23. Pierce spin columns-screw cap (Thermo Scientific).
24. Low protein binding collection tube (2.0 mL) (Thermo Scientific).
25. 2 M KCl.
26. Binding buffer: 250 mM KCl, 10 mM HEPES, 1 mM EDTA, 1 mM dithiothreitol, 1 % (v/v) NP40, pH 7.3.
27. RNase A/T1 cocktail (500 U/mL RNase A; 20,000 U/mL RNase T1).
28. RNase buffer: 10 mM Tris–HCl, 1 mM $MgCl_2$, 40 mM NaCl, pH 7.2.
29. Coomassie Stain.
30. Laboratory Equipment: tanks and power supplies for running agarose and SDS-PAGE gels, rotator, sample shaker, 37 °C water bath, refrigerated microcentrifuge, and nanodrop, spectrophotometer, or equivalent.

3 Methods

The following method explains the steps to synthesize Selenoprotein S SECIS and SECIS-End RNAs by in vitro transcription, to prepare the RNA and adipic acid dihydrazide agarose beads for protein purification, and to isolate and identify proteins binding to the distal 3′ UTR of Selenoprotein S. Before starting, it is important to note that the prevention of RNase contamination is paramount. Many labs use diethylpyrocarbonate-treated water; however, we find that a proper technique can eliminate this need. Frequent changing of gloves, use of nuclease-free water, and sterile pipette tips with filters are essential for reducing the chance of RNase contamination. All the steps are performed at room temperature unless otherwise specified.

3.1 Preparation of Jurkat E6-1 Cell Cytoplasmic Extract

1. Grow Jurkat E6-1 cells in suspension in 50 mL of RPMI 1640 media supplemented with 10% FBS (*see* **Notes 1** and **2**).
2. When cells reach a density of 1–2 × 10^6 cells/mL, pellet cells by centrifuging at 1500 × *g* for 5 min.
3. Remove the supernatant and resuspend the cell pellet in 1 mL of cold PBS. Transfer the cells to a 1.5 mL microcentrifuge tube and centrifuge at 1500 × *g* for 5 min at 4 °C.
4. Remove the supernatant and determine the packed cell volume (PCV) of your cells. To do this, take multiple microcentrifuge tubes and add known volumes of water as standards (50–200 μL). Compare the volumes of water to the cell pellets to approximate PCV. For 10^7 Jurkat E6-1 cells, a typical PCV is approximately 80–100 μL.
5. Resuspend the cells by adding 5× PCV of cytoplasmic extraction buffer. Mix by pipetting. Incubate on ice for 3 min.
6. Centrifuge cells at 1000 × *g* for 5 min at 4 °C.
7. Using a micropipette, carefully transfer the supernatant to a new 1.5 mL microcentrifuge tube. This is the cytoplasmic fraction (*see* **Note 3**).
8. Centrifuge the cytoplasmic fraction at 20,000 × *g* for 10 min at 4 °C to remove any remaining cellular debris.
9. Transfer the supernatant to a new microcentrifuge tube and add glycerol to a final concentration of 20%. Mix gently by pipetting.
10. Measure the protein concentration of the cytosolic fraction. Expect approximately 1.5 mg of cytosolic extract from 10^7 Jurkat E6-1 cells.

3.2 Plasmid Linearization

The pGEM3Zf (+) plasmid is a commonly used cloning vector containing the T7 and SP6 promoters. The SECIS and SECIS-End regions of the Selenoprotein S 3′ UTR described above were cloned into the pGEM3Zf (+) plasmid under the T7 promoter using the HindIII and NotI restriction sites found in the polylinker. When comparing the proteins binding to the two RNA ligands, we are interested in proteins that bind only to the SECIS-End RNA. Other deletions or mutant RNAs could be used in place of the SECIS RNA (*see* **Note 4**).

1. Linearize 10 μg of the SECIS and SECIS-End plasmids using the NotI restriction enzyme (*see* **Note 5**). This cuts the plasmid at only one site in the polylinker portion of the pGEMZf (+) plasmid (*see* **Note 6**).
2. Confirm linearization of your plasmid by running the linearized and non-linearized plasmids on a gel. Add 2 μL of 6× loading dye to 300 ng of cut and uncut plasmid in a final volume of 12 μL. Run the samples on a 1% agarose gel in 1×

TAE buffer at 100 V for 25 min. Linearized plasmid will generally run as a single band and uncut plasmid will run as two bands. The upper band corresponds to a relaxed plasmid and the lower band is the supercoiled plasmid. Linear plasmid usually runs between the relaxed and supercoiled plasmid. If there is still uncut plasmid in your linearized sample, add more restriction enzyme to complete the digestion and repeat this step.

3. Clean up the cut DNA by adding an equal volume of phenol: chloroform:isoamyl alcohol to the reaction mix. Vortex the samples vigorously and centrifuge for 5 min at 12,000 × *g*. As the sample is spinning, prepare the spin columns (Micro Bio-Spin Columns with Bio-Gel 30 in Tris Buffer) according to the manufacturer's instructions.
4. Add the aqueous (upper) phase of your sample to the spin column. The maximum volume for the columns is 70 μL. For larger volumes, the sample can be divided between two or more spin columns. Finish processing the sample by following the manufacturer protocol.
5. Measure the optical density of your cut DNA at 260 nM and 280 nM using a nanodrop or spectrophotometer. The 260:280 ratio for DNA should be between 1.8 and 2.0 (*see* **Note** 7).

3.3 In Vitro Transcription of RNA

RNA is made for both the SECIS and SECIS-End constructs using the Promega RiboMAX T7 kits. The reaction mixes are set up at room temperature and it is important to make sure that any precipitates in the 5× T7 buffer are completely dissolved.

1. The Promega RiboMAX T7 kit contains 100 mM stocks of each rNTP. Make an rNTP mixture containing 25 mM for each of UTP, GTP, CTP, and ATP.
2. The reaction mix consists of 20 μL 5× T7 buffer, 30 μL 25 mM rNTPs, 10 μg linearized DNA template (up to 40 μL), and 10 μL T7 enzyme. The final volume is brought to 100 μL with nuclease-free water. Vortex the reaction gently and pulse down briefly using a microcentrifuge (*see* **Note 8**).
3. Incubate the mixtures in a 37 °C water bath for 2 h.
4. Add 10 μL (1 μL for each μg of DNA template) of the DNase I provided in the Promega RiboMAX T7 kit and incubate at 37 °C for 20 min.
5. Add 110 μL of phenol:chloroform:isoamyl alcohol to the reaction mix and vortex vigorously before centrifuging at 12,000 × *g* for 5 min. As the samples are spinning, prepare the spin columns (Micro Bio-Spin Columns with Bio-Gel 30 in Tris Buffer) according to the manufacturer's protocol.

6. Add the aqueous phase of the sample to the spin column and finish processing the RNA following the manufacturer's protocol.
7. Measure the RNA at 260 nM and 280 nM using a nanodrop or spectrophotometer. The 260:280 ratio for high-quality RNA is between 2.0 and 2.2 (*see* **Note 9**). The yield for a 100 μL transcription reaction should be between 1.5 and 2 mg of RNA (*see* **Note 10**).
8. Add 300 ng of each RNA to a new 1.5 mL microcentrifuge tube. Add 2 μL of 6× loading dye and bring the final volume to 12 μL with nuclease-free water.
9. Heat the RNA at 65 °C for 5 min.
10. Run the RNA on a 2% agarose gel at 100 V for 25 min. Confirm that the size of your RNA is full length and of the correct size (*see* **Notes 11** and **12**).

3.4 Preparation of Adipic Acid Dihydrazide Agarose Beads and Sodium Periodate Treatment of RNA

1. Transfer 250 μL of beads (500 μL of the 50% slurry) to a 1.5 mL microcentifuge tube for each RNA sample. Pellet the beads gently by centrifuging at 1000 × *g* for 1 min. Remove the supernatant and add 1 mL of 0.1 M NaOAc, pH 5.5. Repeat this wash step five times.
2. After the final wash, resuspend the beads in 400 μL of 0.1 M NaOAc, pH 5.5.
3. Dilute 100 μg of each RNA in 150 μL of nuclease-free water (wrap the tubes in aluminum foil). Add an equal volume of 2× coupling buffer. Incubate for 1 h in the dark.
4. After treatment, precipitate the RNA. Add 3× volume of 100% ethanol, 1:10 volume of 3 M NaOAc, pH 5.5, and 2 μL of glycogen (5 mg/mL). Invert the tubes to mix and place them on dry ice for 30 min.
5. After precipitation, centrifuge at 12,000 × *g* for 30 min at 4 °C. The RNA pellet will appear as a white or glassy residue at the bottom of the tube.
6. Carefully remove the supernatant with a pipette and wash the pellet with 1 mL of 75% ethanol. Centrifuge at 12,000 × *g* for 10 min at 4 °C. Remove the supernatant and repeat the 75% ethanol wash. After the final wash, carefully remove the supernatant and let the pellet air dry for 5–10 min.
7. Resuspend the RNA pellet in 60 μL of 0.1 M NaOAc, pH 5.5.
8. Add 60 μL of washed bead slurry to the treated RNA.
9. Incubate at 4 °C overnight rotating slowly to allow the RNA to attach to the beads (*see* **Note 13**).

3.5 Incubation of Lysates with RNA and Beads

1. Transfer the mixture of beads and RNA to a Pierce spin column. Put the spin column into a collection tube. Centrifuge at 1000 × *g* for 1 min. Save the flowthrough.
2. Wash the beads four times with 500 μL of 2 M KCl. After each wash, centrifuge at 1000 × *g* for 1 min and collect the flowthrough.
3. Measure the absorbance at 260 nm of each of the flowthroughs to determine how much RNA bound to the adipic acid dihydrazide beads. Ideally, 80% of the RNA will have bound (*see* **Note 14**).
4. Wash the beads twice with 500 μL of cytoplasmic extraction buffer.
5. The Pierce spin column-screw cap kit includes small plastic plugs that fit into the bottom of the spin columns. Place the plugs securely into the bottoms of the spin columns that contain the washed beads (*see* **Note 15**).
6. Add 500 μg of Jurkat E6-1 cytosolic extract at a 1 μg/μL concentration (diluted in cytosolic extraction buffer). Adjust the final concentration of KCl from 60 mM to 120 mM and the concentration of NP40 from 0.075% to 1% (*see* **Note 16**).
7. Incubate the beads with the protein lysates by rotating for 1 h (*see* **Notes 17** and **18**).
8. After incubation, centrifuge the samples at 1000 × *g* for 2 min. Save the flowthrough as this will contain the proteins that did not bind to the RNA/bead complex. The unbound fractions can be used later to optimize the binding of candidate proteins (*see* **Note 19**).
9. Wash the beads four times with binding buffer. During each wash, place the plugs into the spin columns, and rotate the columns slowly for 5 min. Then remove the wash by centrifuging at 1000 × *g* for 1 min (*see* **Note 20**).
10. Add 10 μL of RNase A/T1 cocktail (Ambion) to 110 μL of RNase buffer.
11. Making sure that the plugs are securely in place, resuspend the beads in 40 μL of the RNase A/T1 cocktail and buffer solution. Incubate at 37 °C for 20 min shaking at 600 RPM.
12. Remove the plugs and place the spin columns into new 1.5 mL microcentrifuge tubes. Centrifuge for 5 min at 1000 × *g* to collect the eluate, which contains the proteins that were bound to the RNA ligands. Eluates can be stored at −70 °C (*see* **Note 21**).

3.6 Coomassie Staining of gel and Identification of Bands of Interest

1. Take 15 μL of the eluate and run it on a 12% SDS-PAGE gel at 200 V for 60 min (*see* **Note 22**).

2. Place the gel in a staining tray and submerge it in Coomassie Blue stain for 15 min. Bands should slowly start to appear (*see* **Note 23**).
3. After staining, destain using distilled water. As the excess stain begins to leave the gel, the protein bands will become more apparent.
4. By putting the gel over a lightbox, it is easy to look for differential bands between the two RNA ligands. These are the bands that will be analyzed by mass spectrometry (*see* **Note 24**).

3.7 Analysis of Mass Spectrometry Results

The resulting gel from our experiment had four bands, ranging from 35 kDa to 70 kDa, that were present in the SECIS-end eluate, but not the SECIS eluate. Analysis of these bands by mass spectrometry revealed that multiple different proteins were present in each band. To narrow down the number of proteins of interest, we set up two criteria. The first criterion was that the protein had to be a known RNA binding protein. The second criterion was that the protein has to be present only in the SECIS-end lane and absent in the SECIS lane based on spectral counts from the mass spectrometry analysis. Confirmation of the mass spectrometry results was accomplished by running a Western blot on the eluates and probing using an antibody against the candidate proteins.

4 Notes

1. Although we used Jurkat E6-1 cells, other cell lines can be used for the purification of RBPs.
2. Because normal tissue culture media is lacking in selenium, the RPMI 1640 media, in which Jurkat E6-1 cells were grown, was supplemented with 60 nM sodium selenite.
3. Pilot experiments were performed using whole cell lysates, nuclear, and cytosolic fractions of Jurkat E6-1 cells for purification. We found that using the cytosolic fraction of Jurkat E6-1 lysates was the most optimal for the isolation of RBPs. The method of lysis and subcellular fraction can be changed to suit your needs.
4. When deciding what RNA ligands to use, keep in mind that short RNA ligands are likely to bind fewer nonspecific RBPs. Using small RNA fragments will reduce the number of proteins that appear in the final Coomassie stained SDS-PAGE gel.
5. The volume of restriction enzyme should not exceed 10% of the total reaction volume to avoid nonspecific cleavage of the plasmid.
6. Plasmids are designed in such a way that restriction sites in the polylinker regions are only present once in the entire plasmid.

However, the sequence that is cloned into the plasmid may contain restriction sites. It is important to make sure that your cloned sequence does not contain a site that will be recognized by the restriction enzyme you choose to use.

7. If the 260:280 ratio is below 1.8, repurify the DNA by repeating the phenol:chloroform:isoamyl alcohol step.
8. Reactions should be set up at room temperature and care should be taken to make sure all reagents are properly resuspended. If the precipitate from the 5× T7 buffer is not completely dissolved, the RNA yield will be low.
9. If the 260:280 ratio for RNA is below 2.0, repurify the RNA using the phenol:chloroform:isoamyl alcohol and Micro Bio-Spin Columns.
10. High RNA yield from the transcription reaction is important. In our experience, low RNA yield is often caused by the rNTPs degrading. Making a fresh rNTP mix can solve this problem. Other factors that can decrease RNA yield are if the 5× T7 reaction buffer has precipitated or if the reaction mix is not set up at room temperature.
11. Because RNA is single stranded, it will run faster than DNA markers. While visualizing the RNA on a gel, you may observe multiple bands, which suggests that the RNA is highly structured and resistant to heating at 65 °C. To melt the structure, heat the RNA at 95 °C for 2 min before running it on a 2% agarose gel.
12. RNA should run in a crisp, single band. If there is a smear beneath one of your bands, there may be RNase contamination. In this case, make all new buffers and clean the work area before linearizing the plasmid again and repeating the transcription step. It is absolutely critical that the RNA is full length to ensure good purification of RBPs.
13. The mix of beads and RNA typically will not move when the tubes are rotating. Because of this, make sure that the beads are suspended in the solution and rotate the tubes as slowly as possible. This will keep the beads from settling and allow the RNA to bind.
14. If a low percentage of RNA binds, the amount of agarose beads in the reaction should be increased. If almost all of the RNA bound, the beads are not saturated with RNA. In this case, more RNA should be added during the RNA coupling step.
15. Before inserting the plugs, be sure to inspect them. If the plug is chipped or worn down, do not use it, as leaks are likely to occur.
16. The concentrations of salt, detergent, and protein are all factors that can be varied to decrease nonspecific RNA binding proteins from appearing in the eluted fraction.

17. The speed of the rotation should be fast enough to ensure that the mixture in the tubes is moving back and forth.
18. The binding can also be performed at 4 °C overnight.
19. The unbound fractions can be used to determine if your protein of interest is binding to the beads. For the Selenoprotein S 3′ UTR, we used UV-crosslinking to show that our band of interest disappeared after the binding step. This also allowed us to optimize binding conditions to increase binding of our protein while at the same time decreasing the binding of nonspecific RNA binding proteins. Not all proteins will crosslink to the RNA, so other methods such as RNA electrophoretic mobility shift assays (REMSAs) can be used.
20. Increasing the volume and time of the washes can remove nonspecific proteins from the beads.
21. All 40 μL of the RNase solution will not be recovered. Expect approximately 30–35 μL of recovered eluate.
22. Depending on the size of the proteins of interest, you may need to change the gel percentage and the amount of time you run the gel to increase resolution of your bands.
23. Coomassie staining is often not sensitive enough to detect low abundance proteins. In this case, silver stain can be used. However, it is important to make sure that the silver stain protocol is compatible with mass spectrometry.
24. Ideally, there will be multiple differential bands present on the gel. However, if there is an abundance of nonspecific RBPs that bind to both RNA ligands, the differential bands will be swamped out and both the lanes on the gel will look identical. In this case, there are multiple steps that can be optimized. Changing the salt and detergent concentration of the binding and wash buffers or adjusting the ratio of RNA ligand to protein lysate can help decrease nonspecific RBP binding. The protein lysates can also be precleared with the negative control RNA ligand. By incubating your protein lysates with the negative control RNA, the nonspecific RBPs can be removed from the protein lysate. The precleared lysate can then be used for RNase-assisted affinity chromatography.

Acknowledgments

We wish to thank Alexis Polce and Dr. Vivek Narayan for critical review of the manuscript. This work was supported by NIH grants # R01DK07859 and R01DK107426.

References

1. Copeland PR, Driscoll DM (1999) Purification, redox sensitivity, and RNA binding properties of SECIS-binding protein 2, a protein involved in selenoprotein biosynthesis. J Biol Chem 274 (36):25,447–25,454
2. Copeland PR, Fletcher JE, Carlson BA et al (2000) A novel RNA binding protein, SBP2, is required for the translation of mammalian selenoprotein mRNAs. EMBO J 19(2):306–314
3. Budiman M, Bubenik J, Miniard A et al (2009) Eukaryotic initiation factor 4a3 is a selenium-regulated RNA -binding protein that selectively inhibits selenocysteine incorporation. Mol Cell 35:479–489
4. Miniard A, Middleton L, Budiman NE et al (2010) Nucleolin binds to a subset of selenoprotein mRNAs and regulates their expression. Nucleic Acids Res 38:4807–4820
5. Mellacheruvu D, Wright Z, Couzens A et al (2013) The CRAPome: a contaminant repository for affinity purification-mass spectrometry data. Nat Methods 10:730–736
6. Michlewski G, Caceres JF (2010) RNase-assisted RNA chromatography. RNA 16 (8):1673–1678
7. Bubenik J, Miniard A, Driscoll D (2013) Alternative transcripts and 3′ UTR elements govern the incorporation of selenocysteine into selenoprotein S. PLoS One 8(4):e62102

Chapter 6

Specific Chemical Approaches for Studying Mammalian Ribosomes Complexed with Ligands Involved in Selenoprotein Synthesis

Olga Kossinova, Alexey Malygin, Alain Krol, and Galina Karpova

Abstract

Chemical approaches are very powerful tools for investigating the molecular structure and architecture of large ribonucleoprotein complexes involving ribosomes and other components of the translation system. Application of RNA nucleotide-specific and cross-linking reagents of a broad specificity range allows the researcher to obtain information on the sites of ligand binding to the ribosome and to each other as well as on the RNA rearrangements caused by the binding. Here, we describe specific chemical approaches including chemical probing and site-directed or bifunctional reagent-mediated cross-linking, which have been used for exploring the mechanism of selenocysteine insertion into a polypeptide chain by mammalian ribosomes.

Key words Selenocysteine, Selenoprotein, Ribosome, RNA, Cross-linking, Chemical modification, Chemical probing

1 Introduction

Selenoproteins constitute a class of various oxidation-reduction proteins containing the amino acid selenocysteine (Sec) [1, 2]. Sec incorporation into selenoproteins is attracting a lot of interest because it is achieved by recoding of an internal, in-frame UGA codon otherwise read as a stop codon. A network of protein-protein and RNA-protein interactions enables the ribosome not to stop at the UGA Sec codon. This network comprises a variety of dedicated proteins and RNAs, among which are the SECIS stem-loop in the 3′UTR of selenoprotein mRNAs and the SECIS binding protein 2 (SBP2) that recruits the specialized translation elongation factor EFSec bound to the Sec-tRNA^{Sec} [3–8]. Other factors such as ribosomal protein eL30, nucleolin, and translation factor eIF4a3 have also been suggested to play a role [9–11]. The molecular interactions leading to synthesis of selenoproteins in eukaryotes have been thoroughly studied in vitro using various cross-

Laurent Chavatte (ed.), *Selenoproteins: Methods and Protocols*, Methods in Molecular Biology, vol. 1661,
DOI 10.1007/978-1-4939-7258-6_6, © Springer Science+Business Media LLC 2018

linking assays and different approaches based on chemical modifications of proteins and RNAs. Contacts between the SECIS and SBP2, ribosomal protein eL30, nucleolin, and eIF4a3 have been identified by means of direct UV-induced cross-linking of the isolated radiolabeled SECIS probe to components of cellular extracts [9–12]. Attempts to reveal the interactions of the SECIS with several ribosomal proteins in the course of the selenocysteine insertion on the mammalian ribosome have been performed with the use of a minimal selenoprotein mRNA whose SECIS-element bore either statistically introduced 4-thiouridines or aminoallyl-containing uridines bearing perfluorophenylazido groups [13]. The SBP2 binding site on the ribosome, and more precisely on the 28S rRNA, as well as ribosomal regions that undergo conformational rearrangements induced by the SBP2 binding have been identified by chemical modification and cross-linking approaches [14, 15]. Thus, chemical approaches, including site-directed cross-linking, are highly informative to study the selenocysteine insertion machinery, especially in the absence of X-ray or cryo-EM structure models of ribosomes complexed with SBP2 or SECIS.

This chapter presents protocols based on site-directed cross-linking and specific chemical modification, which we routinely use in the studies of the translation process in mammals, including the mechanism of selenocysteine insertion into growing polypeptide chain. In particular, cross-linking approaches utilizing photoactivatable RNA derivatives or bifunctional reagents have been applied to obtain information on the architecture of ribosomal complexes with the above ligands [13, 15]. Changes in the 28S rRNA structure occurring in response to the SBP2 binding have been established by chemical probing of RNA with the use of specific probes [15]. Such probes were hydroxyl radicals, which give general information on the accessibility of the backbone at nucleotide resolution, and base-specific reagents, kethoxal, and dimethyl sulfate (DMS), which provide insight into the base pairing status since they modify bases in single strands. All these reagents are complementary in ensuring data on the RNA structure in complexes under investigation and they have been applied in parallel experiments to get the whole picture of the ribosomal complex architecture.

2 Materials

Prepare all solutions, buffers, and reaction mixtures using Milli-Q water (18 MΩ cm at 25 °C) and analytical grade reagents. Keep all the solutions and buffers at 4 °C, unless otherwise indicated. The protocols described here assume the use of standard laboratory equipment: water bath or heating block adapted to 30, 37, 42, 80, and 90 °C, refrigerated centrifuge adapted for 1.5 ml microtubes, vortex mixer, apparatuses for electrophoresis and western-blotting, gel dryer, counters of ^{32}P radioactivity, gel imaging system with appropriate software, gradient fractions collector with optical density detector, UV cross-linker as well as routine chemicals such as EtOH, acetone, HOAc, chloroform, phenol, *N,N,N′,N′*-tetramethylethylenediamine (TEMED) SDS, $MgCl_2$, EDTA, urea, NaOAc, ammonium persulfate, etc.

2.1 Ribosomal Complex Assembly from Purified Components

1. 5× Buffer A: 0.6 M Hepes-KOH, 0.5 M KCl, 10 mM $MgCl_2$, 5 mM DTT, pH 7.5.
2. Ribosomes: purified 60S or 80S ribosomes, dissolved in Milli-Q water or Buffer A, usually 5–10 μM in stock (*see* **Note 1**).
3. SBP2: recombinant Talon-resin purified SBP2, bearing a C-terminal His-Tag, usually 5–10 μM in Buffer A with 20% glycerol.
4. Ultracentrifugation in sucrose gradient: sucrose solution, usually 45–55% (w/v); gradient former.
5. Refrigerated ultracentrifuge (Beckman) with SW41 rotor.
6. Precipitation of ribosomal complex: trichloroacetic acid (TCA) 100% (w/v).
7. Sodium dodecyl sulfate-polyacrylamide gel electrophoresis (SDS-PAGE).
8. Gel loading buffer 1 (GLB1): 0.125 M Tris–HCl, 20% glycerol; 4% SDS; 2% 2-mercaptoethanol; 0.04% bromophenol blue, pH 6.8.
9. 30% acrylamide-bisacrylamide (29:1).
10. Resolving gel buffer: 1.5 M Tris–HCl, pH 8.8.
11. Stacking gel buffer: 0.5 M Tris–HCl, pH 6.8.
12. Shaking platform.
13. Coomassie BB R250 staining solution: 0.25% (w/v) Coomassie BB R250, 20% EtOH, 7% HOAc).
14. Destaining solution: 20% EtOH and 7% HOAc.

2.2 SBP2-Ribosome Diepoxybutane Cross-linking

2.2.1 Diepoxybutane Cross-linking Reaction

1. Diepoxybutane (*see* **Note 2**).
2. 5× Buffer A, 1 M Tris–HCl, pH 7.5.

2.2.2 Analysis of the Cross-link Distribution Between Ribosomal Subunits

1. Dissociation of ribosomes into subunits: 3 M KCl.
2. 5× Buffer B: 0.6 M Hepes-KOH, 2.5 M KCl, 10 mM $MgCl_2$, 5 mM DTT, pH 7.5.
3. Ultracentrifugation in sucrose gradient in Buffer B: *see* Subheading 2.1, **items 4 and 5**.
4. Nitrocellulose membrane for western blotting.
5. PBS tablets.

6. 5% (w/v) fat-free milk, 0.01% Triton X100 in 1× PBS.
7. 0.5% (w/v) fat-free milk, 0.01% Triton X100 in 1× PBS.
8. Rabbit polyclonal anti-SBP2 antibodies.
9. Anti-rabbit HRP-conjugated secondary antibody.
10. X-ray film or Hyperfilm ECL or Chemiluminescence imager (BioRad).

2.2.3 Analysis of the Cross-link Distribution Between Ribosomal RNA and Proteins

1. Analysis of the cross-link distribution between ribosomal proteins: SDS-PAGE, *see* Subheading 2.1, **items 7–14**; dot-blot analysis, *see* Subheading 2.2.2.
2. Separation of ribosomal complex into rRNA and total proteins by ultracentrifugation in sucrose gradient (*see* Subheading 2.1) in Buffer C.
3. 5× Buffer C: 0.1 M Tris–HCl, 10 mM EDTA, 0.5% SDS, pH 7.5.
4. 10 mg/ml RNase A.

2.2.4 SBP2-Cross-linked rRNA Sample Preparation for Reverse Transcription Analysis

1. 1 mg/ml Proteinase K.
2. Dry ice bath.

2.2.5 Primer Labeling

1. DNA oligonucleotide primer 10 μM; [γ-^{32}P]-ATP, 3000 Ci/mmol (Perkin Elmer).
2. T4 polynucleotide kinase (PNK) 10 U/μl.
3. 10× T4 PNK buffer.
4. Lead containers to store radioactive samples; MicroSpin G25 columns (GE Healthcare); protective shields, goggles and gloves.

2.2.6 Reverse Transcription Reaction on Cross-linked rRNA

1. AMV reverse transcriptase 10 U/μl (RT).
2. 5× AMV RT buffer.
3. [^{32}P]-labeled primer.
4. 2 mM dNTP mix.
5. 1 mM ddATP, ddCTP, ddGTP, ddTTP.
6. Gel loading buffer 2 (GLB2): 0.1% bromophenol blue and 0.1% xylene cyanol in deionized formamide.

2.2.7 Separation of the Reverse Transcription Products in Sequencing Gel and Their Imaging

1. 40% acrylamide-bisacrylamide (19:1).
2. 10× TBE buffer: 1 M Tris, 0.9 M boric acid, 0.01 M EDTA, pH 8.9.
3. Whatmann 3MM paper.
4. Transparent wrap.

2.3 Chemical Probing of SBP2-Ribosome Complex

1. Solution I: 12 mM $Fe(NH_4)_2(SO_4)_2$, 62 mM ascorbic acid, 25 mM EDTA-KOH, 0.6% H_2O_2, pH 7.5.
2. 0.1 M thiourea.
3. DMS (*see* **Note 3**).
4. 2-mercaptoethanol.
5. Kethoxal solution (GE Healthcare).
6. 0.5 M potassium borate.

2.4 Ribosomal Complex Assembly and Cross-linking in Cell-Free Protein Synthesizing System

2.4.1 Synthesis of Photoreactive RNA Derivative by Splint-Aided Ligation

1. 5× T7 RNA polymerase buffer: 0.6 M Hepes-KOH, 100 mM $MgCl_2$, 10 mM spermidine and 0.2 M DTT, pH 7.5.
2. 0.1 M rNTPs.
3. DNA template: PCR product or linearized plasmid at 10 mg/ml.
4. RNasin 40 U/μl.
5. 10 mg/ml T7 RNA polymerase.
6. 0.1 M GMP.
7. 3000 Ci/mmol [α-^{32}P]-ATP.
8. 0.1 M 4-thiouridine (s4UTP) or other photoreactive NTP derivatives.
9. Mini spin G15 columns.
10. 50 μM DNA oligonucleotide for splint-aided ligation.
11. 10 U/μl T4 DNA ligase.
12. 10× T4 DNA ligase buffer.

2.4.2 Photoreactive RNA Derivative Purification

1. 40% acrylamide-bisacrylamide (19:1).
2. 10× TBE buffer: 1 M Tris, 0.9 M boric acid, 0.01 M EDTA, pH 8.0.
3. Elution buffer: 0.3 M NaOAc, 0.1% SDS, 0.5 mM EDTA, pH 5.5.

2.4.3 Ribosomal Complex Assembly and Cross-linking Procedure

1. 5× Buffer D:100 mM Hepes-KOH, 500 mM KOAc, 10 mM $Mg(OAc)_2$, 1 mM spermidine, 50 mM sodium creatine phosphate, 5 mM ATP, 2.5 mM GTP, pH 7.5.
2. Rabbit reticulocyte lysate, treated with micrococcal nuclease (Promega).
3. RNasin 40 U/μl.
4. Amino acids mix, 2.5 mM.
5. 10× translational inhibitor (2 mM GMPNP final concentration).
6. 12.5 μM recombinant SBP2.
7. 96-well plate.
8. UV-filter to cut $\lambda < 310$ nm.

9. UV cross-linker (Stratalinker, Stratagene or UV SpotCure, UVP).
10. 2-mercaptoethanol.

2.4.4 Isolation of Ribosomal Complexes

See Subheadings 2.1, **items 1–6** and 2.2.2.

2.4.5 Analysis of Cross-link Distribution Between Ribosomal Subunits

1. Dissociation of ribosomes into subunits: 5× Buffer B; 0.1 M puromycin; 3 M KCl.
2. Ultracentrifugation in sucrose gradient: *see* Subheading 2.1, **item 2**.

2.4.6 Analysis of Cross-link Distribution Between Ribosomal Proteins

See Subheadings 2.1, **items 4 and 5** and 2.2.3.

2.4.7 Analysis of Cross-link Distribution Between Ribosomal RNAs

1. RNA sample preparation: *see* Subheading 2.2.4.
2. Chemicals for gel electrophoresis: *see* Subheading 2.2.7; ethidium bromide stain.

3 Methods

3.1 Ribosomal Complex Assembly from Purified Components

3.1.1 Ribosomal Complex Formation

1. Assemble the mixture of 0.5 μM 60S ribosomal subunits or 80S ribosomes and 1.5 μM recombinant SBP2 in 50 μl of Buffer A.
2. Incubate the mixture at room temperature for 30 min.

3.1.2 Ribosomal Complex Purification and Analysis of SBP2 Content Therein

1. Prepare a 10–30% sucrose gradient in Buffer A.
2. Load the incubated mixture onto the gradient and centrifuge at 90,000 × *g*, 4 °C, for 17 h.
3. Fractionate the gradient into 1.5 ml tubes, measure the OD_{260} of each fraction to detect those containing the ribosomal particles.
4. To 1 volume of the pooled ribosome-containing fractions, add 0.25 volume of ice-cold 100% (w/v) TCA, incubate for 20 min on ice, spin tubes at 12,000 × *g*, 4 °C, for 10 min, carefully remove the supernatant and wash the pellet with 200 μl of ice-cold 70% EtOH (*see* **Note 4**).
5. Dissolve the dry pellet in GLB1, heat the sample at 90 °C for 5 min to denature proteins, and then subject it to a 12% SDS-PAGE.
6. Stain the gel with Coomassie BB R250 for 1 h on a shaking platform and destain with the destaining solution.
7. Analyze band intensities using QuantityOne software and compare the intensity of the band corresponding to SBP2 (the

slowest migrating band) to the intensities of bands in the upper group of ribosomal proteins.

3.2 SBP2-Ribosome Diepoxybutane Cross-linking

3.2.1 Diepoxybutane Cross-linking Reaction

1. Bind 80S ribosomes to SBP2 in 50 μl of Buffer A (Subheading 3.1.1).
2. Add 2 μl of freshly prepared 12.5% diepoxybutane solution in Buffer A (*see* **Note 5**).
3. Incubate at 37 °C for 45 min (*see* **Note 6**).
4. Add 2.5 μl of 1 M Tris–HCl, 7.5, to quench the cross-linking reaction.

3.2.2 Analysis of Cross-link Distribution Between Ribosomal Subunits

1. Add 8.5 μl of 3 M KCl to quenched diepoxybutane cross-linking mixture assembled as described in Subheading 3.2.1 (*see* **Note 7**).
2. Incubate at 37 °C for 20 min.
3. Separate subunits and uncross-linked SBP2 by ultracentrifugation in 10–30% sucrose gradient in Buffer B as described in Subheading 3.1.2, **steps 1–3**.
4. Add on ice $MgCl_2$ up to 20 mM to fractions corresponding to 60S and 40S subunits.
5. Add 1 volume of ice-cold absolute EtOH to each fraction and incubate on ice for 20 min.
6. Spin at 12,000 × *g*, 4 °C, for 20 min.
7. Remove the supernatant and wash the pellet with 200 μl ice-cold 70% EtOH.
8. Dry the pellet and dissolve it in 20 μl of Buffer A, save 18 μl on ice for further analysis.
9. Use the remaining 2 μl of each fraction for dot-blotting onto nitrocellulose membrane.
10. Block the membrane for 1 h in 5% fat-free milk in PBS.
11. Incubate the membrane with rabbit polyclonal anti-SBP2 antibodies (1/2500 dilution) for 1 h in 0.5% fat-free milk in PBS (*see* **Note 8**).
12. Wash the membrane for 10 min in 0.5% fat-free milk in PBS, repeat three times.
13. Incubate the membrane with anti-rabbit HRP-conjugated secondary antibody (1/10,000 dilution) for 1 h in 0.5% fat-free milk in PBS.
14. Wash the membrane for 10 min in 0.5% fat-free milk in PBS, repeat three times.
15. Wash the membrane for 5 min in PBS.
16. Reveal the signal with ECL Plus kit, exposing the treated membrane to either X-ray film or ChemiDoc MP.

3.2.3 Analysis of Cross-link Distribution Between Ribosomal RNA and Proteins

Prepare again a mixture containing cross-linked ribosomal complex as described in Subheading 3.2.1 and divide it into two equal parts.

1. Analysis of cross-linked proteins.
   (a) Add $MgCl_2$ to the first part (*see* **step 4**) up to 100 mM final concentration, place the tube on ice.
   (b) Add two volumes of glacial acetic acid divided into six equal portions during 1 h.
   (c) Incubate on ice for 40 min.
   (d) Pellet the precipitated rRNA by centrifugation at 12,000 × *g*, 4 °C, for 10 min.
   (e) Take the supernatant and add six volumes of acetone.
   (f) Pellet proteins by centrifugation at 12,000 × *g*, 4 °C, for 10 min.
   (g) Discard the supernatant, dry the pellet, and dissolve it in minimal amount of water.
   (h) Analyze the cross-links between SBP2 and ribosomal proteins either by 6% SDS-PAGE with subsequent Coomassie BB R250 staining (as described in Subheading 3.1.2, **steps 5–7**) or by dot-blot analysis (as described in Subheading 3.2.2, **steps 10–16**) (*see* **Note 8**).
2. Analysis of cross-linked ribosomal RNA.
   (a) Supplement the second part (*see* **step 1**) with 0.1% SDS, 5 mM EDTA and 0.05 mg/ml proteinase K and incubate the mixture at 37 °C for 30 min.
   (b) Add an equal volume of phenol/chloroform mix (50 μl) and vortex thoroughly.
   (c) Centrifuge at 12,000 × *g* for 5 min.
   (d) Take the upper phase (water phase) and discard the lower phase (phenol phase).
   (e) Add an equal volume of chloroform (50 μl) and vortex thoroughly.
   (f) Centrifuge at 12,000 × *g* for 5 min.
   (g) Take the upper (water) phase (water phase) and add 1/10 of volume of 3 M NaOAc, pH 5.5, and 3 volumes of absolute EtOH.
   (h) Incubate for 10 min in dry ice.
   (i) Pellet the rRNA by centrifugation at 12,000 × *g*, 4 °C, for 10 min.
   (j) Dissolve the rRNA in water up to the concentration of approximately 1 μM.
   (k) Analyze the cross-links between SBP2 and rRNAs by dot-blot analysis (as described in Subheading 3.2.2, **steps 10–16**) (*see* **Note 8**).

3. Analysis of cross-linked ribosomal RNA in 60S subunits
   (a) Add SDS and EDTA-NaOH, pH 7.5, up to 0.5% and 5 mM, respectively, to the remaining 18 μl of the solution containing cross-linked 60S subunits obtained in Subheading 3.2.2, **step 8**.
   (b) Incubate at 37 °C for 20 min.
   (c) Prepare a 5–20% sucrose gradient in Buffer C using a gradient former.
   (d) Load the complex onto the gradient.
   (e) Centrifuge the gradient at 125,000 × *g*, 17 h, 4 °C.
   (f) Fractionate the gradient, measure the OD_{260} of each fraction to detect those containing the different rRNAs.
   (g) Precipitate the rRNA in the fractions of interest by adding 1/10 of volume of 3 M NaOAc, pH 5.5, and 3 volumes of absolute EtOH.
   (h) Pellet the rRNA by centrifugation at 12,000 × *g*, 4 °C, for 10 min.
   (i) Remove the supernatant and dry the pellet.
   (j) Dissolve the pellet in 10 μl of water; add 1 μl of RNase A solution.
   (k) Incubate at 37 °C for 20 min.
   (l) Analyze the cross-links between SBP2 and rRNA by dot-blot analysis (as described in Subheading 3.2.2, **steps 10–16**) (*see* **Note 8**).

3.2.4 Sample Preparation of SBP2-rRNA Cross-links for Reverse Transcription Analysis

1. Carry out SBP2-ribosome cross-linking by diepoxybutane as described in Subheading 3.2.1.
2. Isolate the RNA from cross-linked 80S ribosomes as described in Subheading 3.2.3, **step 2**(**a–j**).
3. Analyze rRNA nucleotides cross-linked to SBP2 by diepoxybutane by reverse transcription (described below in Subheadings 3.2.5–3.2.7) (*see* **Note 9**).

3.2.5 Primer 5′ Labeling

1. Assemble the labeling reaction: mix 6 μl of water, 1 μl of primer, 1 μl of [γ-^{32}P]-ATP, 1 μl of 10× T4 PNK buffer, and 1 μl of T4 PNK.
2. Incubate at 37 °C for 30 min.
3. Purify the primer from unincorporated [γ-^{32}P]-ATP using MicroSpin G25 columns (follow the manufacturer's instructions).
4. After purification, measure primer activity (aiming at 150,000–200,000 cpm/μl) and adjust with water if needed.

3.2.6 Reverse Transcription

1. Anneal the primer: mix 1 μl of labeled primer with 1 μl of previously modified RNA (1 μM) and 3 μl of water in an Eppendorf tube, heat at 90 °C for 1 min and immediately place on ice (*see* **Note 10**).
2. Keeping the tube on ice, add 2 μl of dNTP mix, 2 μl of 5× AMV RT buffer, and 1 U of AMV reverse transcriptase.
3. Prepare four tubes (labeled A, G, C, and T) with sequencing reaction mixtures each containing, besides the above-mentioned components, 0.5 μl of one of the ddNTPs (e.g., ddATP in the A tube, ddCTP in the C, etc.) (*see* **Note 11**).
4. Incubate at 42 °C for 30 min.
5. After incubation, quickly spin the tubes to collect any condensate.
6. Precipitate the reactions with ethanol as described in Subheading 3.2.3, **step 3(g–i)**.
7. Dissolve the pellets in 4 μl of GLB2.
8. Store the tubes at −20 °C or proceed with sequencing gel.

3.2.7 Sequencing Gel Fractionation and Gel Analysis

1. Prepare an 8% polyacrylamide gel, containing 8 M urea and TBE buffer.
2. Prerun the gel at 50 V/cm until temperature reaches 45 °C.
3. Denature the samples obtained in Subheading 3.2.4: heat the samples at 90 °C for 1 min and immediately place on ice.
4. Load 1–2 μl of each sample onto the gel.
5. Run the gel, keeping the temperature at 45–50 °C (*see* **Note 12**).
6. After the run is completed, place the gel on Whatmann 3MM paper and cover it with transparent film.
7. Dry the gel using a gel dryer at 80–90 °C for 40–50 min.
8. Expose the gel to a PhosphorImager screen for 12–48 h, depending on the intensity of the signals obtained.
9. Analyze the gel in the Quantity One software (*see* **Note 13**).

3.3 Chemical Probing of RNA in SBP2-Ribosome Complex

1. Prepare ribosomes complexed with SBP2 as described in Subheading 3.1.1 and perform chemical probing with the reagent of interest as described in Subheadings 3.3.1–3.3.3 (*see* **Notes 14** and **15**).
2. After chemical probing is performed, isolate RNA as described in Subheading 3.2.3, **step 2(a–j)**.
3. Measure the OD_{260} in each sample to normalize the concentration of RNA if needed.
4. Analyze the RNA modifications by reverse transcription as described in Subheadings 3.2.5–3.2.7 (*see* **Note 16**). Typical

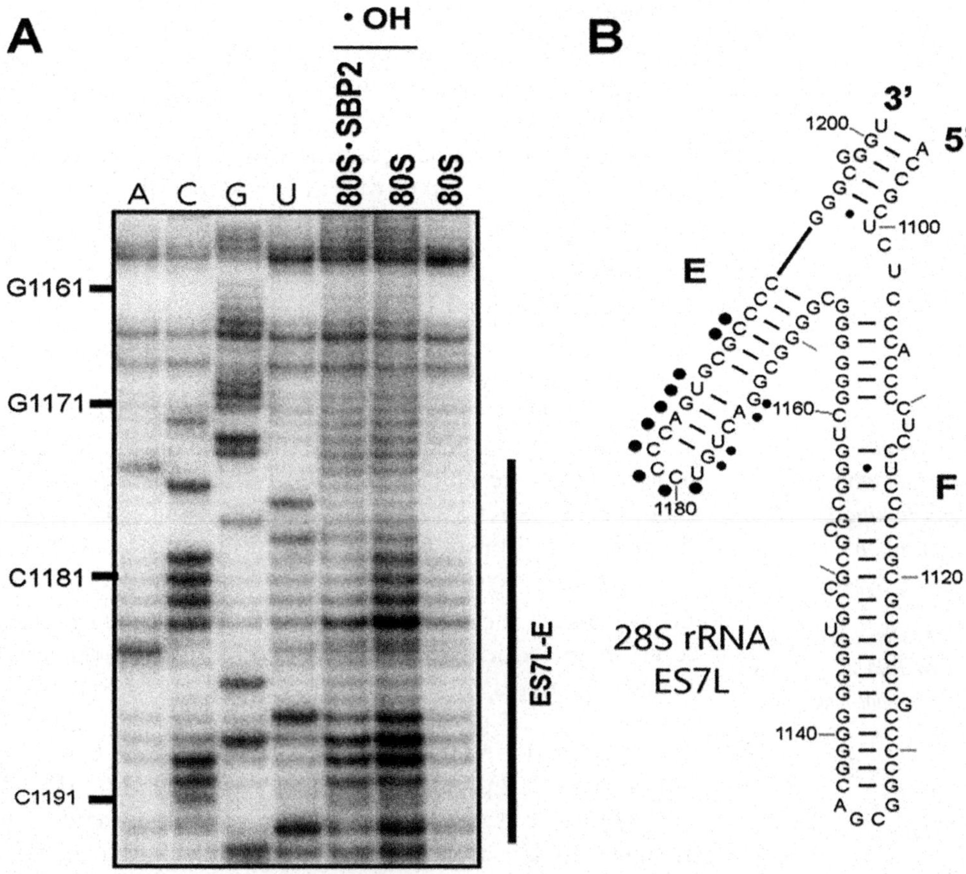

Fig. 1 Direct hydroxyl radical probing of the human 28S rRNA in 80S·CTSBP2 complex. (**a**) Typical autoradiogram of a sequencing gel after fractionation of products obtained by reverse transcription of 28S rRNA isolated from 80S·CTSBP2 complex (*lane 80S·CTSBP2*) and free 80S ribosomes (*lane 80S*) treated with hydroxyl radicals (•OH); *U, G, C, A* are sequencing lanes. (**b**) The identified protection sites mapped on the secondary structure of the 28S rRNA expansion segment ES7L helices E and F [16]; protections observed in the 80S·CTSBP2 complex are shown as *closed circles*

results obtained with hydroxyl radical probes (*see* below in Subheading 3.3.1) are shown in Fig. 1.

3.3.1 Hydroxyl Radical Probing

1. Add 2 μl of the freshly prepared solution I to 20 μl of ribosomal complex (*see* **Note 17**).
2. Incubate the reaction mixture for 2 min at room temperature.
3. Quench the reaction by the addition of 20 μl of 0.1 M thiourea, mix thoroughly, and incubate for 10 min at room temperature.

3.3.2 DMS Probing

1. Prepare a DMS working solution by diluting 7.5 μl of DMS stock in 192.5 μl of absolute EtOH (3.75% DMS).
2. Add 2 μl of the solution to 50 μl of ribosomal complex (0.15% DMS).

3. Incubate on ice for 1 h.
4. Quench the reaction by adding 2 μl of 14.4 M 2-mercaptoethanol and mix thoroughly.

3.3.3 Kethoxal Probing

1. Prepare kethoxal working solution by mixing 5 μl of kethoxal stock solution in 45 μl of water (10% kethoxal).
2. Add 2 μl of the solution to 50 μl of ribosomal complex (0.4% kethoxal).
3. Incubate on ice for 1 h.
4. Add 2.5 μl of 0.5 M potassium borate up to a final concentration of 25 mM (*see* **Note 18**).

3.4 Ribosomal Complex Assembly and Cross-linking in Cell-Free Protein Synthesizing System

3.4.1 Synthesis of the Photoreactive RNA Derivative by Splint-Aided Reaction

1. To synthesize 5′ RNA, assemble the T7 transcription reaction in a final volume of 100 μl of T7 buffer: mix 52 μl of water, 20 μl of T7 buffer 5×, 4 μl of 100 mM rNTPs, 10 μl of DNA template (PCR product), 1 μl of RNasin, and 1 μl of T7 RNA polymerase (*see* **Note 19**). Follow to Subheading 3.4.1, **step 3**.
2. To synthesize the ^{32}P-labeled SECIS 3′ RNA bearing photoreactive nucleotide derivatives, assemble the T7 transcription reaction in a final volume of 100 μl: mix 40 μl of water, 20 μl of T7 buffer 5×, 4 μl of 100 mM CTP, 0.4 μl of 100 mM ATP, 10 μl of [α-^{32}P]-ATP (100 μCi), 8 μl of 100 mM GMP, 2 μl of 100 mM GTP, 2 μl of 100 mM UTP, 0.5 μl of 100 mM s4UTP (or UTP derivative bearing perfluorophenylazido group attached to C5 atom via allylamino linker), 10 μl of DNA template (linearized plasmid, 10 μg/μl), 1 μl of RNasin, and 1 μl of T7 RNA polymerase. Incubate the mixture at 37 °C for 1 h and add 2 μl of 100 mM ATP (*see* **Note 20**). Follow to Subheading 3.4.1, **step 3**.
3. Incubate the above mixtures at 37 °C until a white pyrophosphate precipitate is formed (2–12 h) (*see* **Note 21**).
4. Desalt the resulting RNA on Sephadex G15 mini-spin columns according to the manufacturer's instructions.
5. Estimate the RNA concentration.
6. Mix 150 pmol of 5′ RNA, 150 pmol of splint oligonucleotide and 120 pmol of ^{32}P-labeled SECIS 3′ RNA in a final volume of 70 μl (*see* **Notes 22** and **23**).
7. Incubate at 80 °C for 5 min, place immediately on ice, and keep on ice for 5 min.
8. Add 20 μl of T4 DNA ligase buffer × 5, 1 μl of RNasin, and 10 μl of T4 DNA ligase.
9. Incubate at 16 °C for 12–16 h.
10. Extract proteins with phenol and precipitate RNA with ethanol.

3.4.2 RNA Purification

1. Dissolve the RNA pellet in GLB2, denature at 90 °C for 1 min, and place on ice.
2. Run a 10% polyacrylamide, 8 M urea gel in TBE buffer (stop the run when xylene cyanol dye reaches the bottom of the glass plates).
3. Cover the gel with wrap, expose to X-ray film for 5–10 min, use intensifying screen if necessary. Cut the band corresponding to the ligated full-length RNA product.
4. Put the gel pieces containing the ligated RNA products into a 1.5 ml tube, add elution buffer to cover the gel pieces. Elute for 2–4 h on a shaking platform at room temperature.
5. Take the supernatant, check the efficiency of elution with a Geiger counter, comparing the radioactivity amounts in the gel pieces and supernatant. Repeat the elution step with a new volume of elution buffer if necessary (*see* **Note 24**).
6. Precipitate RNA with 3 volumes of ethanol, wash the pellet with 70% ethanol and dry.

3.4.3 Ribosomal Complex Assembly and Cross-linking Procedure

1. Dissolve the photoreactive RNA derivative (Fig. 2) obtained in Subheading 3.4.2 to 5 pmol/μl in Buffer D and incubate at 37 °C for 10 min (*see* **Note 25**).
2. Assemble the ribosomal complex in rabbit reticulocyte lysate (RRL) by mixing the following components on ice: 50 μl of micrococcal nuclease-treated RRL, 10 μl of translational inhibitor, 18 μl of Buffer D (or Buffer D without GTP, if GMPPNP is used as the inhibitor), 0.25 μl of RNasin, 10 μl of amino acid mixture, and 2 μl of SBP2 (*see* **Note 26**).
3. Incubate at 30 °C for 5 min, add 3 μl of the photoreactive RNA derivative, and incubate at 30 °C for 5 min.

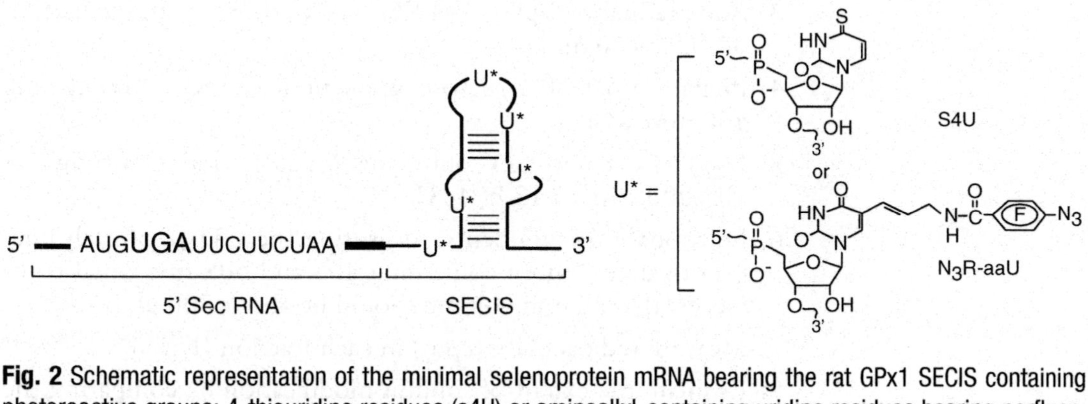

Fig. 2 Schematic representation of the minimal selenoprotein mRNA bearing the rat GPx1 SECIS containing photoreactive groups: 4-thiouridine residues (s4U) or aminoallyl-containing uridine residues bearing perfluorophenylazido group (N3RaaU, where R is perfluorophenyl residue)

4. Divide the mixture in 25 μl aliquots and place each aliquot in a well of a 96-well plate.
5. Place the plate on ice and cover it with UV filter to avoid direct UV cross-links.
6. Irradiate for 2 min (*see* **Note 27**).
7. Add 2-mercaptoethanol up to 0.17% final concentration and incubate on ice for 10 min.

3.4.4 Isolation of Ribosomal Complexes

1. Prepare a 10–30% sucrose gradient in Buffer E.
2. Load the irradiated complex onto the gradient and centrifuge at 90,000 × *g* for 17 h, 4 °C.
3. Fractionate the gradients measuring the OD_{260} of each fraction to detect those containing 48S and 80S ribosomal complexes; the fraction volume should be ~300 μl.
4. Measure radioactivity (cpm) in each fraction.
5. Estimate the extent of RNA binding to ribosome (mol of RNA/mol of ribosome) by comparing the radioactivity amounts in 80S- or 48S-containing fractions and in top fractions containing unbound RNA.
6. Precipitate the complexes with ethanol as described in Subheading 3.2.2, **steps 4–8**.
7. To validate the presence of SBP2 in the assembled complexes, dissolve the pellets in 50 μl of water, and take 10 μl for SDS-PAGE followed by western blotting analysis (as described in Subheading 3.2.2, **steps 10–16**).

3.4.5 Analysis of Cross-link Distribution Between Ribosomal Subunits

1. Dissolve the pellets corresponding to 80S complexes in 100 μl of Buffer A, place on ice, add puromycin up to 0.5 mM and ribosomal subunits (30 pmol of 40S and 60S subunits) as carrier (*see* **Note 28**).
2. Keep on ice for 10 min.
3. Add gradually 12 μl of 3 M KCl, vortex thoroughly, incubate at 37 °C for 30 min.
4. Prepare a 10–30% sucrose gradient in Buffer A containing 450 mM KCl.
5. Load the complexes onto the gradient and centrifuge at 107,000 × *g* for 17 h, 4 °C.
6. Fractionate the gradients, measuring the OD_{260} of each fraction to detect those containing 40S and 60S ribosomal complexes; the fraction volume should be about 300 μl.
7. Measure radioactivity (cpm) in each fraction (Fig. 3).
8. Estimate the extent of subunit modification by comparing the radioactivity amounts in subunit-containing fractions and in top fractions containing uncross-linked RNA.

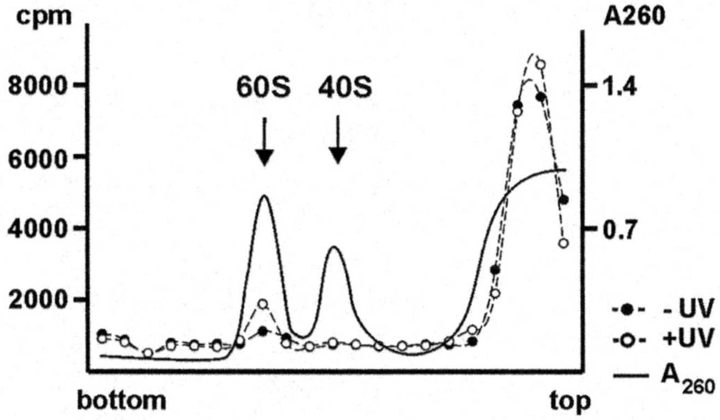

Fig. 3 Typical sucrose gradient sedimentation profile of the 60S and 40S ribosomal subunits isolated from the irradiated 80S complex formed in RRL in the presence of 5 mM anisomycin. The minimal mRNA carried a SECIS derivatized with statistically distributed s4U residues

3.4.6 Analysis of Cross-link Distribution Between Ribosomal Proteins

1. Prepare ribosomal complexes as described in Subheadings 3.4.3 and 3.4.4.
2. Take 20 μl of complex, add SDS and EDTA-NaOH, pH 7.5, up to 0.1% and 0.5 mM, respectively.
3. Incubate at 37 °C for 10 min to dissociate ribosomal particles into rRNA and proteins.
4. Add 1 μl of RNase A (10 mg/ml), incubate at 37 °C for 1 h.
5. Take 10 μl of the mixture and add 60 μl of acetone, keep on ice.
6. Add 1 μl of Proteinase K (10 mg/ml) to the remaining 10 μl of mixture, incubate at 37 °C for 30 min, and then add 60 μl of acetone.
7. Pellet the above two mixtures by centrifugation at 12,000 × *g*, 4 °C, for 10 min.
8. Dissolve the pellets in GLB1, incubate at 90 °C for 5 min.
9. Run 14% SDS-PAGE.
10. Fix the gel in destaining solution for 15 min, dry and expose to PhosphorImager plate for 12–98 h, depending on signal efficiency.
11. Analyze the signals by densitometry using the QuantityOne software.

3.4.7 Analysis of Cross-link Distribution Between Ribosomal RNAs

1. Prepare ribosomal complexes as described in Subheadings 3.4.3 and 3.4.4.
2. Take 20 μl of complex and add 40 μl of elution buffer.
3. Incubate at 37 °C for 15 min.

4. Add 3 µl of Proteinase K (10 mg/ml), incubate at 37 °C for 30 min.
5. Extract the RNA as described previously in Subheading 3.2.3, **step 2**(**b–i**).
6. Dissolve the RNA pellet in GLB2, incubate at 90 °C for 2 min, place on ice.
7. Run 6% polyacrylamide, 8 M urea gel in TBE buffer (stop when xylene cyanol dye reaches the bottom of the glass plates).
8. Stain the gel with ethidium bromide solution or toluidine O blue solution, visualize in ChemiDoc system.
9. Dry the gel and expose on a PhosphorImager screen for 12–98 h, depending on signal efficiency.
10. Analyze the signals by densitometry using the QuantityOne software.

4 Notes

1. We usually dissolve human ribosomal subunits purified by sucrose gradient centrifugation in Milli-Q water. Prior to use, the ribosomal subunits should be renatured by the addition of 1/4 V of 5× Buffer (e.g., Buffer A) with subsequent incubation at 37 °C for 10 min. To assemble 80S ribosome, renatured subunits are mixed in an equimolar ratio and incubated at 37 °C for 5 min.
2. The diepoxybutane stock solution should be opened under inert gas atmosphere and diluted right before use.
3. DMS is highly mutagenic; it is toxic through inhalation and can be absorbed through skin. All manipulations with DMS should be carried out in a ventilated hood. Tubes and tips that were in contact with DMS should be placed in 5 M NaOH and disposed as recommended by institutional safety department.
4. The pellet obtained after TCA precipitation of ribosomes is usually not compact and partly stays on the wall of the tube; therefore care should be taken not to lose ribosomes.
5. Diepoxybutane is a cross-linker specific to amino and thiol groups. Therefore, buffers used to form ribosomal complexes and to store proteins should not contain compounds with primary amino groups (e.g., Tris is not a good choice). Diepoxybutane, being highly reactive toward –NH2 and –SH groups in proteins and N7 atom of guanine in RNA, produces protein-protein and RNA-protein cross-links forming vicinal diol linkages between the respective groups. The linkages can be cleaved by periodate oxidation.

6. Diepoxybutane cross-linking conditions must be subject to optimization for each ribosomal complex. Prior to main experiments, one needs to optimize the concentration of the cross-linker and the cross-linking time. Under optimal conditions, ribosomal subunits obtained by dissociation of 80S ribosomes cross-linked to SBP2 sediment in much the same way as uncross-linked subunits in sucrose gradient.
7. Include a control ribosomal complex with no cross-linking reagent to ensure that dissociation of 80S ribosomes into subunits induced by high-salt was efficient. This control will reveal the locations of peaks corresponding to the 40S and 60S subunits and allows detection of potential subunit multimers arising from excessive cross-linking.
8. Anti-SBP2 antibodies should be checked beforehand for cross-reactivity with ribosomal proteins.
9. Diepoxybutane cross-linking sites are studied here with a reverse transcription assay. Reverse transcriptase can encounter a number of stop or pause sites in rRNAs; to distinguish those caused by the actual diepoxybutane modifications, include a control ribosomal complex lacking diepoxybutane.
10. For better annealing of GC-rich primers, DMSO can be added to the annealing mix up to 5% (v/v).
11. The dNTP/ddNTP ratio here is 1/8, which is optimal to get readable sequencing lanes and to read 150–200 nucleotides of RNA. The ratio may be adjusted to get longer reads.
12. Gel temperature higher than 50 °C may cause breaking of the sequencing glass plates. Check with the hand the sequencing equipment for temperature limit and use power supply with automatic temperature control.
13. Since diepoxybutane reacts with N7G that is not involved in Watson-Crick base-pairing, reverse transcriptase will not be arrested at the modified G. However, such a nucleotide is able to retard the transcription process, causing a noticeable pause [17].
14. Include a control ribosomal complex that will not be subjected to the chemical treatment.
15. Although the concentration of the reagent is given here, it is a good practice to carry out preliminary experiments on RNA modification, varying the concentration of the reagent at a fixed concentration of RNA with subsequent reverse transcription assay, and thus to choose optimal concentration of the reagent. Low level of modification as well as overmodification should be avoided.
16. Reverse transcriptase stops occur 3′ to the actual DMS or kethoxal modification or hydroxyl radical-mediated cleavage sites.

17. Avoid Tris-, glycerol-, alcohol-containing buffers in hydroxyl radical-mediated cleavage reactions since these chemicals were shown to quench hydroxyl radicals [18].
18. Since kethoxal modification is reversible, all solutions in the subsequent steps (RNA extraction, reverse transcription) should contain 25 mM potassium borate to stabilize kethoxal-guanine adducts.
19. To ensure sufficient level of splint-aided RNA ligation, it is important to obtain 5′ RNA with homogenous 3′ end. Several protocols have been described [19, 20].
20. Taking into account that s4U insertion efficiency is about four times lower than that of UTP [21], one can assume that no more than 2 of the 32 uridine residues of the SECIS 3′ RNA are substituted by s4U residues.
21. Magnesium pyrophosphate precipitate can be dissolved by adding EDTA to chelate Mg^{2+}.
22. Start at least with a tenfold excess of each individual RNA, since the usual yield of ligated RNA is about 5–10% because of the low efficiency of the reaction and losses of RNA at subsequent purification steps.
23. A detailed description of RNA ligation optimization is given in [22].
24. RNA elution efficiency is around 70–75%.
25. Prior to cross-linking experiments, several validating experiments are needed: (a) toe-printing assay to show that the RNA is phased correctly; (b) mass-spectrometry, western blotting, tRNA-post-labeling—to demonstrate that assembled complexes contain all necessary components; (c) nitrocellulose filter binding assay to compare the binding efficiency of unmodified RNA and photoactivatable RNA derivative.
26. Rabbit reticulocyte lysate should be thawed on ice and all solutions should be added on ice. Each batch of RRL can survive two freeze-thaw cycles. 80S ribosome concentration in RRL is around 0.5–1 pmol of ribosomes/5 μl of RRL.
27. Prepare a control complex not subjected to the irradiation step to ensure that the bands observed in the SDS-PAGE analysis (*see* Subheading 3.4.6) of the irradiated sample indeed arose from cross-linking.
28. Incubation with puromycin causes premature peptide chain release, thus helping to dissociate 80S ribosomal complexes containing fragments of endogenous mRNAs which may be present in the pellet together with complexes under investigation. This improves resolution of the ribosomal subunit peaks in the sucrose gradient sedimentation profile.

Acknowledgments

This work was supported by the ARCUS and Supramolecular Chemistry programs (to G.K. and A.K.), the CNRS Laboratoire International Associé LIA NUCPROT (to A.K.), and the Russian Foundation for Basic Research (grant 12- 04-93111-CNRSL_a to G.K.).

References

1. Papp LV, Lu J, Holgrem A, Khanna KK (2007) From selenium to selenoproteins: synthesis, identity, and their role in human health. Antioxid Redox Signal 9:775–806
2. Lescure A, Rederstorff M, Krol A, Guicheney P, Allamand V (2009) Selenoprotein function and muscle disease. Biochim Biophys Acta 90:1569–1574
3. Berry MJ, Banu L, Chen Y, Mandel S, Kieffer J, Harney J, Larsen P (1991) Recognition of UGA as a selenocysteine codon in type I deiodinase requires sequences in the 3′ untranslated region. Nature 353:273–276
4. Fagegaltier D, Hubert N, Yamada K, Mizutani T, Carbon P, Krol A (2000) Characterization of mSelB, a novel mammalian elongation factor for selenoprotein translation. EMBO J 19:4796–4805
5. Tujebajeva RM, Copeland PR, XM X, Carlson BA, Harney JW, Driscoll DM, Hatfield DL, Berry MJ (2000) Decoding apparatus for eukaryotic selenocysteine insertion. EMBO Rep 1:158–163
6. Copeland PR, Fletcher JE, Carlson BA, Hatfield DL, Driscoll DM (2000) A novel RNA binding protein, SBP2, is required for the translation of mammalian selenoprotein mRNAs. EMBO J 19:306–314
7. Allmang C, Wurth L, Krol A (2009) The selenium to selenoprotein pathway in eukaryotes: more molecular partners than anticipated. Biochim Biophys Acta 1790:1415–1423
8. Bulteau AL, Chavatte L (2015) Update on selenoprotein biosynthesis. Antioxid Redox Signal 23:775–794
9. Chavatte L, Brown BA, Driscoll DM (2005) Ribosomal protein L30 is a component of the UGA-selenocysteine recoding machinery in eukaryotes. Nat Struct Mol Biol 12:408–416
10. Miniard AC, Middleton LM, Budiman ME, Gerber CA, Driscoll DM (2010) Nucleolin binds to a subset of selenoproteins mRNAs and regulates their expression. Nucleic Acids Res 38:4807–4820
11. Budiman ME, Bubenik JL, Miniard AC, Middleton LM, Gerber CA, Cash A, Driscoll DM (2009) Eukaryotic initiation factor 4a3 is a selenium-regulated RNA-binding protein that selectively inhibits selenocysteine incorporation. Mol Cell 35:479–489
12. Copeland PR, Driscoll DM (1999) Purification, redox sensitivity, and RNA binding properties of SECIS-binding protein 2, a protein involved in selenoprotein biosynthesis. J Biol Chem 274:25447–25454
13. Kossinova OA, Malygin AA, Krol A, Karpova GG (2013) A novel insight into the mechanism of mammalian selenoprotein synthesis. RNA 19:1147–1158
14. Caban K, Copeland PR (2012) Selenocysteine insertion sequence binding protein 2 alters conformational dynamics of residues involved in tRNA accommodation in 80S ribosomes. J Biol Chem 287:10664–10673
15. Kossinova OA, Malygin AA, Krol A, Karpova GG (2014) The SBP2 protein central to selenoprotein synthesis contacts the human ribosome at expansion segment 7L of the 28S rRNA. RNA 20:1046–1056
16. Anger A, Armache JP, Berninghausen O, Habeck M, Subklewe M, Wilson D, Beckman R (2013) Structures of the human and Drosophila 80S ribosome. Nature 497:80–85
17. Yanshina DD, Bulygin KN, Malygin AA, Karpova GG (2015) Hydroxylated histidine of human ribosomal protein uL2 is involved in maintaining the local structure of 28S rRNA in the ribosomal peptidyl transferase center. FEBS J 282:1554–1566
18. Tullius TD, Dombroski BA, Churchill ME, Kam L (1987) Hydroxyl radical footprinting: a high-resolution method for mapping protein–DNA contacts. Methods Enzymol 155:537–558
19. Kao C, Zheng M, Rüdisser S (1999) A simple and efficient method to reduce nontemplated nucleotide addition at the 3′ terminus of RNAs transcribed by T7 RNA polymerase. RNA 5:1268–1272

20. Wichlacz A, Legiewicz M, Ciesiolka J (2004) Generating in vitro transcripts with homogenous 3′ ends using trans-acting antigenomic delta ribozyme. Nucleic Acids Res 32:e39
21. Dereuil YL, Expert-Bezançon A, Favre A (1991) Conformation and structural fluctuations of a 218 nucleotides long rRNA fragment: 4-thiouridine as an intrinsic photolabelling probe. Nucleic Acids Res 19:3653–3660
22. Frelander MJ, Turunen JJ (2005) RNA ligation using T4 DNA ligase. In: Hartmann RK, Bindereif A, Schön A, Westhof E (eds) Handbook of RNA biochemistry. Wiley, Hoboken, NJ, pp 36–53. ISBN 3-527-30826-1

Chapter 7

In Vitro Translation Assays for Selenocysteine Insertion

Mark H. Pinkerton and Paul R. Copeland

Abstract

The molecular characterization of the protein and RNA factors that are required for Sec incorporation in mammals has been largely carried out using in vitro translation systems specifically modified for this purpose. This chapter outlines the various systems and modifications that have been used to decipher the mechanism of Sec incorporation.

Key words Rabbit reticulocyte lysate, Wheat germ lysate, SECISBP2, eEFSec, Translation, mRNA, 3′ UTR

1 Introduction

In vitro translation systems have proved to be a valuable tool for the study of the molecular mechanisms required for protein synthesis, and they have allowed for rapid and independent evaluation of gene products, mutant proteins, and more. Specifically, the development of the rabbit reticulocyte lysate (RRL) [1] and wheat germ lysate (WGL) [2] systems for in vitro translation has been central to these types of analyses. In the field of selenocysteine (Sec) incorporation, a major difficulty developed in determining the core mechanism involved in eukaryotic Sec incorporation. The current knowledge of Sec incorporation consists of a recoding event at an in-frame UGA codon in the presence of a conserved RNA hairpin loop motif known as the Sec insertion sequence (SECIS), which is present in the 3′ untranslated region (UTR) of all selenoprotein mRNAs. The SECIS is bound by SECIS binding protein 2 (SBP2), which in turn interacts with a Sec elongation factor (eEFSec) to deliver a Sec-$tRNA^{Sec}$ to the ribosome. These core components must be present in order for the translation of a selenoprotein mRNA to occur.

SBP2 was initially characterized as a SECIS binding protein [3, 4], but determining its essentiality in the Sec incorporation reaction was greatly facilitated by the fact that rabbit reticulocyte lysate is naturally devoid of its activity [5]. Such was not the case for eEFSec

Laurent Chavatte (ed.), *Selenoproteins: Methods and Protocols*, Methods in Molecular Biology, vol. 1661, DOI 10.1007/978-1-4939-7258-6_7, © Springer Science+Business Media LLC 2018

[6], so the plant-based wheat germ system, which is naturally lacking all selenocysteine-related factors due to the absence of Sec in plants, was utilized to test its function [7]. Thus, these in vitro systems where purified factors are used to reconstitute Sec insertion proved to be pivotal in the identification and validation of the known core Sec insertion requirements. Here, we will discuss methods and techniques using RRL and WGL to study Sec insertion in vitro and purification of the core Sec insertion components.

2 Materials

All the solutions are prepared using ultrapure water (prepared by purifying deionized water) and are stored at room temperature unless otherwise noted. Proteins are aliquoted and stored at −80 °C and kept on ice after thawing unless otherwise noted.

2.1 Recombinant SBP2 Purification

1. *Escherichia coli* BL21 bacteria transformed with a plasmid harboring an Xpress/His-tagged C-terminal domain of SBP2 (e.g., pTRCHis-TOPO; *see* **Note 1**).
2. Luria Broth.
3. 100 μg/mL ampicillin (Sigma) stock.
4. 1 M Isopropyl-β-D-thiogalactopyranoside (IPTG; Fisher). Store at −20 °C and keep on ice until use.
5. PBS 1×: 137 mM NaCl, 2.7 mM KCl, 4.3 mM Na_2HPO_4, 1.47 mM KH_2PO_4, pH 8.
6. SBP2 Buffer A: 1× PBS with 850 mM NaCl, 1% Tween 20, 30 mM Imidazole. Filter through a 0.22 μm filter and store at 4 °C.
7. 100 mM PMSF in ethanol stored at 4 °C.
8. Sonicator or other equipment/materials for cell lysis method.
9. Metal Chelate Affinity Chromatography Column, store at 4 °C.
10. 0.1 M $NiSO_4$ solution, filter through a 0.22 μm filter and store at 4 °C.
11. Heparin Chromatography Column, store at 4 °C.
12. Heparin Buffer A: 1× PBS containing 300 mM NaCl, filter through a 0.22 μm filter and store at 4 °C.
13. Heparin Buffer B: 1× PBS containing 1 M NaCl, filter through a 0.22 μm filter and store at 4 °C.
14. Storage Buffer: 1× PBS with 10% glycerol (v/v) 0.1% tween-20 (v/v) and 2 mM DTT.
15. 50 kDa MWCO centrifugal filter unit.
16. Standard SDS-PAGE reagents for evaluation.

2.2 Recombinant Flag-eEFSec Preparation

1. *Escherichia coli* BL21 bacteria transformed with 5'Flag-eEFSec TA cloned into pTRCHis-TOPO.
2. 2 L autoclaved L.B. medium split into two 4 L flasks. Weigh, add ultrapure water, adjust pH, and autoclave.
3. 100 μg/mL ampicillin stock. Prepare with autoclaved ultrapure water, store at 4 °C.
4. 1 M IPTG. Prepare with autoclaved ultrapure water, aliquot and store at −20 °C, keep on ice until use.
5. EDTA prepared at pH 8.0.
6. 10× EFS Buffer: 200 mM Tris–HCl, 200 mM KCl, 1 mM EDTA, pH 7.5.
7. EFS Buffer A: 100 mL 10× EFS buffer plus 29.22 g NaCl (0.5 M), 250 mL glycerol, 10 mL Tween 20 (1%) up to 1 L with water.
8. EFS Buffer B: 100 mL 10× EFS buffer, 250 mL glycerol up to 1 L with water.
9. Anti-Flag antibody conjugated to magnetic beads (Sigma).
10. 3× FLAG peptide (lyophilized powder).
11. Standard SDS-PAGE reagents for evaluation .

2.3 Source of Sec-tRNA^{Sec}

There are multiple methods and applications for the purification of labeled [^{75}Se]-Sec-tRNA^{Sec} and unlabeled Sec-tRNA^{Sec} (*see* Chapter 4). Here, we describe the method we use to obtain total aminoacyl tRNA (aa-tRNA) from flash frozen rat or rabbit testicles, which provides a rich source of Sec-tRNA^{Sec} for in vitro translation in a Sec-tRNA^{Sec} -dependent system such as wheat germ lysate. This preparation is also used as a source for further purification of Sec-tRNA^{Sec} by eEFSec affinity chromatography.

1. Phenol, unbuffered, pH 4.0.
2. 5× Buffer T: 50 mM NaOAC, 3.25 M NaCl, 50 mM $MgOAc_2$, 5 mM EDTA, pH 4.0.
3. Testis Lysis Buffer: 20 mM Tris–HCl, 100 mM KCl, 2.5 mM $MgCl_2$, 2 mM DTT, 0.4 mM GTP, 0.25 mM spermidine, 20% glycerol and EDTA-free protease inhibitor .
4. 100% Ethanol stored at −20 °C.
5. 70% Ethanol stored at −20 °C.
6. RNAse-free water.
7. Flash frozen rabbit testicles (Pel-Freeze).

2.4 In Vitro Translation in Rabbit Reticulocyte Lysate

1. Nuclease-Treated Rabbit Reticulocyte Lysate Extract (Promega).
2. Amino Acid mixture, 1 mM.

3. [^{75}Se]-selenite diluted to 5 μM or [^{75}Se]-Sec-tRNA^{Sec}.
4. Capped and polyadenylated mRNA (native selenoprotein or luciferase reporter with in-frame UGA codon) and a SECIS element in the 3′ UTR.
5. Purified C-terminal SBP2 8 μM .
6. RNase-free water.

2.5 Detection of Luciferase Activity

1. White, flat-bottom 96-well microplates.
2. Microplate Luminometer with reagent injector (Berthold).
3. Luciferase assay system (Promega).

2.6 In Vitro Translation in Wheat Germ Lysate

1. Wheat germ extract (Promega).
2. Amino Acid mixture, 1 mM.
3. aa-tRNA preparation from rabbit or rat testicles (10 μg/μL) or purified [^{75}Se]-Sec-tRNA^{Sec}.
4. Capped and polyadenylated mRNA (native selenoprotein or luciferase reporter with in-frame UGA codon and a SECIS element in the 3′ UTR).
5. Purified eEFSec, 8 μM.
6. Purified C-terminal SBP2, 8 μM.
7. RNase-free water.

3 Methods

3.1 Recombinant SBP2 Purification

1. Grow overnight culture of *E. coli* harboring plasmid encoding Xpress-His tagged CT-SBP2 in 4 mL of LB with 100 μg/mL ampicillin (4 μL of 100 mg/mL stock). Incubate overnight at 37 °C shaking at 225 rpm.
2. Prepare a 2 L flask with 1 L LB.
3. Inoculate each flask with 2 mL of the overnight culture and add ampicillin to 100 μg/mL.
4. Allow the cells to grow at 37 °C to an OD_{600} of ~1.0. Check the OD_{600} at 3 h to estimate growth. Doubling time is typically 0.5 h. Add 100 μL of 1 M IPTG for each 1 L culture and incubate for 2.5 h at 30 °C.
5. Pellet cells in 1 L bottles by centrifugation at 4000 × *g* for 15 min each.
6. Pour off the supernatant and freeze pellets at −80 °C overnight.
7. Resuspend each pellet in 20 mL of cell resuspension buffer plus 0.5 mM PMSF in original bottles (40 mL total).
8. Transfer to two 50 mL falcon tubes (~20 mL each).

9. Sonicate for 4 × 40 s to lyse cells, keeping extracts on ice and allowing a break of 1 min between sonication to keep it cool. We use a Misonix sonicator set at "8" with a pulse of ~30% per second. Setting for optimal balance of lysis and protein solubility must be determined for each instrument .
10. Spin at 15,000 × *g* for 15 min at 4 °C.
11. Transfer supernatants (~40 mL) to a clean tube. This extract can be used for purification now or can be frozen at −80 °C and thawed when needed.
12. Filter the extract through a 0.22 μm filter immediately before injection onto the column.
13. Charge metal affinity column with Nickel by flushing column with 3 mL of water then 3 mL of 0.1 M nickel sulfate. Rinse column with 5 mL water using a syringe.
14. Attach column to chromatography system pump and equilibrate with at least 10 column volumes of SBP2 Buffer A. This and all subsequent steps should be carried out at 4 °C.
15. Apply the sample to column at 0.5 mL/min or less.
16. Wash column with SBP2 Buffer A until baseline is stabilized.
17. Inject 5 mL of 300 mM imidazole in 1× PBS into loop and inject onto column.
18. Manually collect peak fraction starting when OD_{280} reaches 0.2 on either side of the peak .
19. Inject another 5 mL of imidazole to clean column.
20. Dialyze against Heparin Buffer A for three changes of 1 L >4 h each change.
21. Equilibrate Heparin Column in Heparin Buffer A.
22. Load dialyzed sample onto Heparin column.
23. Elute with linear gradient to 100% Heparin Buffer B.
24. Collect fractions and analyze for CT-SBP2 content by SDS-PAGE. CT-SBP2 migrates at approximately 72 kDa and elutes late in the gradient at ~80% Buffer B.
25. Pool peak fractions and concentrate and buffer exchange with storage buffer in a 50 MWCO Centrifugal filter unit. Add DTT to 2 mM prior to storage.
26. Use SDS-PAGE to quantitate, aliquot, and store at −80 °C.

3.2 Recombinant eEFSec Preparation

1. Grow overnight culture of E coli harboring FLAG-eEFSec in 5 mL of LB with 100/mL ampicillin shake at 225 rpm @ 37 °C.
2. Prepare a 2 L flask with 1 L LB.
3. Inoculate each flask with 2 mL of the overnight culture and add ampicillin to 100 μg/mL.

4. Incubate at 37 °C with shaking at 225 rpm until culture reaches an OD_{600} of 1.0. This usually takes 3.5–4.5 h. Check the OD_{600} at 3.5 h to estimate growth. Induce by adding 100 μL of 1 M IPTG for each 1 L culture and incubate for 16 h at 16 °C.
5. Pellet cells in 1 L bottles by centrifugation at 4000 × *g* for 15 min each.
6. Pour off the supernatant and freeze pellets at −80 °C overnight.
7. Resuspend each pellet (1 L culture) in 40 mL of eEFSec buffer A plus 0.5 mM PMSF in original bottles (40 mL total).
8. Transfer to two 50 mL tubes (~20 mL each).
9. Sonicate for 40 s to lyse cells. Do this a total of four times per tube on ice with 1 min breaks to keep lysate cool.
10. Spin at 15,000 × *g* for 15 min at 4 °C.
11. Transfer supernatants (~40 mL) to a clean tube. Spin again at 15,000 × *g* for 15 min at 4 °C.
12. Transfer supernatants (~40 mL) to a clean tube.
13. This extract can be used for purification now or can be frozen at −80 °C and thawed when needed.
14. For batch purification, transfer 1 mL of anti-FLAG beads to a 50 mL disposable tube and wash beads in eEFSec Buffer A. Perform two washes of 20 mL each .
15. Add 40 mL of lysate to the beads and incubate 2 h in the cold room with end-over-end rotation.
16. Wash beads five times 3–5 min each, with 20 mL of cold EFS Buffer A. Transfer to a clean tube for the final two washes.
17. Elute protein with 100 μL of eEFSec Buffer B containing 110 μg/mL 3× FLAG peptide. Mix 30 min in the cold room. Proceed to gel quantitation and/or store protein at -80C after adding DTT to 2 mM.

3.3 Aminoacyl tRNA Purification from Rabbit Testicles

1. Thaw frozen tissue in lysis buffer using 2 mL of buffer per gram of tissue.
2. Homogenize tissue with a hand blender, keeping mixture cool in an ice bath.
3. Clear lysate by centrifugation at 12,000 × *g* for 30 min at 4 °C, remove the supernatant, and transfer to a clean 50 mL tube.
4. For 20 mL of extract, add 12 mL of water, 8 mL of 5× Buffer T, and 20 mL of phenol.
5. Mix by inverting, do not vortex. There can be significant variability regarding rRNA contamination depending on the extent of agitation.

6. Spin at 12,000 × *g* for 10 min at 4 °C.
7. Transfer the aqueous layer to a clean tube.
8. Add another 20 mL of phenol, vortex and repeat **step 6**.
9. Remove the aqueous layer and place in a fresh tube.
10. Add 2.5 volumes of cold 100% ethanol and chill to −80 °C for 10 min to precipitate.
11. Spin at 12,000 × *g* for 15 min at 4 °C.
12. Wash the pellet with cold 70% ethanol.
13. Discard the supernatant without perturbing the pellet.
14. Spin max speed for 5 min at 4 °C and remove the reminder supernatant with a fine pipette.
15. Dry at room temperature for ~10 min.
16. Resuspend in 10 mM NaOAc water to achieve a concentration of ~10 mg/mL.
17. Evaluate purity and quality by agarose gel electrophoresis stained with ethidium bromide.

3.4 Purification of Sec-tRNA^{Sec} with eEFSec

1. Start with FLAG-eEFSec immobilized on anti-FLAG beads from Subheading 3.2, **step 16**.
2. Add 7.5 mg of a total testis aa-tRNA prep (Subheading 3.3) and 0.5 mM GTP in Testis Lysis Buffer lacking spermidine, glycerol and protease inhibitors (Buffer TB).
3. Incubate at 4 °C for 30 min.
4. Wash five times, 3–5 min each wash, with cold buffer TB. Transfer to a clean falcon tube for the final two washes.
5. Elute protein/tRNA complex with 100 μL 110 μg/mL 3× FLAG peptide and in 1 mL buffer TB. Mix 30 min at 4 °C for 30 min.
6. Transfer the eluent to a fresh tube and use in assays as a purified ternary complex (eEFSec/Sec-tRNA^{Sec}/GTP).
7. For Sec-tRNA^{Sec} purification, add 1/5 volume 5× Buffer T to the eluent and proceed as described in Subheading 3.3.

3.5 In Vitro Translation

1. Prepare the necessary components and thaw them in a room temperature water bath.
2. As soon as components thaw immediately move components to ice. Freeze-thaw of the lysate and CTSBP2 more than two times can result in diminished translation and Sec insertion activity.
3. The components for a typical single reaction for wheat germ lysate or rabbit reticulocyte lysate in vitro translation reactions are found in Tables 1 and 2 respectively.

Table 1
Components contained in a wheat germ lysate in vitro translation reaction modified for selenoprotein synthesis

Components	Volume (μL)
Nuclease-treated wheat germ lysate	6.25
Complete amino acid mix 1 μM	0.25
mRNA (200 ng/μL)	0.5
eEFSec 8 μM	0.5
CT-SBP2 8 μM	0.5
^{35}S Met or [^{75}Se]-Sec-tRNA^{Sec} (*see* **Note 4**)	0.5
aa-tRNA 10 mg/mL	0.5
Potassium acetate (875 mM)	1
Nuclease-free H_2O	3
Total	12.5

Table 2
Components contained in a rabbit reticulocyte lysate in vitro translation reaction modified for selenoprotein synthesis

Components	Volume (μL)
Nuclease-treated rabbit reticulocyte lysate	8
Complete amino acid mix 1 μM	0.25
mRNA (200 ng/μL)	0.5
CT-SBP2 8 μM (2 pmol/rxn)	0.5
[^{75}Se]-selenite 5 μM (*see* **Note 3**)	0.5
Nuclease-free H_2O	2.75
Total	12.5

4. Make a cocktail of common components, and then add the remaining reagents individually. Mix by gentle agitation of the tube. If necessary, briefly centrifuge the reaction to return the reaction at the bottom of the tube or plate.
5. Incubate at 30 °C for 1 h for rabbit reticulocyte lysate or 25 °C for 2 h for wheat germ lysate.
6. Analyze protein products by SDS-PAGE electrophoresis, fixation, and gel drying. It is not advisable to load more than 4 μL of rabbit reticulocyte lysate reactions per lane.
7. Expose dried gel to film or phosphorimager screen and Sec insertion can be observed in proteins of interest (*see* **Note 5**).

4 Notes

1. In our assays, we typically use the C-terminal half of rat SBP2 (amino acids 399-8946) as it is fully functional for all known Sec incorporation events analyzed in vitro [5].
2. For wheat germ lysate, each mRNA requires some fine tuning for concentration that will provide the largest yield, some factors may require the addition of potassium acetate to further optimize yield.
3. [^{75}Se]-selenite can be replaced with ^{35}S Methionine or Cysteine. It is advisable to perform a no mRNA control reaction to account for background bands.
4. [^{75}Se]-Sec-tRNA^{Sec} can be replaced with unlabeled Sec-tRNA-^{Sec} and ^{35}S Methionine or Cysteine. It is advisable to perform a no mRNA control reaction to account for background bands .
5. Free [^{75}Se]-selenite travels with the dye front and will appear if left on the gel, we recommend allowing the dye front to just start to run off the gel or cut off the dye front after fixing the gel.

Acknowledgments

This work was supported by NIH grants GM077073 and HD083616 to PRC.

References

1. Pelham HR, Jackson RJ (1976) An efficient mRNA-dependent translation system from reticulocyte lysates. Eur J Biochem 67:247–256
2. Anderson CW, Straus JW, Dudock BS (1983) Preparation of a cell-free protein-synthesizing system from wheat germ. Methods Enzymol 101:635–644
3. Lesoon A, Mehta A, Singh R, Chisolm GM, Driscoll DM (1997) An RNA-binding protein recognizes a mammalian selenocysteine insertion sequence element required for cotranslational incorporation of selenocysteine. Mol Cell Biol 17:1977–1985
4. Copeland PR, Driscoll DM (1999) Purification, redox sensitivity, and RNA binding properties of SECIS-binding protein 2, a protein involved in selenoprotein biosynthesis. J Biol Chem 274:25447–25454
5. Copeland PR, Fletcher JE, Carlson BA, Hatfield DL, Driscoll DM (2000) A novel RNA binding protein, SBP2, is required for the translation of mammalian selenoprotein mRNAs. EMBO J 19:306–314
6. Kinzy SA, Caban K, Copeland PR (2005) Characterization of the SECIS binding protein 2 complex required for the co-translational insertion of selenocysteine in mammals. Nucleic Acids Res 33:5172–5180
7. Gupta N, Demong LW, Banda S, Copeland PR (2013) Reconstitution of selenocysteine incorporation reveals intrinsic regulation by SECIS elements. J Mol Biol 425:2415–2422

Chapter 8

Studying Selenoprotein mRNA Translation Using RNA-Seq and Ribosome Profiling

Brian K. Dalley, Lisa Baird, and Michael T. Howard

Abstract

Deep sequencing of ribosome protected mRNA footprints, also called ribosome profiling or Ribo-Seq, is a relatively new methodology well suited to address questions regarding the mechanisms and efficiency of protein expression. Specifically, the ability of this technique to quantify ribosome abundance with codon resolution enables experiments aimed at studying many aspects of translation, including gene-specific translational efficiency, translation of regulatory upstream short open reading frames, sites of ribosome pausing, and most importantly for selenoproteins, the efficiency by which UGA codons are redefined to encode selenocysteine. Here, we describe a streamlined protocol that was developed in our lab to process mammalian tissue to produce the requisite matched ribosome profiling and RNA-Seq libraries for deep sequencing.

Key words Codon redefinition, Recoding, Ribo-seq, Ribosome profiling, RNA-Seq, RNA stability, Selenium, Selenocysteine, Translation

1 Introduction

Ribosome profiling is based on the observation that translating ribosomes protect ~30 nucleotides (nts) of the mRNA from digestion with ribonucleases [1, 2]. The method involves isolation of ribosome:mRNA complexes from tissues or cells, digestion of the unprotected mRNA with ribonucleases, and deep sequencing of the ribosome protected mRNA footprints (RPFs). Counting the number of footprints that map to open reading frames reflects ribosome density and provides an approximation of gene-specific protein synthesis rates. Importantly, additional information lies underneath this gene-level summary of translational activity. Changes in ribosome density at specific codons or regions across an individual mRNA can shed light into translational mechanisms. For example, ribosome profiling has revealed alternative translation start sites and previously unknown protein coding sequences as well as evidence for translation of regulatory upstream open reading

Laurent Chavatte (ed.), *Selenoproteins: Methods and Protocols*, Methods in Molecular Biology, vol. 1661,
DOI 10.1007/978-1-4939-7258-6_8, © Springer Science+Business Media LLC 2018

frames in 5′ UTRs [3]. Sites of ribosome pausing appear as regions with increased ribosome density [4, 5]. Sites of translation initiation and termination can be mapped as locations where the ribosome density abruptly increases or decreases, respectively. This last feature is of particular interest for the analysis of selenoprotein expression since selenoprotein biosynthesis requires redefinition of a UGA codon to encode for selenocysteine (Sec) rather than terminate translation. This noncanonical translation event is in competition with termination and thus presents a limiting step in selenoprotein synthesis, which is subject to gene-specific regulation [6, 7].

Measuring the efficiency of Sec incorporation during selenoprotein expression from mRNAs produced from their native genomic configuration in vivo has not previously been possible due to the inability to reliably monitor an unstable termination product [8] relative to the full-length Sec containing protein. However, this question can be addressed in vivo using ribosome profiling to monitor the change in ribosome density before and after UGA-Sec codons [6], with RPFs located 3′ of the UGA-Sec codon (3′ RPFs) being proportional to the number of ribosomes that have successfully incorporated Sec and are translating downstream codons.

Herein, we present a streamlined method (Fig. 1), based on the ribosome profiling technology originally published by the Weissman laboratory [9] that is applicable to mammalian tissue. It should be noted that this is a still developing approach to study translation with several technical and bioinformatic components that can lead to misinterpretation of the data. While we address some of these issues in the Notes section, it is important—as it is with any method—to develop independent experimental approaches to validate the results obtained by ribosome profiling.

2 Materials

Prepare all solutions with RNase-free water, work in a clean environment, and wear gloves to prevent unintended RNase contamination.

2.1 Buffers

1. 5× Polysome buffer: 50 mM Tris–HCl, 1500 mM KCl, 50 mM $MgCl_2$, pH 7.5. Prepare 20 ml.
2. Lysis buffer (prepare 2 ml for each sample): 0.4 ml of 5× polysome buffer, 0.2 ml of 10% Triton X-100, 20 μl of 100 mM 1,4-Dithiothreitol, 20 μl of 2 U/μl DNaseI (RNase-free), 16 μl of 25 mg/ml cycloheximide, 1.344 ml RNase-free H_2O (*see* **Note 1**).
3. Sucrose Buffer: 1.2 ml of 5× polysome buffer, 3.9 ml of 77% sucrose solution, 6 μl of 20 U/μl SUPERase In, 20 μl of 1 M

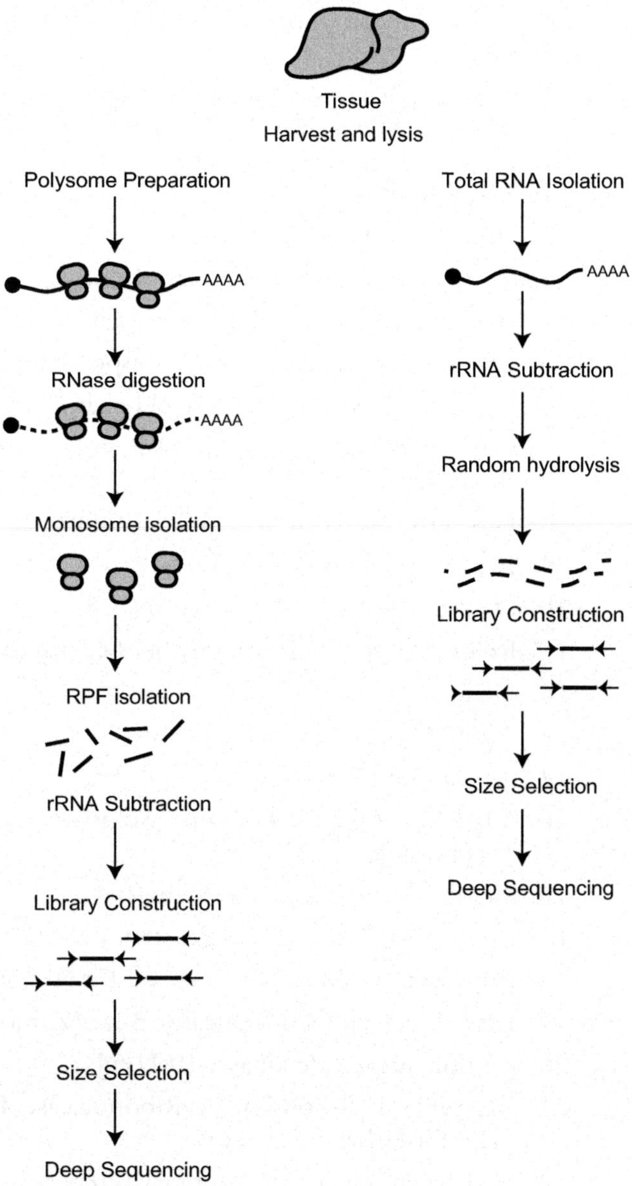

Fig. 1 Overview of ribosome profiling

$MgCl_2$, 30 μl of 100 mM 1,4-Dithiothreitol, 48 μl of 25 mg/ ml cycloheximide, 796 μl of RNase-free H_2O (*see* **Note 2**).

4. Denaturing 15% polyacrylamide gel, 4.8 g Urea, 3.75 ml 40% Acrylamide/Bis 19:1, 1 ml 10× TBE buffer, 2.5 ml RNase-free H_2O, 4 μl TEMED, 33 μl 10% APS.
5. DEPC RNase-free H_2O: Prepare by adding 2 ml of DEPC (Diethylpyrocarbonate) to 2 l of H_2O. Shake the bottle vigorously. Let the bottle sit overnight at room temperature and then autoclave the next day.

6. DNA Dilution Buffer, 10 mM Tris–HCl, 0.05% Tween-20, pH 8.0.
7. T4 Polynucleotide Kinase Buffer (10×), 700 mM Tris–HCl, 100 mM $MgCl_2$, 5 mM DTT, pH 7.6.

2.2 Enzymes and Other Commercial Reagents

1. RNase 1, 100 U/μl.
2. DNase 1, 2 U/μl, RNase free.
3. SUPERase In (ThermoFisher Scientific).
4. miRNeasy Mini Kit (Qiagen).
5. 2.0 ml sterile screw cap vial (Biospec Products) (*see* **Note 3**).
6. 2.3 mm diameter chrome steel beads (Biospec Products).
7. Ultra centrifuge tubes, thickwall, polycarbonate, 11 × 34 mm (Beckman).
8. 10× TBE buffer: 890 mM Tris-Borate, 20 mM EDTA.
9. 40% Acrylamide/Bis 19:1.
10. DEPC (Diethylpyrocarbonate) (Amresco).
11. RnaseZAP spray (ThermoFisher Scientific).
12. Urea.
13. TEMED.
14. Ammonium Persulfate.
15. 20/100 Oligo Length Standard (Integrated DNA Technologies).
16. Sucrose, BioUltra grade (Sigma).
17. SYBR Gold Nucleic acid gel stain (ThermoFisher Scientific).
18. Ribo-Zero Gold rRNA Removal Kit (Illumina).
19. RNA Clean and Concentrator-5 kit (Zymo Research).
20. T4 polynucleotide kinase, 10 U/μl.
21. RNaseOUT Recombinant Ribonuclease Inhibitor, 40 U/μl (ThermoFisher Scientific).
22. NEBNext Small RNA Library Prep Set for Illumina (New England Biolabs).
23. QIAquick PCR Purification Kit (Qiagen).
24. Qubit ds DNA Assay Kit (ThermoFisher Scientific).
25. 3% Agarose Dye-Free Marker F Pippin Gel Cassette (Sage Science).
26. High Sensitivity D1000 ScreenTape (Agilent Technologies).
27. Kapa Library Quantification Kit for Illumina (Kapa BioSystems).

2.3 Equipment

1. Cell homogenizer, Mini-Beadbeater-8 (Biospec Products) (*see* **Note 4**).
2. Biopulverizer, capacity 0.1–1 g (Biospec Products).

3. Magnetic stand.
4. TL-100 Ultracentrifuge (Beckman).
5. TLA100.2 rotor (Beckman).
6. Mini-Protean Tetra Cell gel apparatus (BioRad).
7. PCR Thermocycler.
8. Heating block.
9. SpeedVac Concentrator (ThermoFisher Scientific).
10. Vortex.
11. Qubit Fluorometer (ThermoFisher Scientific).
12. Pippin Prep (Sage Science).
13. 2200 TapeStation Instrument (Agilent Technologies).
14. CFX Connect Real-Time PCR Detection System (Bio-Rad Laboratories).

3 Methods

The method described here is for mouse tissue with specific information provided for liver and skeletal muscle. We have successfully applied this method to several different tissues types and consequently believed this method can be broadly applied to other mammalian tissues. As ribosomes move at ~6 codons/s and nucleases digest RNA very quickly, do not let samples "rest" between steps unless indicated.

3.1 Tissue Lysis and Purification of RPFs

1. Following harvest, tissues should be immediately cut into approximately ~1 cm square portions and plunged into liquid nitrogen. Tissue may be stored for up to several months at −80 °C prior to processing for ribosome profiling.
2. For each sample, pipette 1.5 ml Lysis buffer into a 2.0 ml sterile screw cap vial containing 3 × 2.3 mm diameter chrome steel balls. Place on ice (*see* **Note 5**).
3. Slice off approximately 60–80 mg piece of tissue, while still frozen, and place immediately into the tube containing lysis buffer from **step 2**. If processing skeletal muscle tissue we recommend first "flattening" the frozen tissue in a Biopulverizer. This step is not needed for liver tissue (*see* **Note 6**).
4. Homogenize on highest setting in the Mini-Beadbeater-8, 2 × 30 s (*see* **Note 7**).
5. Immediately add 10 μl of RNase 1 and incubate for 30 min (skeletal muscle) or 60 min (liver) at RT with gentle rocking (*see* **Note 8**).
6. Prepare Sucrose Buffer during the RNase 1 digestion.

7. Add 50 μl of SUPERase In to the reaction tube to inhibit RNA digestion. Vortex gently to mix.
8. Centrifuge 12,000 × *g* in a microfuge for 10 min at 4 °C. Remove the supernatant to a new 1.5 ml tube.
9. For each sample, pipette 220 μl of lysate supernatant from **step 7** into 6× ultra centrifuge tubes. Underlay with 660 μl of Sucrose Buffer using a 1 cc syringe and 18-gauge needle. Place tubes in a TLA 100.2 rotor (*see* **Note 9**).
10. Centrifuge tubes at 200,000 × *g* for 3.5 h in TL-100 ultracentrifuge.
11. Carefully remove all supernatant. The ribosome pellet will be translucent but visible.
12. Resuspend the six ribosome pellets in a total volume of 1.4 ml Qiazol (*see* **Note 10**).
13. Proceed to **step 1** of Subheading 3.3.

3.2 Tissue Lysis and Processing of Total RNA for RNA-Seq

1. For each sample, pipette 1.4 ml Qiazol into a 2.0 ml screw cap centrifuge tube containing 3 × 2.3 mm diameter chrome steel balls. Place on ice.
2. From the same tissue sample used in Subheading 3.1, **step 3**, slice off approximately 50 mg piece of tissue, while still frozen, and place immediately into the tube containing Qiazol from **step 1**.
3. Homogenize on highest setting in the Mini-Beadbeater-8, 2 × 30 s.
4. Proceed to **step 1** of Subheading 3.3.

3.3 RNA Purification

1. Split the samples from Subheading 3.1, **step 13** or Subheading 3.2, **step 4** into 2 × 1.5 ml centrifuge tubes (700 μl each). Incubate at RT for 5 min.
2. Add 140 μl of chloroform to each tube and shake vigorously by hand for 15 s.
3. Incubate at RT for 2–3 min.
4. Centrifuge at 12,000 × *g* for 15 min at 4 °C.
5. Remove upper phase to a new centrifuge tube (*see* **Note 11**).
6. Add 1.5 volumes of EtOH and pipette up and down to mix thoroughly.
7. Add 700 μl onto a miRNeasy Mini spin column.
8. Centrifuge 8000 × *g* for 15 s. Discard the flow through and repeat with the remainder of the sample.
9. Add 700 μl Buffer RWT (miRNeasy Mini kit). Centrifuge 8000 × *g* for 15 s. Discard the flow-through.
10. Add 500 μl Buffer RPE (miRNeasy Mini kit). Centrifuge 8000 × *g* for 15 s. Discard the flow-through.

11. Repeat **step 10**.
12. Place the spin column in a new collection tube and centrifuge at 8000 × *g* for 1 min.
13. Place the spin column in a new 1.5 ml centrifuge tube, add 30 μl of H_2O and centrifuge at 8000 × *g* for 1 min to collect purified RNA (*see* **Note 12**).
14. Estimate the concentration by the analysis of the optical density at 260 nm. Using this approach we typically obtain between 3 and 10 μg of RNA per sample. If necessary, concentrate samples in a SpeedVac concentrator (no heat) such that the concentration is at least 0.1 μg/μl. Electrophoresis of ~500 ng of RNA on a denaturing 15% polyacrylamide gel is also recommended for the RPF samples to check the extent of RNA digestion, as well as an initial time point experiment taken through this entire experiment to determine the optimal time of digestion (*see* **Note 13**). Next proceed to library construction, Subheading 3.4.

3.4 Library Construction for Total RNA and RFPs

3.4.1 Prepare Ribo-Zero Magnetic Beads

1. For each sample, add 90 μl of Ribo-Zero Magnetic Beads to a 1.5 ml microfuge tube.
2. Place the tubes on a magnetic stand for 2 min.
3. Remove and discard the supernatant.
4. Add 90 μl of RNase-free H_2O to each tube.
5. Vortex the tubes to resuspend the magnetic beads.
6. Briefly spin tubes by centrifugation and place the tubes on a magnetic stand for 2 min.
7. Remove and discard the supernatant.
8. Repeat the H_2O wash one additional time for a total of two washes.
9. Add 24 μl of Ribo-Zero Mag Beads Resuspension Sln to each tube.
10. Vortex the tubes to resuspend the magnetic beads.
11. Add 0.4 μl of Riboguard RNase Inhibitor to each tube and mix by gently pipetting the solution up and down 5× with a pipette set at 25 μl.
12. Store the beads at RT until used in Subheading 3.4.2, **step 8**.

3.4.2 Ribo-Zero Treatment of RNA

1. Add 25–500 ng of RPFs or total RNA (*see* **Note 14**) to a new 1.5 ml microfuge tube and adjust the volume to 11.2 μl with water.
2. Add 1.6 μl of Ribo-Zero rRNA Reaction Buffer to each tube.
3. Add 3.2 μl of Ribo-Zero Removal Mix-Gold to each tube for a total volume of 16 μl.

4. Place the tubes on an Eppendorf MixMate or suitable mixer and mix at 1400 rpm for 30 s.
5. Briefly centrifuge tubes to bring contents to bottom of tubes.
6. Transfer the tubes to a heat block set at 68 °C and incubate for 10 min.
7. Remove the tubes from the heat block and incubate at RT for 5 min.
8. Transfer the samples incubated with rRNA Removal Mix to the tubes containing Magnetic Beads (from Subheading 3.4.1, **step 12**) using a pipettor set at 16 μl and immediately pipette up and down 10×.
9. Place the tubes on an Eppendorf MixMate or suitable mixer and mix samples at 1400 rpm for 30 s.
10. Incubate the tubes at RT for 5 min.
11. Following brief centrifugation, transfer the tubes to a magnetic stand for 2 min.
12. Transfer the supernatant (approximately 36 μl) to a new 1.5 ml microfuge tube containing 64 μl of water. Avoid the transfer of beads as these will be bound to rRNA.
13. Proceed to Subheading 3.4.3.

3.4.3 Purification of Ribo-Zero Treated Total RNA and RPFs

1. Using a Zymo Research RNA Clean and Concentrator-5 kit, add RNA Binding Buffer to the Ribo-Zero-treated samples as described below:
   (a) Add 100 μl of RNA Binding Buffer to each Ribo-Zero-Treated Total RNA sample for a total volume of 200 μl.
   (b) Add 200 μl of RNA Binding Buffer to each RPF RNA sample for a total volume of 300 μl.
2. Vortex the tubes for 5 s to mix and then briefly spin by centrifugation.
3. Add EtOH to the Ribo-Zero-treated samples as described below:
   (a) Add 100 μl of 100% EtOH to each Ribo-Zero-Treated Total RNA sample for a final volume of 300 μl.
   (b) Add 450 μl of 100% EtOH to each Ribo-Zero-Treated RPF RNA sample for a final volume of 750 μl.
4. Vortex the tubes for 5 s to mix.
5. Load each sample onto a Zymo Research RNA Clean and Concentrator-5 spin column.
6. Centrifuge the column at 12,000 × *g* for 30 s.
7. Remove and discard the flow-through.
8. Add 400 μl of Zymo Research RNA Prep Buffer to each column.

9. Centrifuge the column at 12,000 × *g* for 30 s.
10. Remove and discard the flow-through.
11. Add 700 μl of Zymo Research RNA Wash Buffer to each column.
12. Centrifuge the column at 12,000 × *g* for 30 s.
13. Remove and discard the flow-through.
14. Add 400 μl of Zymo Research RNA Wash Buffer to each column.
15. Centrifuge the column at 12,000 × *g* for 2 min.
16. Transfer the column to a new 1.5 ml microfuge tube.
17. Add 8 μl of water to the column and incubate at RT for 1 min.
18. Centrifuge the column at 12,000 × *g* for 30 s to elute the RNA.
19. Immediately proceed to Subheading 3.4.4.

3.4.4 Fragmentation of Total RNA Samples

This step is for Total RNA samples only and should not be applied to RPF samples.

1. Transfer a 7 μl volume of each Ribo-Zero-treated Total RNA sample to a new PCR tube.
2. Add 3 μl of T4 Polynucleotide Kinase Buffer (10×) to each Total RNA Sample (*see* **Note 15**).
3. Place the tubes on an Eppendorf MixMate and mix at 2000 rpm for 30 s or gently vortex. Briefly spin in a centrifuge.
4. Transfer the tubes to a thermal cycler with a heated lid set at 100 °C and run the following profile:
   - 94 °C Pre-heat hold.
   - 94 °C for 8 min.
   - 4 °C Hold.
5. Following the 8 min incubation at 94 °C, immediately transfer the tubes to an ice bath for 1 min.
6. Immediately proceed to Subheading 3.4.5, **step 1b**.

3.4.5 T4 Polynucleotide Kinase Treatment of rRNA-Depleted RPF and Total RNA

1. Prepare T4 polynucleotide kinase (PNK) master mix:
   (a) Prepare PNK Master Mix for RPF Samples by combining the following volumes of reagents per reaction and add 43 μl of master mix to each 7 μl volume of RPF sample for a final volume of 50 μl. Mix by vortex and briefly spin down. Prepare 10% overage when processing multiple samples.
   - 30.6 μl of water.
   - 6 μl of T4 Polynucleotide Kinase Reaction Buffer (10×).

- 6 μl of 10 mM ATP.
- 2.4 μl of T4 Polynucleotide Kinase 10 U/μl.
- 1 μl RNaseOUT.

(b) PNK Master Mix Preparation for Fragmented Total RNA Samples: Prepare PNK Master Mix for Fragmented Total RNA Samples by combining the following volumes of reagents per reaction and add 40 μl of reaction mix to each 10 μl volume of fragmented, total RNA sample for a final volume of 50 μl. Mix by vortex and briefly spin down. Prepare 10% overage when processing multiple samples.

- 30.6 μl of water.
- 6 μl of 10 mM ATP.
- 2.4 μl of T4 Polynucleotide Kinase 10 U/μl.
- 1 μl RNaseOUT.

2. Place the tubes on an Eppendorf MixMate and mix at 2000 rpm for 30 s. Briefly spin by centrifugation.
3. Transfer the tubes to a thermal cycler with a heated lid set at 42 °C and run the following profile:
   - 37 °C Pre-heat Hold.
   - 37 °C for 60 min.
   - 4 °C Hold.
4. Immediately proceed to Subheading 3.4.6.

3.4.6 Purify T4 PNK-Treated RNA on a Zymo RNA Clean and Concentrator-5 Column

1. Transfer each 50 μl T4 PNK-treated sample to a 1.5 ml microfuge tube containing 50 μl of water.
2. Add 200 μl of RNA Binding Buffer from a Zymo Research RNA Clean and Concentrator-5 kit to the diluted T4 PNK-treated sample and mix by vortexing.
3. Add 450 μl of 100% EtOH and mix well by inversion.
4. Load each sample (approximately 750 μl) onto a Zymo RNA Clean and Concentrator-5 spin column and precede with the buffers provided in the Zymo RNA Clean and Concentrator-5 kit as described below.
5. Centrifuge the column at 12,000 × *g* for 30 s.
6. Remove and discard the flow-through.
7. Add 400 μl of RNA Prep Buffer to each column and centrifuge for 30 s at 12,000 × *g*.
8. Remove and discard the flow-through.
9. Add 700 μl of Zymo Research RNA Wash Buffer to each column and centrifuge for 30 s at 12,000 × *g*.
10. Remove and discard the flow-through.

11. Add 400 μl of Zymo Research RNA Wash Buffer to the column and centrifuge for 2 min at 12,000 × *g*.
12. Transfer the column to a new 1.5 ml microfuge tube.
13. Add 7 μl of water to each column and incubate for 1 min at RT.
14. Centrifuge the column for 1 min at 12,000 × *g* to elute the RNA.
15. Proceed to Subheading 3.4.7.

3.4.7 Ligate the Multiplex 3′ SR Adapter

1. The remaining steps of library preparation will use reagents from the NEBNext Small RNA Library Prep Set for Illumina.
2. Combine in a thin-wall PCR tube 1.0 μl of 3′ SR Adaptor with 6 μl of purified RNA from Subheading 3.4.6, **step 14**, for a total volume of 7 μl.
3. Place the tubes on an Eppendorf MixMate and mix at 2000 rpm for 30 s. Briefly spin in a centrifuge.
4. Transfer the samples to a thermal cycler with a heated lid set at 100 °C and run the following profile to denature the adapter.
   - 70 °C Pre-Heat Hold.
   - 70 °C for 2 min.
   - 4 °C Hold.
5. Following the 2 min incubation at 70 °C, immediately transfer the reaction tubes to an ice water bath.
6. Prepare a 3′ Ligation Master Mix by combining the reagents below. Mix by vortex and briefly spin down by centrifugation. Prepare 10% overage if you are preparing master mix for multiple samples.
   - 10 μl of 3′ Ligation Reaction Buffer (2×).
   - 3 μl 3′ Ligation Enzyme Mix.
7. Add 13 μl of the master mix to each reaction tube from **step 5** for a total volume of 20 μl.
8. Place the tubes on an Eppendorf MixMate and mix at 2000 rpm for 30 s.
9. Transfer the tubes to a thermal cycler without heated lid and run the following profile:
   - 25 °C Pre-Heat Hold.
   - 25 °C for 1 h.
   - 4 °C Hold.
10. Immediately proceed to Subheading 3.4.8.

3.4.8 Hybridize the Reverse Transcription Primer

1. Prepare RT Primer Master Mix by combining the following volumes of reagents per reaction. Mix by vortex and briefly spin down by centrifugation. Prepare 10% overage when processing multiple samples.
   - 4.5 μl of water.
   - 1.0 μl SR RT Primer for Illumina.
2. Add 5.5 μl of RT Primer Master Mix to each tube from the 3′ Adapter Ligation Reaction for a total volume of 25.5 μl.
3. Place the tubes on an Eppendorf MixMate and mix at 2000 rpm for 30 s. Briefly spin in a centrifuge.
4. Transfer the tubes to a thermal cycler with a heated lid set at 80 °C and run the following profile:
   - 75 °C Pre-heat Hold.
   - 75 °C for 5 min.
   - 37 °C for 15 min.
   - 25 °C for 15 min.
5. Immediately proceed to Subheading 3.4.9.

3.4.9 Ligate the 5′ SR Adaptor

1. For each sample, aliquot 1.1 × N μl of the 5′ SR Adaptor into a new PCR tube where N equals the number of reactions that you will be processing.
2. Place the tubes in a preheated thermal cycler and run the following protocol. At the end of the 2 min incubation, immediately transfer the tubes to an ice bath. Use the denatured adapter within 30 min.
   - 70 °C Pre-Heat Hold.
   - 70 °C for 2 min.
   - 4 °C Hold.
3. Prepare 5′ Ligation Master Mix by combining the following volumes of reagents for each reaction. Following the addition of the Ligase Enzyme, pipette the master mix up and down three to four times. Prepare 10% overage when processing multiple samples.
   - 1.0 μl of 5′ SR Adapter for Illumina (denatured).
   - 1.0 μl of 5′ Ligation Reaction Buffer (10×).
   - 2.5 μl of 5′ Ligase Enzyme Mix.
4. Briefly vortex the master mix and spin contents to the bottom of the tube by brief centrifugation.
5. Add 4.5 μl of 5′ Ligation Master Mix to each reaction tube from Subheading 3.4.8, **step 5** for a total volume of 30 μl.
6. Place the reaction tubes on an Eppendorf MixMate and mix at 2000 rpm for 30 s followed by brief centrifugation.

7. Transfer the reaction tubes to a thermal cycler without heated lid and run the following profile:
   - 25 °C Hold.
   - 25 °C for 1 h.
   - 4 °C Hold.
8. Immediately proceed to Subheading 3.4.10.

3.4.10 Perform Reverse Transcription

1. Prepare Reverse Transcription Master Mix using the following volumes of reagents for each reaction. Prepare 10% overage when processing multiple samples.
   - 8.0 μl of First Strand Synthesis Reaction Buffer.
   - 1.0 μl of Murine RNase Inhibitor.
   - 1.0 μl of Protoscript II Reverse Transcriptase.
2. Briefly vortex the master mix and then spin contents down by centrifugation.
3. Add 10 μl of Reverse Transcription Master Mix to each reaction tube containing 3′ and 5′ Adapter Ligated RNA from Subheading 3.4.9, **step 8**, for a total volume of 40 μl.
4. Place the reaction tubes on an Eppendorf MixMate and mix at 2000 rpm for 30 s. Briefly centrifuge.
5. Transfer the tubes to a thermal cycler with a heated lid set at 55 °C and run the profile:
   - 50 °C Pre-Heat Hold.
   - 50 °C for 1 h.
   - 4 °C Hold.
6. Following completion of the reverse transcription protocol, immediately proceed to Subheading 3.4.11.

3.4.11 PCR Amplification

1. Prepare a PCR Master Mix with the following volumes of reagents for each reaction. Prepare 10% overage when processing multiple samples.
   - 50 μl of LongAmp Taq 2× Master Mix.
   - 2.5 μl of SR Primer for Illumina.
   - 5 μl of water.
2. Vortex the PCR Master Mix, then briefly spin contents down by centrifuge.
3. Add 57.5 μl of Master Mix to each sample from Subheading 3.4.10, **step 6**.
4. Add 2.5 μl of one of the RNA Index Primers to each tube for a final volume of 100 μl. Samples in the same multiplex group for sequencing must contain unique PCR Primer indices.

5. Place the reaction tubes on an Eppendorf MixMate and mix at 2000 rpm for 30 s followed by brief centrifugation.
6. Perform PCR amplification with heated lid set at 100 °C and run the following profile:
   - 94 °C for 30 s.
   - 12 cycles of:
     - 94 °C for 15 s.
     - 62 °C for 30 s.
     - 70 °C for 15 s.
   - 70 °C for 5 min.
   - 4 °C Hold.
7. Following completion of PCR, immediately proceed to Subheading 3.4.12.

3.4.12 Qiagen Cleanup of PCR Amplified Library

1. Transfer the PCR reaction (100 μl) to a 1.5 ml microfuge tube.
2. Add 5 volumes (500 μl) of Buffer PB from the QIAquick PCR Purification Kit to the sample and mix by vortexing followed by brief centrifugation.
3. Place a QIAquick PCR Purification Kit column in a 2 ml collection tube.
4. Transfer the sample containing Buffer PB to a QIAquick column.
5. Centrifuge the QIAquick tube for 60 s at 12,000 × *g* to bind DNA to the column.
6. Remove tubes from the centrifuge and discard the flow-through.
7. Add 700 μl of Buffer PE (with EtOH added) to each column.
8. Place QIAquick tubes in a centrifuge and spin at 12,000 × *g* for 1 min.
9. Remove tubes from the centrifuge and discard the flow-through.
10. Centrifuge the tubes one additional time at 12,000 × *g* for 5 min.
11. Remove QIAquick column from the 2 ml collection tube and position the column in a new 1.5 ml microfuge tube.
12. Add 32 μl of Qiagen Buffer EB to the center of each column and incubate for 1 min at RT.
13. Centrifuge the QIAquick columns for 1 min at 12,000 × *g* to elute and collect the amplified DNA in the 1.5 ml microfuge tube.

14. Measure the concentration of the eluted DNA by Qubit HS DNA Assay.
15. If you do not immediately proceed to size selection of the library on the Pippin Prep, the sample can be safely stored at −20 °C.

3.5 Library Size Selection and Normalization

3.5.1 Size Selection Using a Sage Science Pippin Prep

Calibration should be performed on a daily basis or whenever the cassette type is changed.

1. Using the Protocol Editor Tab on a Sage Science Pippin Prep, set up a size selection protocol with the following run parameters:
   - Cassette: 3% DF Marker F.
   - Start: 105.
   - End: 170.
   - Target: 138.
   - Ref Lane: Same value as the lane number.
   - Range Flag: Broad.
   - Internal Standards.
2. Allow DNA Marker F to come to Room Temp for 30 min.
3. Combine 20 μl of purified library with 10 μl of water in a new 1.5 ml microfuge tube.
4. Vortex the DNA Marker F and add 10 μl of the marker to each sample for a total volume of 40 μl.
5. Load 40 μl of the Library (containing DNA Marker F loading solution) into each well of a 3% Agarose Dye-Free Pippin Gel Cassette.
6. Run the size selection protocol using the parameters described above.
7. Following completion of the run, set a pipettor to 60 μl, and remove the volume of buffer present in the elution module and transfer it to a 1.5 ml microfuge tube. This volume will contain the size-selected library. Repeat this process for the remaining elution modules.
8. Measure the concentration of the eluted DNA using a Qubit ds DNA HS Assay Kit.
9. Run an aliquot of the eluted library on an Agilent 2200 TapeStation High Sensitivity D1000 ScreenTape to verify the size distribution of the library. Due to the presence of Pippin Running Buffer, the eluted library will resolve at an apparent size distribution approximately 20–30 bp larger than expected. This anomaly can be corrected with purification of the size-selected library using a Qiagen QIAQuick PCR Purification Kit if desired.

3.5.2 Normalization and Pooling of Size Selected Libraries Using a KapaBiosystems Illumina Library Quantification Kit

1. For each library being assayed, dispense 1.5 ml of DNA Dilution Buffer into three microcentrifuge tubes.
2. Dispense 1 μl of the size-selected library into each tube of DNA Dilution Buffer for a final dilution of 1:1500. Vortex the diluted libraries thoroughly and centrifuge for 1 min at 12,000 × *g*.
3. Prepare Kapa Biosystems qPCR Master Mix with Primer PreMix as described and mix thoroughly by vortex. Dispense 12 μl of qPCR Master Mix into a 96-well plate for each reaction.
4. Assay DNA standards in triplicate reactions. Dispense the following volumes of reagents into each well of a 96-well plate being used for assay of DNA Standards:
   - 12 μl of KAPA SYBR FAST qPCR Master Mix containing Primer Premix.
   - 4 μl of water.
   - 4 μl of DNA Standard.
5. Assay the 1:1500 diluted library (prepared in triplicate dilutions) by combining the following volumes of reagents into each well of a 96-well plate:
   - 12 μl of KAPA SYBR FAST qPCR Master Mix containing Primer Premix.
   - 4 μl of water.
   - 4 μl of 1:1500 diluted library.
6. Seal the 96-well plate with a Bio-Rad Microseal "B" Seal and centrifuge the plate for 1 min at 1000 rpm.
7. Transfer the 96-well plate to a Bio-Rad CFX Connect Real-Time System and run the following profile:
   - 95 °C for 5 min.
   - 36 cycles of:
     - 95 °C for 30 s.
     - 60 °C for 45 s.
8. Following the completion of the run, prepare a copy of the data in Microsoft Excel format. Calculate the molarity of each sequencing library by averaging the triplicate qPCR values and comparing those values to the concentrations of the known DNA Standards.
9. Normalize the concentrations of the undiluted libraries by the addition of 10 mM Tris–HCl, 0.01% Tween 20, pH 8.0.
10. Pool equal volumes of each normalized library that will be sequenced in the same multiplex group into a microcentrifuge tube.

3.6 Illumina Sequencing

1. Pooled sequencing libraries are chemically denatured and adjusted to a final concentration of 21 pM.
2. Denatured libraries are applied to an Illumina HiSeq SR v4 flowcell using an Illumina cBot. Hybridized molecules are clonally amplified and annealed to sequencing primers with reagents from an Illumina HiSeq SR Cluster Kit v4.
3. Transfer the flowcell to an Illumina HiSeq 2500 instrument and perform a 50-cycle sequence run with an additional 7 base index read using a HiSeq SBS Kit v4.

3.7 Preliminary Bioinformatic Analysis

For those just starting with bioinformatic analysis of ribosome profiling data sets, we highly recommend using RiboGalaxy: http://ribogalaxy.ucc.ie [10]. Results from our own ribosome profiling analyses are briefly outlined as follows. From a single sequencing lane we typically obtain ~300,000 sequence reads. Adapter sequences are trimmed from sequences using fastx_clipper from the Hannon Lab FASTX Toolkit (http://hannonlab.cshl.edu/fastx_toolkit/index.html) discarding all adapter only sequences and N sequences while keeping sequences 18 nts in length or greater. After clipping we typically have between 220,000 and 250,000 sequence reads. We find Bowtie 1 to be useful for aligning trimmed sequence reads to various sequence features. First, we align sequence reads against rRNA sequences from the organism being interrogated. From mouse liver ribosome profiles processed as described above, ~20% of reads align to rRNA. Of the remaining sequence reads, approximately 40–50% align to RefSeq mRNAs (uniquely aligning to a RefSeq database containing a single representative mRNA for each gene, allowing for two mismatches). We typically discard RPFs with predicted A-sites that map in the first 15 codons of each mRNA as this region is subject to drug-induced accumulation of ribosomes [11]. A-sites typically map to approximately the center of the footprint but may be precisely determined by meta-analysis of all genes (the position of footprints is determined relative to the start and stop codons of all mRNAs and analyzed collectively). The 5′ end of ribosome footprints should increase dramatically just upstream of the collective start codon. By determining the average distance from the 5′ end of these sequence reads to the start codons, the P-site positions can be determined. Conversely, by looking at the average distance from where the footprints drop in abundance near the stop codons, A-site positions may be inferred. The triplet codon phasing of the footprints downstream of the start codons and upstream of the stop codons should also be apparent in such a meta-analysis. We also typically align to a Bowtie index composed of only selenoprotein mRNAs and identify ribosomes upstream and downstream of the UGA according to the annotated position of the UGA relative to the A-site position of each footprint. Raw counts can be normalized as reads per kilobase per million mapped reads for comparison between experiments. Alternatively, we have also found EdgeR (https://bioconductor.org/packages/release/bioc/html/edgeR.html) to be a useful tool to determine alternative normalization factors and test differential expression of ribosome footprints and RNA.

4 Notes

1. 5× Polysome Buffer may be made ahead of time and stored in aliquots at −20 °C. 100 mM DTT should be stored at −20 °C in single use aliquots. Cycloheximide should be resuspended in DMSO and stored at 4 °C. Components of Lysis Buffer should be assembled immediately before use on ice.
2. Prepare 77% Sucrose solution first. In a 50 ml Conical tube, weigh 5 g of Sucrose. Add 3 ml H_2O and vortex. Place the tube in a beaker with warm H_2O. Vortex every few minutes until sucrose is in solution. Using a 50 ml tube (rather than a 15 ml tube for instance) aids in this process because the bottom of the bigger tube allows for better mixing. Next, assemble all the ingredients of Sucrose Buffer in a 15 ml tube on ice. Add the Sucrose solution last and mix by vortexing.
3. Heavy-duty homogenization 2.0 ml tubes from Biospec Products are recommended because they withstand the forces exerted by the steel beads better than standard polypropylene tubes.
4. The Mini Bead Beater-8 we use is no longer manufactured but BioSpec now sells a similar product. Other homogenizers should be suitable.
5. Do not use tubes that have been subjected to cold temperatures (i.e., −20 °C or −80 °C) because they have a higher risk of cracking during the homogenization process. Do not use a volume of lysis buffer less than 1.5 ml when homogenizing to minimize the possibility of cracking the tube during the homogenization process.
6. Optional: For disruption of fibrous tissue first prechill the biopulverizer in liquid nitrogen. Take care to be sure that all parts of the unit are well chilled. Place piece of frozen skeletal muscle tissue in the pulverizer and give three sharp hits with the mallet provided with the unit on the pestle. Turn the pestle slightly between each hit. The tissue should now be flat and pancake like in appearance. Use a clean RNAse-free set of forceps to remove the tissue from the device and place it into the tube containing the lysis buffer. Proceed with homogenization.

7. Work as quickly as possible to complete homogenization after adding the tissue to the lysis buffer.
8. A Labquake® rocker works well for this step but any piece of equipment that provides gentle mixing is acceptable.
9. Slowly dispense the Sucrose Buffer solution at the bottom of the centrifuge tube using the syringe and needle. The RNA lysate will float on the top of the expelled Sucrose solution.
10. Add 233 μl of Qiazol to the first centrifuge tube and resuspend the pellet by repeated pipetting. Let the tube sit. Add 233 μl to each of the subsequent tubes, one by one, and pipette repeatedly to resuspend each pellet. Go back to the first tube and repeatedly pipette and then remove the liquid to a new tube. Repeat this step for each tube in turn and pool all of the Qiazol into a single tube with the final volume being approximately 1.4 ml.
11. Generally this volume is 350 μl and the subsequent volume of EtOH to add is 525 μl.
12. An optional second elution with 30 μl H_2O can also be applied. The concentration of RNA in the second elution will be low but having an increased total yield of RNA may be useful for validation experiments.
13. Dilute ~500 ng of RNA in a total volume of 5–10 μl containing 1× TBE Urea dye. Also dilute 0.5 μl of the 20/100 Oligo Length Standard and use it as a size marker on the gel. A band running the size of RPF fragments (~30 bp) should be apparent just slightly larger than the second to smallest band in the 20/100 Oligo Length Standard. The Mini-Protean Tetra cell system with 1.0 mm gels works well for acrylamide gel electrophoresis. First, prepare 10 ml of a 15% denaturing acrylamide gel solution (see the buffers section) without APS and TEMED. Gentle warming of the solution helps to dissolve the Urea but allow the solution to return to room temperature before pouring the gel. Clean the glass plates for the gel well with detergent and H_2O followed by a treatment with RNAseZAP and a final rinse with DEPC-treated water. Assemble plates and pour gel after adding APS and TEMED to the gel solution. Electrophorese at 200 V constant voltage as per gel rig manufacturer's recommendations. Stain the gel, after electrophoresis, in 50 ml of 1× TBE containing 5 μl SYBR Gold nucleic acid stain, for 2 min. We also strongly recommend doing an initial ribosome profiling experiment of several different time points of RNase 1 digestion to determine the minimal RNase digestion time to reduce footprints size to ~30 nts. Incomplete digestion can lead to slightly larger footprint size (*see* Fig. 2), whereas over digestion may lead to ribosome degradation and increased rRNA contamination.

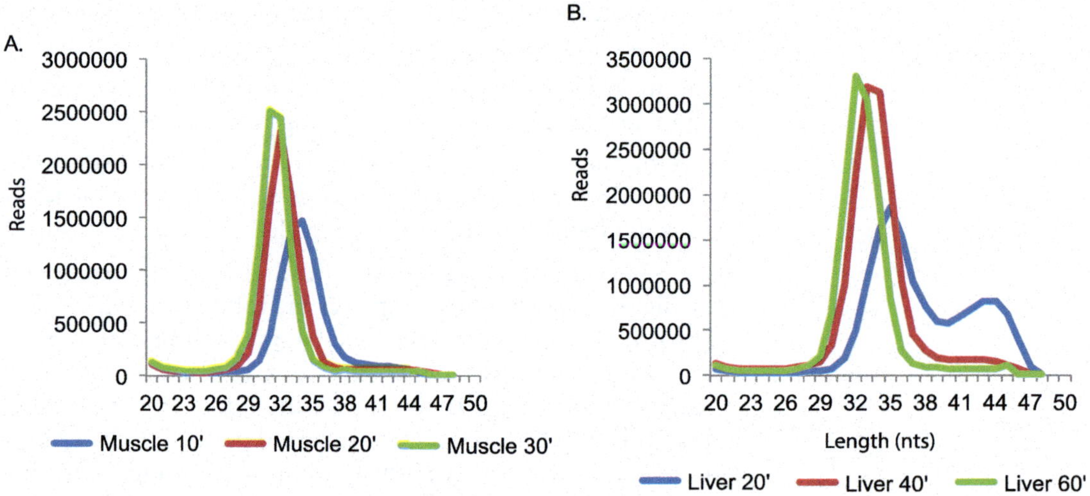

Fig. 2 RNase 1 time course of muscle and liver tissue for ribosome profiling. (**a**) Quadriceps muscle extracts (70 mg of tissue) were incubated for 10, 20, and 30 min with RNase1. Shown are the numbers of sequence reads according to size. (**b**) Same as (**a**) for liver extracts incubated with RNase 1 for 20, 40, and 60 min

14. The use of inadequate amounts of RNA may lead to less even coverage across mRNAs with extensive peaks separated by regions showing little to no coverage.
15. RNA fragmentation occurs when the RNA is heated in the presence of Magnesium in the T4 PNK buffer.

Acknowledgments

The authors would like to acknowledge our many selenium collaborators for thoughtful discussions and funding from the National Institutes of Health, GM114291 (MTH) and ES022716 (MTH).

References

1. Steitz JA (1969) Polypeptide chain initiation: nucleotide sequences of the three ribosomal binding sites in bacteriophage R17 RNA. Nature 224(5223):957–964
2. Wolin SL, Walter P (1988) Ribosome pausing and stacking during translation of a eukaryotic mRNA. EMBO J 7(11):3559–3569
3. Ingolia NT, Brar GA, Stern-Ginossar N, Harris MS, Talhouarne GJ, Jackson SE, Wills MR, Weissman JS (2014) Ribosome profiling reveals pervasive translation outside of annotated protein-coding genes. Cell Rep 8 (5):1365–1379. doi:10.1016/j.celrep.2014.07.045
4. Ingolia NT, Lareau LF, Weissman JS (2011) Ribosome profiling of mouse embryonic stem cells reveals the complexity and dynamics of mammalian proteomes. Cell 147(4):789–802. doi: S0092-8674(11)01192-5 [pii] 10.1016/j.cell.2011.10.002
5. Shalgi R, Hurt JA, Krykbaeva I, Taipale M, Lindquist S, Burge CB (2013) Widespread regulation of translation by elongation pausing in heat shock. Mol Cell 49(3):439–452. doi:10.1016/j.molcel.2012.11.028
6. Howard MT, Carlson BA, Anderson CB, Hatfield DL (2013) Translational redefinition of UGA codons is regulated by selenium

availability. J Biol Chem 288 (27):19401–19413. doi: M113.481051 [pii] 10.1074/jbc.M113.481051

7. Shetty SP, Shah R, Copeland PR (2014) Regulation of selenocysteine incorporation into the selenium transport protein, selenoprotein P. J Biol Chem 289(36):25317–25326. doi:10.1074/jbc.M114.590430
8. Lin HC, Ho SC, Chen YY, Khoo KH, Hsu PH, Yen HC (2015) CRL2 aids elimination of truncated selenoproteins produced by failed UGA/Sec decoding. Science 349 (6243):91–95. doi:10.1126/science.aab0515
9. Ingolia NT, Ghaemmaghami S, Newman JR, Weissman JS (2009) Genome-wide analysis in vivo of translation with nucleotide resolution using ribosome profiling. Science 324 (5924):218–223. doi: 1168978 [pii] 10.1126/science.1168978
10. Michel AM, Mullan JP, Velayudhan V, O'Connor PB, Donohue CA, Baranov PV (2016) RiboGalaxy: a browser based platform for the alignment, analysis and visualization of ribosome profiling data. RNA Biol 13 (3):316–319. doi:10.1080/15476286.2016.1141862
11. Gerashchenko MV, Gladyshev VN (2014) Translation inhibitors cause abnormalities in ribosome profiling experiments. Nucleic Acids Res 42(17):e134. doi:10.1093/nar/gku671

Chapter 9

Modification of Selenoprotein mRNAs by Cap Tri-methylation

Anne-Sophie Gribling-Burrer, Gilbert Eriani, and Christine Allmang

Abstract

Several selenoprotein mRNAs undergo 5′ cap maturation events whereby their classical monomethylated m^7G cap becomes trimethylated ($m_3^{2,2,7}G$) by the trimethylguanosine synthase 1 (Tgs1). Here, we describe immunoprecipitation methods for the detection of endogenous $m_3^{2,2,7}G$-capped selenoprotein mRNAs from total cell extracts or after polysome fractionation of cytoplasmic extracts. We have also developed a method for the in vitro cap hypermethylation of selenoprotein mRNA transcripts using purified Tgs1 enzyme.

Key words Selenoprotein mRNAs, $m_3^{2,2,7}G$ cap, TMG cap, Cap hypermethylation, Cap immunoprecipitation, Tgs1

1 Introduction

Selenoprotein mRNAs constitute an interesting class of mRNAs. Not only are they subjected to translational recoding of an in frame UGA_{Sec} codon but they also undergo distinctive mRNA biogenesis and maturation pathways [1–3]. The assembly and biogenesis mechanism of selenoprotein mRNPs resembles in many aspects that of the spliceosomal uridyl-rich small nuclear RNAs (UsnRNAs) and some of the small-nucleolar RNAs (snoRNAs) [2, 4]. Recently, we characterized a new mechanism of maturation of the 5′-end cap of selenoprotein mRNAs [5]. The 5′ cap has multiple functions and is involved in mRNA processing, translation, and degradation [6–11]. Mammalian mRNAs, synthetized by RNA polymerase II (polII), are generally characterized by the presence of a 7-methylguanosine (m^7G) cap structure at their 5-end [12]. We showed that several selenoprotein mRNAs bear a trimethylated $m_3^{2,2,7}G$ cap (or TMG cap) [5]. The trimethylated cap is synthesized by the trimethylguanosine synthase 1 (Tgs1) by the same mechanism as for the cap of UsnRNAs and snoRNAs. This modification is required for the expression of selenoprotein GPx1 in vivo [5]. We also showed that

Laurent Chavatte (ed.), *Selenoproteins: Methods and Protocols*, Methods in Molecular Biology, vol. 1661,
DOI 10.1007/978-1-4939-7258-6_9, © Springer Science+Business Media LLC 2018

the hypermethylated-capped selenoprotein mRNAs are associated with polysomes in the cytoplasm, suggesting that they are actively translated [5]. The detection of the $m_3^{2,2,7}$G-capped RNAs can be performed by immunoprecipitation experiments using a highly specific anti-$m_3^{2,2,7}$G cap R1131 serum (or anti-TMG serum) that was demonstrated not to recognize monomethylated caps [13, 14]. This experiment, called TMG-IP, has been used successfully to characterize the TMG-capped snRNAs and snoRNAs. Here, we present an adaptation of this assay coupled to real-time quantitative PCR for the detection of endogenous $m_3^{2,2,7}$G-capped selenoprotein mRNAs in vivo from cultured cells (Fig. 1). The method can also be applied to recover $m_3^{2,2,7}$G-capped selenoprotein mRNAs after polysome fractionation of cytoplasmic mRNAs. This allows us to estimate the ratio of ribosome bound compared to free $m_3^{2,2,7}$G-capped selenoprotein mRNAs (Fig. 2).

Finally, we have also developed a new method for the in vitro cap hypermethylation of a selenoprotein mRNA transcript using purified Tgs1 enzyme. Tgs1 catalyzes two successive methyl group transfers, from *S*-adenosyl-l-methionine (AdoMet) to the exocyclic nitrogen N2 of the m^7G cap to generate the $m_3^{2,2,7}$G cap. The enzyme is specific for m^7G-capped RNA and m^7GTP, which represents the minimal substrate, while nonmethylated 5′-cap RNA or GTP are not modified [15]. Two biological forms of Tgs1 have been reported, a full-length cytoplasmic isoform and a shorter nuclear isoform ($Tgs1_{577-851}$), both exhibit methyltransferase

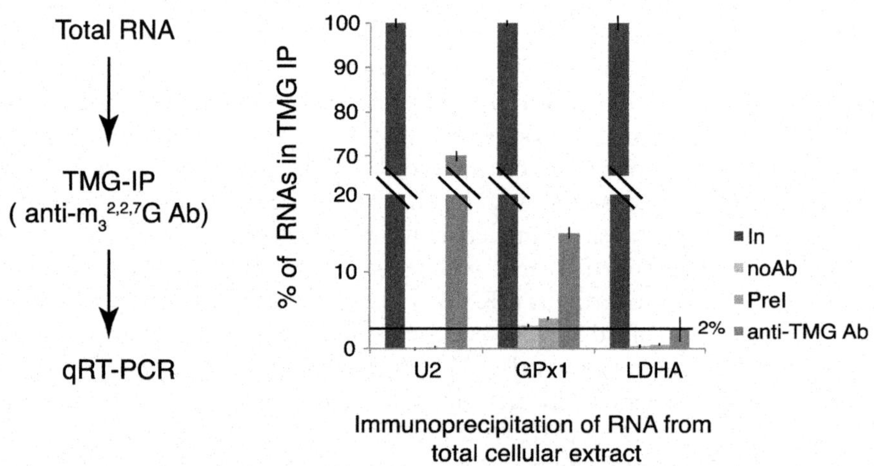

Fig. 1 Example of anti-$m_3^{2,2,7}$G cap immunoprecipiation experiment performed on total RNA. RNA extracted from HEK293FT cell extracts was immunoprecipitated with anti-TMG serum (anti-$m_3^{2,2,7}$G Ab, R1131). Bound RNA was analyzed by qRT-PCR. (In) input 10% of total RNA; (noAb) control without antibodies; (PreI) preimmune antibodies; (anti-TMG Ab) with anti-$m_3^{2,2,7}$G antibodies. The graph represents the percentage of RNAs in the IP compared with the input RNA. Error bars represent standard deviation of an average of three independent experiments. Nonspecific binding observed in the absence of antibodies or for housekeeping mRNAs typically ranges between 1 and 2% (*horizontal line*). Typical results are presented for GPx1selenoprotein mRNA, the non-selenoprotein m^7G capped LDHA mRNA and U2 UsnRNA, which serves as a positive control for cap hypermethylation

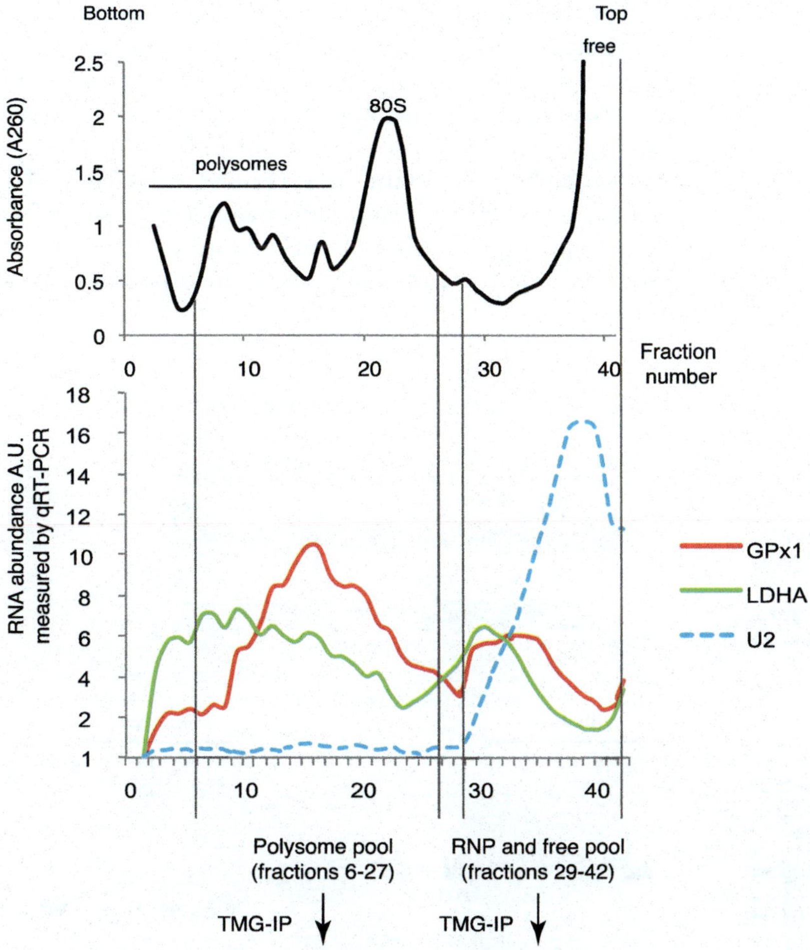

% of RNA in TMG IP

	Polysome pool	RNP pool
U2	7.6 +/- 0.4	102 +/- 25
GPx1	8.5 +/- 0.7	1.4 +/- 0.16
LDHA	0.7 +/- 0.3	0.15 +/- 0.01

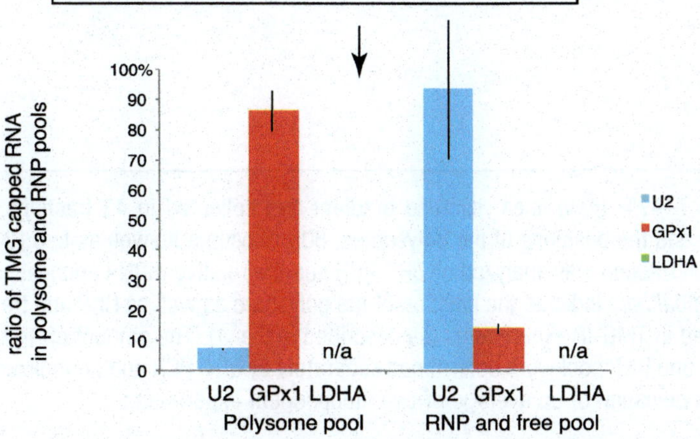

Fig. 2 Polysome distribution of endogenous HEK293 selenoprotein mRNAs and ratio of TMG-capped mRNAs in ribosome bound and free polysome fractions. Cytoplasmic extracts from HEK293FT cell were fractionated onto

activity [16]. Furthermore, the crystal structure of an active truncated form Tgs1$_{618\text{-}853}$ has been solved [17, 18]. We purified Tgs1$_{577\text{-}851}$ and Tgs1$_{618\text{-}853}$ (Fig. 3) and used both successfully for in vitro methylation of RNA transcripts. Methylation activity tests for Tgs1 have been reported so far only on m^{7}GTP. The method presented here allows the conversion of a m^{7}G capped to a m$_3$^{2,2,7}G capped-mRNA in a quantitative manner and can apply to any RNA transcript (Fig. 4). It represents an interesting alternative to methods based on the co-transcriptional incorporation of cap analogs.

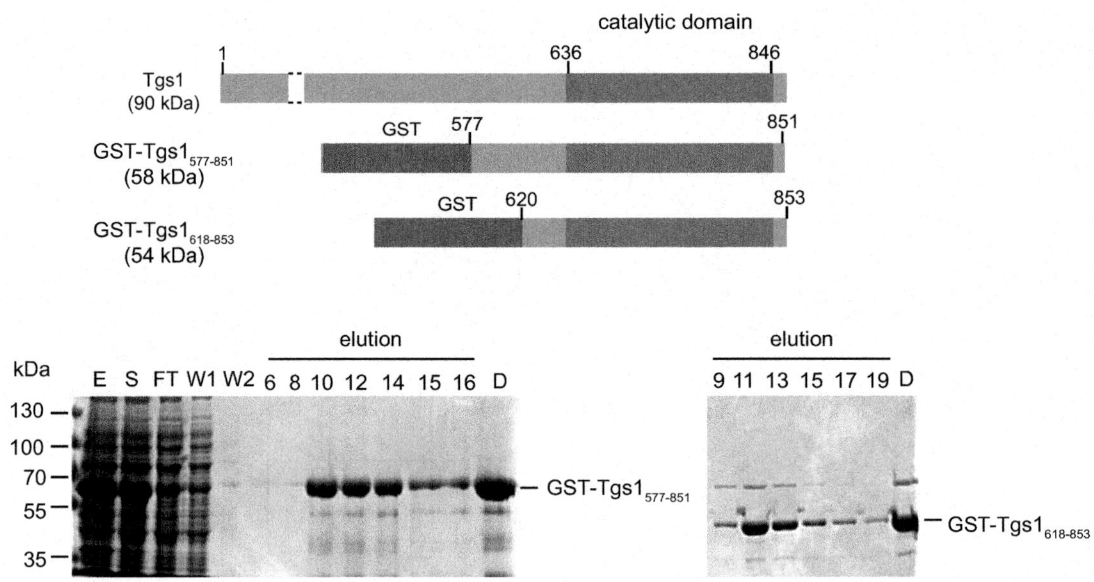

Fig. 3 Purification of the human Trimethyl guanosine synthase (Tgs1). The two truncated forms of the enzyme Tgs1$_{577\text{-}851}$ and Tgs1$_{618\text{-}853}$ overexpressed in *E. coli* BL21 (DE3) RIL are represented. Both contain the functional catalytic dimethyl transferase domain and are fused with a Glutathion S-transferase domain (GST). Proteins have been purified on Glutathione Sepharose 4B resin. The purification steps are analyzed on 1% SDS-10% PAGE. *E* total cell extract, *S* supernatant, *FT* flow through, *W1* and *W2* washing steps, *D* protein after dialysis

Fig. 2 (continued) 7–47% (w/v) linear sucrose gradient and collected in 43 fractions. Typical absorbance profiles are shown and the positions of the polysomes, 80S ribosome as well as free RNA are indicated. The RNA content of each fraction was analyzed by qRT-PCR and the relative mRNA abundance was represented in arbitrary units. *Vertical bars* indicate the position of the polysome as well as RNP and free fractions that were pooled and analyzed in TMG-IP experiments (as described in Fig. 1). The amount of RNA immunoprecipitated from the polysome and RNP pool were determined separately by qRT-PCR and normalized to 100%. Error bars represent standard deviation of an average of two independent experiments

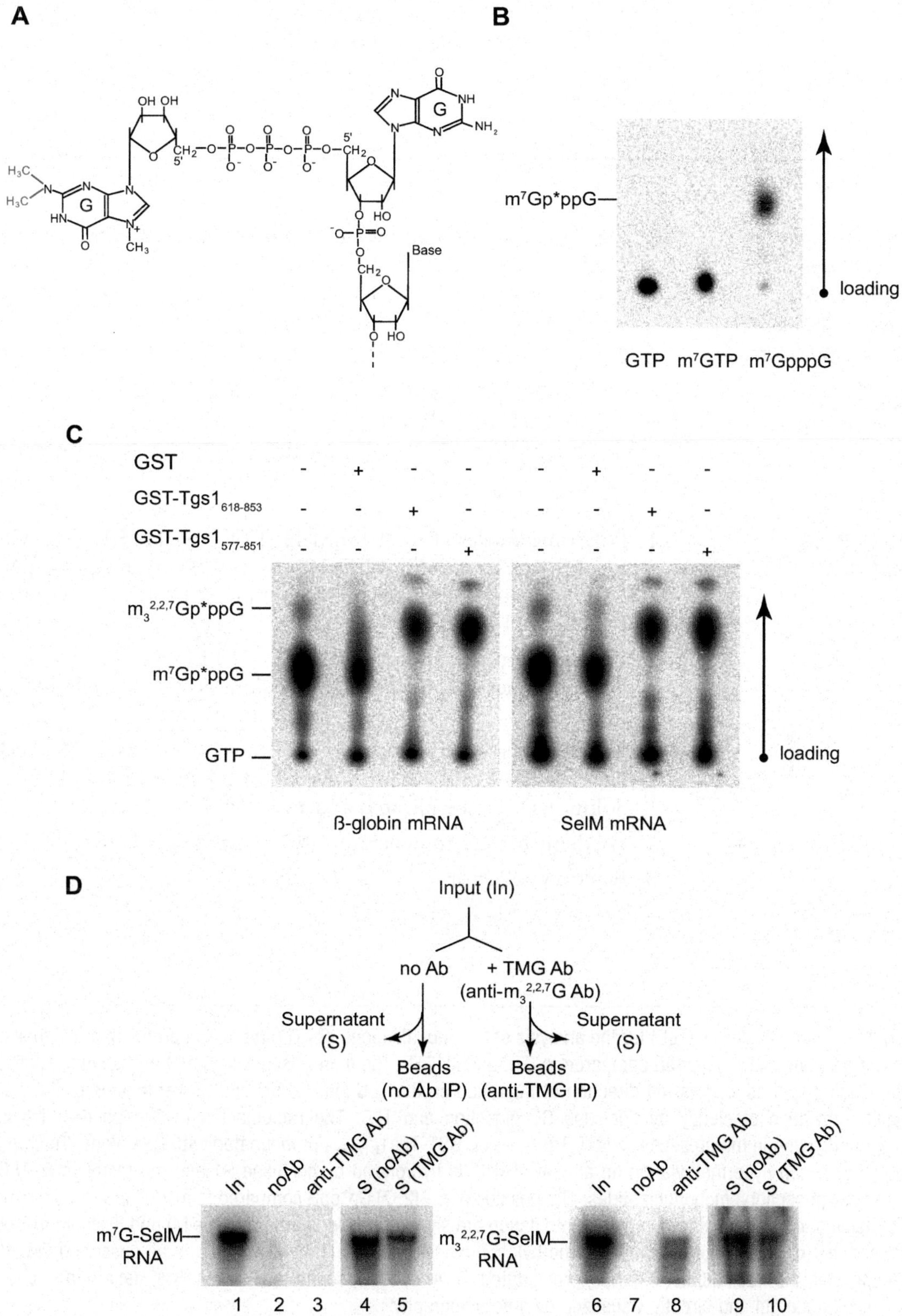

Fig. 4 In vitro cap hypermethylation of mRNA transcripts. (**a**) Structure of the $m_3^{2,2,7}G$ cap. The methyl groups added by Tgs1 to the exocyclic N2 of the m^7G cap are represented in *red*. (**b**, **c**) Thin-layer Chromatography

Thus, the methods presented here allow the in vivo detection of endogenous hypermethylated capped selenoprotein mRNAs or the generation of in vitro $m_3^{2,2,7}$G-capped mRNAs for molecular biology studies.

2 Material

2.1 Immuno-precipitation of Endogenous Hypermethylated-Capped Selenoprotein mRNAs

2.1.1 Cell Culture

1. DMEM cell culture medium: Dulbecco's Modified Eagle Media (DMEM), 10% Foetal Calf Serum (FCS), 1% penicillin-streptomycin, 1% geneticin, 10 nM sodium selenite (*see* **Note 1**).
2. PBS: 137 mM NaCl, 26.8 mM KCl, 40.2 mM Na_2HPO_4, 17.6 mM KH_2PO_4, pH 7.4.
3. Trypsin/EDTA 0.25%.
4. Sterile 25 and 75 cm^2 culture flasks. 10 cm petri dishes for cell culture.

2.1.2 Polysome Fractionation

1. Polysome Buffer (PB): 10 mM HEPES-NaOH, 50 mM KCl, 5 mM $MgCl_2$, 1 mM DTT, 100 units RNasin/mL, 400 μM VRC (Vanadyl Ribonucleotide Complex), protease inhibitors cocktail, pH 7.5.
2. PBD denaturing buffer: 10 mM HEPES-NaOH, 10 mM potassium acetate, 0.5 mM magnesium acetate, 5 mM DTT, pH 7.5.
3. 7 and 47% sucrose solutions: 7%/47% sucrose, 25 mM Tris–HCl, 50 mM KCl, 5 mM $MgCl_2$, 1 mM DTT, pH 7.4. Solutions must be filtrated and autoclaved (*see* **Note 2**).
4. Gradient Master to generate 7–47% sucrose gradients.
5. Fraction collector.

Fig. 4 (continued) assays (TLC) for the analysis of cap modifications. (**b**) TLC migration profile of radiolabeled monomethylated cap m^7GpppG compared to GTP and m^7GTP. The buffer used is 100 mM ammonium sulfate. The loading spot as well as the direction of migration on the TLC plate is shown. (**c**) Verification of the cap hypermethylation efficiency by nuclease S1 digestion and TLC. The radiolabeled m^7G capped-mRNA is hypermethylated in the presence of GST-Tgs$1_{577\text{-}851}$, GST-Tgs$1_{618\text{-}853}$ or incubated with GST alone. To verify the level of cap hypermethylation an aliquot of RNA is hydrolyzed by nuclease S1 and the resulting 5′ TLC separates phosphoryl mononucleotides. The position of m_3 [2, 7] Gp*ppG compared to m^7Gp*ppG is indicated. Examples of efficient in vitro cap hypermethylation are shown for selenoprotein M (SelM) and ß-globin mRNA transcripts. (**d**) Verification of cap hypermethylation by TMG-IP. The TMG-IP method is as described Fig. 1, except that the radiolabeled immunoprecipitated RNA can be separated on an 8% denaturing urea-polyacrylamide gel and directly visualized by autoradiography

2.1.3 Trimethyl-guanosine-Capped mRNA Immunoprecipitation (TMG-IP)

1. Polyclonal anti-m_3G (TMG) serum (Synaptic Systems).
2. 50% protein-A Sepharose beads: Swell lyophilized protein-A Sepharose beads according to the manufacturer's instructions for 30 min to 1 h in PBS at room temperature (*see* **Note 3**). Always sediment and recover beads by centrifugation at 3000 × *g*. Wash gently 3× with 3 volumes of PBS. Always recover beads by centrifugation at 3000 × *g*. Suspend the beads in 1 volume of PBS (stock of beads 50%). Prepare 2.5× the volume of beads necessary per condition to be tested.
3. Purified BSA (20 mg/mL), and total yeast tRNA (20 mg/mL) to saturate protein A-Sepharose beads.
4. NT2 buffer: 50 mM Tris–HCl, 150 mM NaCl, 1 mM $MgCl_2$, 0.05% NP40, pH 7.5 [19].
5. NT2/RNasin: NT2, 100 units/mL.
6. NT2/VRC/DTT buffer: NT2, 400 μM VRC, 1 mM DTT.
7. Rotating wheel.
8. qRT-PCR apparatus, Reverse Transcriptase and qPCR Master Mix.

2.2 mRNA Cap Hypermethylation In Vitro

2.2.1 Production and Purification of Tgs1

1. *E. coli* BL21 (DE3) RIL.
2. Luria Bertani broth: 10 g/L tryptone, 10 g/L NaCl, 5 g/L Yeast Extract.
3. Glutathione Sepharose 4B resin.
4. Branson Annemasse utrason apparatus.
5. Lysis buffer: 50 mM Tris–HCl, 500 mM NaCl, 2 mM EDTA, 2 mM DTT, pH 7.5.
6. Washing buffer: 50 mM Tris–HCl, 1 M NaCl, 2 mM DTT, pH 7.5.
7. Loading blue: 400 mM Tris–HCl, pH 8.8, 4% SDS, 4% ß-mercaptoethanol, 30% glycerol, 0.05% bromophenol blue.
8. Staining solution: 0.1% Coomassie blue, in 50% methanol, 0.1% acetic acid.
9. Destaining solution: 7% acetic acid, 5% ethanol.
10. Dialysis buffer: 20 mM Tris–HCl, 150 mM NaCl, 2 mM DTT, 50% glycerol, pH 7.5.
11. Bradford reagent.

2.2.2 RNA Purification, Gel Electrophoresis, and Thin-Layer Chromatography Assay (TLC)

1. TMSDT buffer: 280 mM Tris–HCl, 140 mM $MgCl_2$, 35 mM DTT (dithiothreitol), 0.07% Triton-X100, 7 mM spermidine, pH 8.1.
2. T7 RNA polymerase and Pyrophosphatase.

3. Long thin polyacrylamide denaturing sequencing gels apparatus.
4. [α^{32}P] guanosine 5′-triphosphate (GTP) (111 TBq, 3000 Ci/mmol 370 MBq, 10 mCi/mL).
5. TBE: 89 mM Tris-Base, 89 mM boric acid, 2.5 mM EDTA.
6. Denaturing dye: 95% formamide, 0.025% bromophenol blue, 0.025% xylene cyanol.
7. Elution buffer: 3 mM NaCl, 0.5 mM EDTA, 10 mM Tris–HCl, pH 7.5.
8. Electroelution device from Whatman Schleicher & Schuell also called Elutrap for the electroelution of RNA samples contained in polyacrylamide bands [20].
9. Scriptcap m^{7}G capping system (Epicentre Biotechnologies) to cap the 5′ end of mRNAs. The kit contains the Vaccinia virus Capping Enzyme (VCE) which synthesize the Cap 0 in a 3-step reaction.
10. Phosphorimager equipment with Phosphor Imaging Plates to detect radiolabeled RNAs.
11. Liquid scintillation counter to measure radioactive RNA labeling.
12. S1nuclease buffer: 200 mM sodium acetate, 1.5 M NaCl, 10 mM $ZnSO_4$, pH 4.5.
13. Thin-layer chromatography (TLC) plates are purchased to Macherey-Nagel (Polygram CEL300 PEI). TLC plates are developed in vertical TLC tanks (27 × 26.5 × 7 cm).

3 Methods

3.1 Immuno-precipitation of Endogenous Hypermethylated-Capped Selenoprotein mRNAs

3.1.1 Cell Extract Preparation

1. Cultivate HEK293FT in DMEM cell culture medium at 37 °C in 5% CO_2 in the presence of 10 nM Na_2SeO_3 (*see* **Notes 1** and **4**).
2. Initially thaw out 3 × 10^6 cells from frozen stocks in 8 mL of DMEM cell culture medium supplemented with 30% of FCS in 25 cm^2 culture flasks.
3. After 24 h, when cells reach confluence, remove culture medium and rinse cells gently with 5 mL of sterile PBS to eliminate residual serum. Add 1 mL of Trypsin/EDTA to disrupt cell. Incubate for 30–60 s at room temperature. Dilute cells 10 times (approximately 10^6 cells) in 10 mL of DMEM cell culture medium and transfer them to 75 cm^2 culture flasks for maintenance.
4. To prepare cell extracts, transfer 10^6 cells to 10 cm petri dishes and growth them to 80% confluence.

5. Wash cells with 3 mL (or 3 volumes of) PBS, recover cells by scraping them in 1 mL PBS. Centrifuge cells for 10 min at 1000 × *g*. At this stage, total RNA (*see* **step 6**) as well as total or cytoplasmic cell extracts can be prepared (*see* **step** 7).
6. Direct RNA extraction can be performed using TriReagent (MRC) according to the manufacturer's instructions and directly be subjected to TMG-IP (Fig. 1).
7. For total cell extract preparation resuspend the cell pellet in 1 volume of Polysome Buffer (PB) and incubate 5 min on ice. Freeze cells in liquid Nitrogen and store at −80 °C (*see* **Note 5**).

3.1.2 Polysome Fractionation

1. To prepare cytoplasmic extracts for polysome fractionation collect 2×10^6 cells as described in Subheading 3.1.1. Incubate cells in 1 volume of PBS in the presence of 150 μg/mL cycloheximide for 20 min on ice (*see* **Note 6**).
2. Centrifuge cells for 3 min at 250 × *g*. To lyse the cells, resuspend the pellet in 1 volume of PB and 150 μg/mL cycloheximide, incubate for 5 min on ice.
3. Disrupt cells with the help of a syringe and a narrow-gauge 0.4 mm hypodermic needle (27G). After 5–10 strokes check the lysis under the microscope by the addition of Trypan blue solution to an aliquot of cells (*see* **Note** 7).
4. Centrifuge cells for 10 min at 18,000 × *g* and recover the supernatants that correspond to the cytoplasmic extract. Snap-freeze the extracts with liquid nitrogen and store at −80 °C.
5. Prepare a 7–47% linear sucrose gradient in 11 mL PB. This can be performed manually or with the help of a Gradient Master (*see* **Notes 8** and **9**).
6. Add 250 μL 7% sucrose solution to 250 μL of cytoplasmic extracts and load at the top of the 7–47% sucrose gradient.
7. Centrifuge for 2.5 h at 235,000 × *g* using a Beckmann SW41 rotor to separate the polysomes.
8. Collect fractions of 300 μL and monitor gradients by following the absorbance at 254 nm (*see* Fig. 2).
9. Extract RNA from each fraction. After two phenol:chloroform: isoamyl alcohol (25:24:1) extractions, ethanol precipitate the RNA present in the upper aqueous phase in the presence of sodium acetate 0.3 M and 1 μg of glycogen (RNase free). Resuspend the RNA pellet in 30 μL milliQ water. The RNA repartition in each fraction can be analyzed by RT-qPCR (protocol 3.1.4, *see* Fig. 2).
10. Recover RNA from 250 μL of cytoplasmic extract. This will serve as a total cytoplasmic RNA control.

3.1.3 Trimethyl-guanosine-Capped mRNA Immunoprecipitation (TMG-IP)

Trimethyl-capped RNAs are immunoprecipitated using a strictly specific rabbit polyclonal anti-m_3G (TMG) serum (Synaptic Systems), also referred to as R1131 [13, 14]. The experiment should be carried out at 4 °C and incubations are performed on a rotating wheel. All the steps can be performed in microtubes. Washing of the beads is done by gentle centrifugation at 3000 × *g* for 3 min. A volume of 50 μL of protein A-Sepharose beads (50%) is required for one experimental condition, as well as for the preclear of the RNA samples.

1. Saturate protein A-Sepharose beads. Add 1 μL of purified BSA (20 mg/mL) and 1 μL of total yeast tRNA (20 mg/mL) to 50 μL of protein A-Sepharose beads (50%) in PBS. Incubate for 30 min at 4 °C on a rotating wheel.
2. Bind antibodies to protein A-Sepharose beads. Add 10 μL anti-m_3G (TMG) serum to 50 μL of the saturated protein A-Sepharose beads (50%). Complete with PBS up to 500 μL. Incubate for 18 h at 4 °C. In parallel, prepare beads without antibodies and/or beads coupled to rabbit preimmune serum. These beads will serve as control for nonspecific binding analysis.
3. Wash beads three times with 500 μL of PBS and three times with 500 μL NT2/VRC/DTT buffer. Resuspend the beads in 50 μL of NT2/VRC/DTT buffer.
4. Prepare the samples to be immunoprecipitated. TMG-IP can be performed on 200 μg of total RNA (*see* Subheading 3.1.1, **step 6**) or RNA extracted from polysome fractions (*see* Subheading 3.1.2, **step 9**). In this case, pool the RNAs extracted from the ribosome bound fractions (fractions 6–27, Fig. 2) and those extracted from free and ribonucleoprotein (RNP) fractions (fractions 29–42). Prepare also the corresponding total cytoplasmic RNA extracts (*see* Subheading 3.1.2, **step 4**) for TMG-IP analysis. Resuspend RNA samples in 1 mL final of NT2/VRC/DTT. In each case save 10% of the starting RNA material (referred to as "input") for subsequent qRT-PCR analysis.
5. Preclear the RNA extracts to remove proteinA-sepharose non-specific RNA binders from the samples. For that purpose, add 1 mL of the RNA preparation to 50 μL of saturated protein A-Sepharose beads (50%) without antibodies (*see* Subheading 3.1.3, **step 1**). Incubate for 30 min at 4 °C. Centrifuge for 3 min at 3000 × *g* to sediment the beads and recover the supernatant containing the precleared RNA samples.
6. Add 1 mL of preclear RNA samples to 50 μL of anti-TMG coupled protein A-Sepharose beads (*see* Subheading 3.1.3, **step 3**) and incubate for 2 h at 4 °C.

7. Sediment beads by centrifugation for 3 min at 3000 × *g* and remove the supernatant (save unbound material). Wash beads six times with 1 mL NT2 buffer. Transfer beads in a new tube before the last wash to eliminate RNA bound aspecifically to the tubes.
8. Recover the immunoprecipitated RNA. Resuspend the beads in 200 μL of NT2/RNasin. Recover the RNA by two phenol: chloroform:isoamyl alcohol (25:24:1) extractions and ethanol precipitation in the presence of 1 μg of glycogen. Resuspend in 10 μL milliQ water.

3.1.4 qRT-PCR Analysis of Selenoprotein mRNA Recovered in TMG-IPs

1. DNase I treatment of the immunoprecipitated RNAs. Incubate the recovered RNAs with 2.5 units of DNaseI RNase free in 20 μL of DNAse buffer for 15 min at 25 °C (*see* **Note 10**).
2. Reverse transcribe 10 μL of DNase I treated RNAs using 100 pmol of random nonamers (dN9) and AMV-RT or equivalent, according to the manufacturer's instructions.
3. Analyze cDNAs by quantitative RT-PCR (qRT-PCR) (*see* **Note 11**). Analysis of immunoprecipitated $m_3^{2,2,7}$G-capped RNAs from total cell extracts is illustrated in Fig. 1 and the immunoprecipitated $m_3^{2,2,7}$G-capped RNAs from the polysome fractions are shown in Fig. 2. qRT-PCR should be performed according to the MIQE guidelines [21]. Validated oligonucleotides directed against human selenoprotein mRNAs.

3.2 mRNA Cap Hypermethylation In Vitro

3.2.1 Production and Purification of Tgs1

Two active truncated forms of human Tgs1 (Trimethyl guanosine synthase) proteins can be used for hypermethylation assays [16, 17]. Both contain the catalytic dimethyl transferase domain located in carboxy-terminal part of the enzyme plus 18 or 59 amino acid residues from the central domain fused with the GST domain (Glutathion S-transferase) (*see* Fig. 3). The two proteins, named $Tgs1_{577-851}$ and $Tgs1_{618-853}$, are overexpressed in *E. coli* BL21 (DE3) RIL starting from plasmid pGEX-6P containing the corresponding genes. The expression and purification protocol is the following:

1. Inoculate 500 mL of Luria Bertani broth supplemented with 100 μg/mL ampicillin by 0.5 mL of an overnight preculture of BL21 (DE3) RIL transformed with pGEX-6P-$Tgs1_{577-851}$ or pGEX-6P-$Tgs1_{618-853}$. Incubate at 37 °C with shaking (150 rpm).
2. When the OD_{600nm} reaches 0.6, add IPTG to a final concentration of 0.5 M and incubate at 18 °C during 18 h.
3. Harvest the cells by 10 min centrifugation at 5000 × *g* at 4 °C. All the following steps are performed at 4 °C.
4. Resuspended cell pellets in 10 mL of Lysis buffer.

5. Lyse the cells by sonication: six times 30 s at 120 V.
6. Centrifuge for 10 min at 8000 × *g* to eliminate cell debris. Then, the supernatant is centrifuged for 1 h at 105,000 × *g* to remove polyribosomes. Take the supernatant and keep it on ice.
7. Pour 1.5 mL of Glutathione Sepharose 4B resin in a column, wash with 15 mL of Lysis buffer.
8. Add the fraction 105,000 × *g* (**step 6**) on the top of the column. Seal the column with caps and rotate 1 h 30 min at 4 °C on a wheel (at 30 rpm).
9. Wash the column with 5 mL of Lysis buffer and then with 5 mL of Washing buffer.
10. Elute bound proteins with 10 mL of Lysis buffer supplemented with 25 mM glutathione. Collect 500 μL fractions.
11. Analyze the fractions by 1% SDS-10% PAGE. Mix together 7 μL of the fractions with 3 μL of loading blue and migrate at 100 V until the blue reaches the bottom of the gel.
12. Color the gel with staining solution during 30 min. Then place the gel in destaining solution.
13. Pool the fractions of interest and dialyze against 250 mL of Dialysis buffer during 4 h at 4 °C using a dialysis membrane tubing (MWCO 6-8000).
14. Measure concentrations of Tgs1 proteins with Bradford reagent and validate the concentration values by SDS-PAGE with a BSA sample of known concentration. Typically, 500 mL of culture yields 3 mg of purified $Tgs1_{577\text{-}851}$ and 2 mg of $Tgs1_{618\text{-}853}$.

3.2.2 In Vitro RNA Transcription

1. Mix 20 μL of DNA template (containing 10 μg of DNA), 14.3 μL of 7× TMSDT buffer, 10 μL of 50 mM ATP, 10 μL of 50 mM CTP, 10 μL of 50 mM ATP, 10 μL of 50 mM GTP, 10 μL of 50 mM UTP, 1 μL of RNAsin, 19.7 μL of milliQ water, and 5 μL of T7 RNA polymerase (2 mg/mL).
2. Incubate for 1 h at 37 °C, add 2 μL of pyrophosphatase, and incubate for another 30 min at 37 °C. Pyrophosphatase hydrolyzes the formed PPi, which is an inhibitor of the forward reaction, thereby extending further the transcription reaction.
3. Add 2 μL of DNase I RNase free and incubate for 1 h at 37 °C, in order to degrade the DNA template.
4. Add 100 μL of milliQ water and 200 μL of phenol solution saturated with water. Mix vigorously for 10 s and centrifuge for 2 min at 18,000 × *g* at room temperature.
5. Take the supernatant containing the RNA and extract similarly a second time with 200 μL of phenol–chloroform solution saturated with water.

6. Take the supernatant and add 10 μL of 5 M NaCl solution and 600 μL 100% ethanol. Leave at −20 °C for at least one night.
7. Retrieve the precipitated RNA by centrifugation at 18,000 × *g* for 20 min and dissolve in 100 μL of denaturing dye.
8. Heat at 95 °C for 5 min and load on 4% denaturing urea-polyacrylamide gel (400 × 200 × 1 mm) in TBE buffer. Migrate at 30 W for 1 h 30 min and detect RNA by UV shadowing.
9. Cut the full-length RNA band with a scalpel blade and electro-elute the RNA sample in an electroelution device, according to the manufacturer. Then, the RNA solution recovered from the electroelution chamber is precipitated by ethanol as described above (**step 6**) and dissolved in 50 μL milliQ water.
10. Measured concentration of the purified RNA by measure of the absorbance at 260 nm in a spectrophotometer.

3.2.3 Cap 0 Radioactive Synthesis at the 5′ End of mRNAs

1. Efficient capping is achieved using the Vaccinia virus Capping Enzyme (VCE) from Scriptcap m^7G capping kit. The VCE synthesizes the Cap 0 structure usually found on the 5′ end of eukaryotic mRNAs.
2. Capping is performed on 50 μg of purified RNA according to the manufacturer's instructions. However, to radiolabel the RNA, the unlabeled solution of GTP from the kit was replaced by 2.5 μL radioactive [α^{32}P] guanosine 5′-triphosphate (GTP) (25 μCi) (*see* **Note 12**) in order to synthesize an mRNA starting with a radioactive Cap 0.
3. After cap synthesis, the mRNA solution is de-proteinized by phenol and phenol/chloroform extractions and precipitated by ethanol as previously described (previous paragraph, **steps 4–6**). Retrieve the precipitated RNA by centrifugation at 18,000 × *g* for 20 min as described above.
4. Dissolve the dried radioactive RNA pellet in 30 μL of denaturing dye and load on 4% denaturing urea-polyacrylamide gel (400 × 200 × 1 mm) in TBE buffer.
5. Perform migration at 30 W for 1 h 30 min and detect radiolabeled RNA by autoradiography according to standard methods (*see* **Note 13**).
6. Cut the radioactive band corresponding to the Cap 0-capped mRNA and elute the RNA by passive diffusion overnight at 4 °C in 500 μL of Elution buffer.
7. Recover and precipitate the aqueous phase by ethanol precipitation as described above.
8. Centrifuge at 18,000 × *g* for 20 min and dissolve radioactive RNA in 20 μL of milliQ water. Measure the total cpm number by counting aliquot in Cerenkov mode in a liquid scintillation

counter. Typical labeling reaction raises RNA solutions with specific activity ranging between 50,000 and 150,000 cpm/μL of the final solution (*see* **Note 14**).

3.2.4 Cap Hypermethylation Catalyzed by Trimethyl Guanosine Synthase

1. Hypermethylation of the Cap 0 is achieved using the human trimethyl guanosine synthase (Tgs1) purified (*see* Subheading 3.2.1 and Fig. 4a).
2. Add to 2 μL of [^{32}P]-labeled capped-mRNA (≈100,000 cpm), 1 μL of 1 M Tris–HCl, pH 8.1, 2 μL of 50 mM DTT, 1 μL of freshly prepared solution of 1 mM S-adenosyl methionine (SAM), 1 μL RNasin, 12 μL milliQ water, and 1 μL of purified Tgs1 enzyme (6×10^{-5} M).
3. Incubate for 2 h at 37 °C.
4. Add 1 μL glycogen (2 μg/μL), 2 μL 3 M sodium acetate, and 75 μL of 100% ethanol. Precipitate the RNA overnight at −20 °C.
5. Centrifuge at 18,000 × *g* for 15 min and discard the supernatant. Wash the pellet by 100 μL of 80% ethanol and centrifuge at 18,000 × *g* for 5 min. Dry the RNA pellet under vacuum for 5 min and dissolve the radioactive RNA in 5 μL of milliQ water.

At this step, the hypermethylated RNA is ready for additional experiments such as translation assays, in vitro RNP assembly experiments, etc.

3.2.5 Verification of Cap Hypermethylation by TLC Assay

1. To verify the efficiency of cap hypermethylation, the radiolabeled capped-mRNA can be hydrolyzed by nuclease S1 and the resulting 5′ phosphoryl mononucleotides are analyzed by thin-layer chromatography (TLC) as shown in Fig. 4. To this end, add 4 μL of 5× S1nuclease buffer, 11 μL of milliQ water, and 1 μL of nuclease S1 (25 units/mL). Incubate for 20 min at 30 °C.
2. Add 2 μL 0.5 M EDTA, pH 8.0 and incubate 10 min at 70 °C in order to inactivate nuclease S1.
3. Rinse the thin-layer chromatography (TLC) plate with milliQ water for 10 min and allow drying at 37 °C.
4. Load 2 μL of the nuclease S1-treated samples at 2 cm of the bottom of the plate. After drying, place the plate in a chromatography tank felt with a solution of 100 mM ammonium sulfate (final height about 1 cm).
5. Allow chromatographic migration for 90–120 min, then remove the plate and dry it at 37 °C.
6. Detect the radioactivity signal using an Imaging Plate and Bioimager apparatus or an auto radiographic film (*see* **Note 15** and Fig. 4b, c).

3.2.6 Verification of Cap Hypermethylation by TMG-IP

1. To further verify the specificity of cap hypermethylation, a TMG-IP analysis can be performed as described in Subheading 3.1.3 on [^{32}P]-labeled hypermethylatedcapped-mRNA (≈30,000–50,000 cpm).
2. Dissolve the radioactive RNAs recovered from the TMG-IP in 10 μL of formamide denaturing dye, load on 8% denaturing urea-polyacrylamide gel in TBE buffer and detect radiolabeled RNA by autoradiography as described in Subheading 3.2.3 (*see* Fig. 4d).

4 Notes

1. The addition of 10 nM Na_2SeO_3 is required for an optimal synthesis of selenoproteins.
2. We advise to prepare and autoclave a 70% sucrose solution and to prepare and filter all other ingredients separately. Dilute sucrose to 7 and 47% extemporarily. This avoids the difficulty to filter the highly viscous 47% solution.
3. Typically, 1 g of ProteinA-Separose powder swell to 3–4 mL of hydrated gel. For each condition tested 50 μL of beads are needed for the preclear of the extracts and 50 μL for the immunoprecipitation reaction.
4. Any other cell types can be used and cultured according to the manufacturer's protocol.
5. DTT, RNasin, VRC, and protease inhibitors cocktail must be added freshly prior to use.
6. Cycloheximide is used to block translation elongation and therefore to "freeze" translating ribosomes on the mRNAs.
7. The Trypan blue solution is used to monitor cell lysis. The dye is excluded from intact cells, but stains the nuclei of lysed cells. Lysis should be >80–90%.
8. Control gradients can also be run with PBD denaturing buffer.
9. Pipet 5.5 mL of the 7% sucrose into a 14 × 89 mm Ultra-Clear centrifugation tube (Beckman Coulter). Use a syringe to carefully layer 5.5 mL of the 47% sucrose solution under the 7% one, without mixing the two solutions. Avoid the introduction of air bubbles. Add caps to the tubes and follow the instructions of the gradient forming device to generate the gradient by automatic mixing.
10. The 10% input is treated in the same way as immunoprecipitated RNA fractions and analyzed in parallel.
11. We routinely dilute cDNA 10–20 times before qRT-PCR.

12. Radioisotope techniques have the advantages of low cost and very high sensitivity. However, their application presents a considerable health hazard (due to carcinogenicity), and thus requires special care and attention.
13. RNAs are fractionated on long thin polyacrylamide denaturing sequencing gels that are run in a vertical arrangement. This polyacrylamide gel format provides optimal separation of long and short RNA fragments that differ by only a single nucleotide.
14. Counts per minute (cpm) of radiolabeled RNA are measured in Cerenkov mode in a liquid scintillation counter apparatus.
15. Radiolabeled RNAs are detected using Phosphorimager equipment with Imaging Plates, which are reusable and offer enhanced sensitivity compared to auto radiographic films.

Acknowledgments

We thank Thomas Monecke and Ralf Ficner for the gift of pGEX-6P hTgs1$_{618\text{-}853}$ plasmid, Rémy Bordonné for the gift of pGEX-sTgs1$_{577\text{-}851}$ plasmid.

References

1. Allmang C, Wurth L, Krol A (2009) The selenium to selenoprotein pathway in eukaryotes: more molecular partners than anticipated. Biochim Biophys Acta 1790:1415–1423
2. Boulon S, Marmier-Gourrier N, Pradet-Balade B et al (2008) The Hsp90 chaperone controls the biogenesis of L7Ae RNPs through conserved machinery. J Cell Biol 180:579–595
3. Bulteau A, Chavatte L (2015) Update on selenoprotein biosynthesis, Antioxidants & redox signaling. 23:775–94
4. Cléry A, Bourguignon-Igel V, Allmang C et al (2007) An improved definition of the RNA-binding specificity of SECIS-binding protein 2, an essential component of the selenocysteine incorporation machinery. Nucleic Acids Res 35:1868–1884
5. Wurth L, Gribling-Burrer AS, Verheggen C et al (2014) Hypermethylated capped selenoprotein mRNAs in mammals. Nucleic Acids Res 42:8663–8677
6. Topisirovic I, Svitkin YV, Sonenberg N et al (2011) Cap and cap-binding proteins in the control of gene expression, Wiley interdisciplinary reviews. RNA 2:277–298
7. Houseley J, Tollervey D (2009) The many pathways of RNA degradation. Cell 136:763–776
8. Kohler A, Hurt E (2007) Exporting RNA from the nucleus to the cytoplasm. Nat Rev Mol Cell Biol 8:761–773
9. Fortes P, Inada T, Preiss T et al (2000) The yeast nuclear cap binding complex can interact with translation factor eIF4G and mediate translation initiation. Mol Cell 6:191–196
10. Wilusz CJ, Wormington M, Peltz SW (2001) The cap-to-tail guide to mRNA turnover. Nat Rev Mol Cell Biol 2:237–246
11. Izaurralde E, Lewis J, Gamberi C et al (1995) A cap-binding protein complex mediating U snRNA export. Nature 376:709–712
12. Shatkin AJ (1976) Capping of eucaryotic mRNAs. Cell 9:645–653
13. Luhrmann R, Appel B, Bringmann P et al (1982) Isolation and characterization of rabbit anti-m3 2,2,7G antibodies. Nucleic Acids Res 10:7103–7113
14. Tycowski KT, Aab A, Steitz JA (2004) Guide RNAs with 5′ caps and novel box C/D snoRNA-like domains for modification of snRNAs in metazoa. Curr Biol 14:1985–1995
15. Hausmann S, Zheng S, Costanzo M et al (2008) Genetic and biochemical analysis of yeast and human cap trimethylguanosine synthase: functional overlap of 2,2,7-

trimethylguanosine caps, small nuclear ribonucleoprotein components, pre-mRNA splicing factors, and RNA decay pathways. J Biol Chem 283:31706–31718
16. Girard C, Verheggen C, Neel H et al (2008) Characterization of a short isoform of human Tgs1 hypermethylase associating with small nucleolar ribonucleoprotein core proteins and produced by limited proteolytic processing. J Biol Chem 283:2060–2069
17. Monecke T, Dickmanns A, Ficner R (2009) Structural basis for m(7)G-cap hypermethylation of small nuclear, small nucleolar and telomerase RNA by the dimethyltransferase TGS1. Nucleic Acids Res 37:3865–3877
18. Monecke T, Dickmanns A, Strasser A et al (2009) Structure analysis of the conserved methyltransferase domain of human trimethylguanosine synthase TGS1. Acta Crystallographica Section D 65:332–338
19. Keene JD, Komisarow JM, Friedersdorf MB (2006) RIP-Chip: the isolation and identification of mRNAs, microRNAs and protein components of ribonucleoprotein complexes from cell extracts. Nat Protoc 1:302–307
20. Göbel U, Maas R, Clad A (1987) Quantitative electroelution of oligonucleotides and large DNA fragments from gels and purification by electrodialysis. J Biochem Biophys Methods 14:245–260
21. Bustin SA, Benes V, Garson JA et al (2009) The MIQE guidelines: minimum information for publication of quantitative real-time PCR experiments. Clin Chem 55:611 LP–611622

Part III

Selenoprotein Detection and Quantification

Chapter 10

Total Selenium Quantification in Biological Samples by Inductively Coupled Plasma Mass Spectrometry (ICP-MS)

Véronique Vacchina and Jean Dumont

Abstract

Selenium (Se) is an element readily absorbed during the intestinal tract, distributed in the body by means of blood and excreted mainly by urine or feces. Here, we describe the method allowing the determination of the total Se content in biological tissues and fluids by Inductively Coupled Plasma Mass Spectrometry (ICP MS).

Key words Selenium, ICP MS, Biological fluids and tissues

1 Introduction

Most inorganic and low molecular weight selenocompounds, e.g., selenized amino-acids, inorganic selenium, are efficiently absorbed from the intestinal tract [1, 2]. The absorbed selenium (Se) is then transported in the blood mainly bound to proteins [3] and is distributed into different organs (liver, kidney, muscles, heart, etc.) or animal products when appropriate (eggs, milk, etc.). The Se is then excreted from the body by three distinct routes: in urine via the kidney, in feces, and in the expired air via the lung [3]. Moreover, Se levels in hair and nails are also of interest as they are considered to reflect long-term intake [3] providing that Se-containing treatments have not been applied. The evaluation of the total Se content in the biological organs and fluids is therefore of primary importance to evaluate the repartition of the Se in the body.

In all these products, organs, and fluids, the Se concentration is expected to be at the trace or ultra-trace level. Its determination requires therefore the use of sensitive methods. Inductively Coupled Plasma Mass Spectrometry (ICP MS) is now recognized as the major technique for trace level inorganic analysis in biological fluids and tissues due to its sensitivity, specificity, and good reproducibility among others [4–6].

Laurent Chavatte (ed.), *Selenoproteins: Methods and Protocols*, Methods in Molecular Biology, vol. 1661,
DOI 10.1007/978-1-4939-7258-6_10, © Springer Science+Business Media LLC 2018

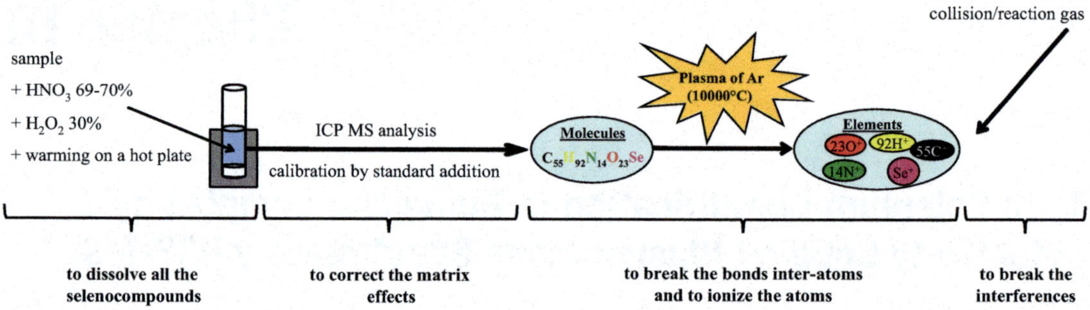

Fig. 1 Summary diagram of the analytical method allowing the determination of total Se in biological fluids and tissues by ICP MS

The aim of the paper is therefore to describe the analytical method allowing the determination of the total Se content in biological fluids, tissues, and products by ICP MS. This method is summarized in Fig. 1.

2 Materials

Prepare all solutions using ultrapure water (18.2 MΩ.cm) and analytical grade reagents. Prepare and store all reagents at room temperature (unless stated otherwise). Diligently follow all waste disposal regulations when disposing waste materials.

2.1 Reagents

1. 69–70% Nitric acid (HNO_3) (Instra).
2. 30% Hydrogen peroxide (H_2O_2) (Instra).
3. HNO_3 2%: dilute 14 mL of HNO_3 69–70% in 486 mL of ultrapure water.
4. Se standard solution: either mono- or multi-elemental with a concentration of at least 100 μg Se/mL.

2.2 Instrumentation

1. Freezer.
2. Hot plate.
3. Grinder.
4. Analytical grade balance.
5. Adjustable volume pipettes.
6. ICP MS equipped with a collision/reaction cell filled with H_2 (*see* **Notes 1–3**).
7. Plastic tubes fitting with the hot plate and with the auto-sampler of the ICP MS.

3 Methods

Carry out all the procedures at room temperature unless stated otherwise.

3.1 Samples Storage and Pretreatment

1. For blood samples, a Se-free anti-coagulant has to be added at the time of the sampling in the sampling tube.
2. Fresh tissues and fluids samples have to be stored at −20 °C.
3. The solid samples such as liver, eggs have to be freeze-dried and grinded prior to analysis. For liquid samples such as blood, milk, urine the analysis can be performed on liquid-state samples.

3.2 Mineralization

1. Precisely weigh empty plastic tubes that are consistent with the hot plate (*see* **Note 4**). Include at least three tubes devoted for blanks and tubes devoted for the Certified Reference Material(s) (CRM) (*see* **Note 5**).
2. Precisely weigh 0.25 g or 500 μL for respectively solid and liquid samples in these tubes.
3. Add 2 mL or 1 mL of 69–70% HNO_3 respectively for solid and liquid samples.
4. Close the tubes and leave them in a fume hood for predigestion over at least 8 h.
5. Add 1 mL or 500 μL of 30% H_2O_2 respectively for solid and liquid samples (*see* **Note 6**).
6. Close the tubes and place them in the hot plate. For liquid samples, the following temperature program is applied: up to 65 °C in 30 min and 4 h at 65 °C. For solid samples, the following temperature program is applied: up to 45 °C in 20 min, 40 min at 45 °C, up to 85 °C in 20 min, and 85 °C during 160 min.
7. Add 10 mL or 20 mL of ultrapure water respectively for liquid and solid samples. Precisely weigh the final mass of the closed tubes.

3.3 ICP MS Analysis

The quantification of the Se in the mineralizates must be performed by the standard addition method (*see* **Notes** 7 and **8**).

1. Prepare a 10 μg Se/mL solution in 2% HNO_3 by (multiple-step) dilution of the mother Se standard solution.
2. Analyze this solution together with a sample to roughly evaluate the concentration in the sample and estimate the spikes to add to build the standard addition curve. The ICP MS conditions are given Table 1.

Table 1
ICP MS conditions

Parameters	Value
Rf power	1500 W
Isotopes monitored	^{76}Se, ^{77}Se, ^{78}Se
Dwell time	100 msec
Collision/reaction cell	Filled with H_2
Carrier gas flow, lens voltage, etc.	Optimized daily

3. The sequence of analysis should be as follows: the blanks, the standard addition curve in the increasing order of the concentrations, the samples, and the CRM when appropriate. Between each sample, the ICP MS introduction system should be rinsed with 2% HNO_3. Around every 10 samples, a new standard addition curve should be analyzed to check the stability of the instrument (*see* **Note 9**).

3.4 Calculations

The validation of the standard addition curve) and of the concentration found for the CRM should be made according to the rules in force in the laboratory.

1. The concentrations in the raw sample are calculated according to the following equation:

$$C_i = \frac{I_i}{S_i} \times D \times \frac{(m_f - m_e)}{(m_s)}$$

with: C_i: concentration for the isotope i

I_i: intensity for the isotope I after corresponding blank subtraction

S_i: slope of the standard addition curve for the isotope i

D: dilution factor when appropriate

m_f: mass of the tube when full

m_e: mass of the tube when empty

m_s: mass of the sample.

2. The three concentrations obtained for the three isotopes of Se scanned must be averaged. Should one of these three concentrations differ by more than 20% from the others, it should be removed from this average calculation (*see* **Note 10**). The average concentration calculated this way corresponds to the concentration in the sample.

4 Notes

1. ICP MS is an elemental mass spectrometry technique. The samples are introduced into a plasma of argon where the molecules are atomized and the elements are ionized. The elements are then transferred into a mass analyzer where they are separated as a function of their mass above charge (*m/z*) ratio. For routine analysis, the mass analyzers used are generally quadrupole with a mass resolution of 1 amu. In other words, elements with *m/z* ratios that differ less than 1 cannot be discriminated in this way. Finally, the elements are detected.
2. Different kinds of spectroscopic interferences exist in ICP MS: (1) the isobaric ones corresponding to elements that have close *m/z* ratios (difference below 1). To correct these interferences, it is possible to apply correction equations; (2) the doubly charged ones associated with elements that can be doubly charged and become an interference for less heavy elements. To minimize this, the optimization of the ICP MS should be made to limit the formation of doubly charged ions; (3) the polyatomic ones. They are due to the formation of, for example, oxydes (or dimer etc.) in the plasma. To overcome these interferences, the use of an ICP MS equipped with a collision/reaction cell has now been democratized (*see* **Note 3**). Anyway, from a general point of view, the evaluation/correction of the interferences can be performed by scanning as many isotopes of the analyte as possible. The main interferences for Se are summarized in Table 2.
3. To break the polyatomic interferences, the use of an ICP MS equipped with a collision/reaction cell is now generalized. In the first case, an inert collision gas such as He is used and the interference is eliminated by Kinectic Discrimination Energy (KED). It is easy to use as no reaction subproducts are generated. In the other case, reactive gases such as H_2, NH_3 are used. They allow the elimination of the interference by modification

Table 2
Main Se interferences in ICP MS with or without H_2 as reaction gas

Isotope	Interferences without reaction gas	Interferences with H_2 as reaction gas
^{76}Se	$^{40}Ar^{36}Ar^+$, $^{40}Ar^{36}S^+$, $^{31}P_2{}^{14}N^+$	/
^{77}Se	$^{40}Ar^{36}Ar^1H^+$, $^{38}Ar_2{}^1H^+$, $^{40}Ar^{37}Cl^+$	/
^{78}Se	$^{40}Ar^{38}Ar^+$, $^{31}P_2{}^{16}O^+$	/
^{80}Se	$^{40}Ar_2{}^+$, $^{79}Br^1H^+$	$^1H^{79}Br$
^{82}Se	$^{40}Ar_2{}^1H_2{}^+$, $^{12}C^{35}Cl_2{}^+$, $^{34}S^{16}O_3{}^+$, $^{81}Br^1H^+$	$^1H^{81}Br$

of the mass of the analyte or the interference. Care must be taken as reaction subproducts may be generated inducing new interferences. In the case of Se, even if different collision/reaction gas may be used, the most frequently used one remains H_2 [7].

4. Close the tubes to weigh them as later on, they will need to be weighed full and therefore preferably closed.
5. The CRMs are used as quality control of the analysis. A list of the commercially available CRMs for Se analysis in biological fluids and tissues is given Table 3. The choice of the appropriate CRM should be made according to two parameters: the matrix and the concentration. Both should be as close as possible of the samples to analyze.

Table 3
List of commercially available CRMs for Se determination in biological fluids and tissues

Name	Matrix	Supplier
DORM-4	Fish proteins	NRC-CNRC
ERM-BB422	Fish muscle	IRMM
1946	Lake superior fish tissue	NIST
1947	Lake Michigan fish tissue	NIST
1566b	Oyster tissue	NIST
ERM-BB184	Bovine muscle	IRMM
BOVM-1	Bovine muscle	NRC-CNRC
BCR-185R	Bovine liver	IRMM
1577c	Bovine liver	NIST
DOLT-5	Dogfish liver	NRC-CNRC
ERM-BB186	Pig kidney	IRMM
LUTS-1	Non-defatted lobster hepatopancreas	NRC-CNRC
TORT-3	Lobster hepatopancreas	NRC-CNRC
EGGS-1	Egg powder	NRC-CNRC
CRM n°13	Human hair	NIES
ERM-BD001	Human hair	IRMM
ERM-BD150 and 151	Skimmed milk powder	IRMM
Seronorm	Freeze-dried blood, urine and serum	SERO
BCR-637, 638 and 639	Human serum	IRMM
909c	Human serum	NIST
2670a	Human urine	NIST
CRM n°18	Human urine	NIES

6. Depending on the matrix, the addition of H_2O_2 may be quite reactive. It is better to let the reaction end during a few minutes before closing the tubes.
7. In addition to spectroscopic interferences already discussed, non-spectroscopic interferences may occur in ICP MS. They come from the matrix (dissolved salts, acid concentration, etc.). They can lead to the suppression or enhancement of the signal of the analyte. They can be corrected by the use of the proper internal standard (absent from the matrix, distinguishable from the analyte, similar physico-chemical properties than the analyte), by dilution (when possible) or by standard addition calibration. This latter is well adapted to the correction of the matrix effects providing that different cautions are taken: the volumes added should be low enough to avoid dilution and the quantity added should be around double the initial signal. As a consequence, before building the calibration curve, the concentration in the sample should be roughly evaluated. The standard addition curve should contain at least four points (blank excluded).
8. In the case of Se, the signal is enhanced in the presence of a high carbon content [8] in the extract. Standard addition quantification is therefore necessary to measure the Se content in the mineralizates of organic samples.
9. Should the signal intensity of a sample be higher than the one of the most concentrated standard, this sample should be reanalyzed after proper dilution. In this case, a new standard addition curve should be made fitting with this dilution factor.
10. Table 2 summarizes the main Se interferences in ICP MS using H_2 as a reaction gas. In Br-containing samples (such as most of the biological samples), ^{80}Se and ^{82}Se are interfered. That is the reason why they are usually not scanned. In case of doubts, ^{78}Se remains the most reliable isotope for this analysis.

References

1. Alexander J (2014) Selenium. In: Handbook on the toxicology of metals, vol 1, 4th edn. Academic Press, San Diego, pp 1175–1208
2. WHO (1987) Selenium. Environmental Health Criteria, Geneva, p 58
3. Reilly C (2006) The biology of selenium. In: Selenium in food and health. Springer, New York, pp 1–206
4. Goullé JP, Saussereau E, Mahieu L et al (2014) Clinical role of ICP MS in clinical toxicology and forensic toxicology: a metallic profile. Bioanalysis 6:2245–2259
5. Vacchina V, Huin V, Hulo S et al (2014) Use of dreid-blood spot and inductively coupled plasma mass spectrometry for multi-element determination in blood. J Trace Elem Med Biol 28:255–259
6. Bakirdere S, Aydin F, Bakirdere EG et al (2011) From mg/kg to pg/kg levels: a story of trace elements determination: a review. Appl Spectrosc Rev 46:38–66

7. Darrouzes J, Bueno M, Lespès G et al (2005) Operational optimisation of ICP MS – octopole collision/reaction cell – MS for applications to ultratrace selenium total and speciation determination. J Anal at Spectrom 20:88–94

8. Larsen EH, Stürup S (1994) Carbon-enhancer inductively coupled plasma mass spectrometric detection of arsenic and selenium and its application to arsenic speciation. J Anal At Spectrom 9:1099–1105

Chapter 11

Quantification of SeMet and SeCys in Biological Fluids and Tissues by Liquid Chromatography Coupled to Inductively Coupled Plasma Mass Spectrometry (HPLC-ICP MS)

Véronique Vacchina, Katarzyna Bierla, Joanna Szpunar, and Ryszard Lobinski

Abstract

Selenium (Se) is an element readily absorbed during the intestinal tract and distributed in the body. In biological fluids, tissues, and animal products, Se is known to be present mainly in the form of a selenoamino-acid (selenomethionine (SeMet) or selenocysteine (SeCys)). Both amino-acids have different biological activities which justifies their discrimination. Here, we describe the method allowing the simultaneous determination of SeMet and SeCys in blood/plasma, animal tissues, milk, and eggs by two-dimensional Liquid Chromatography coupled to Inductively Coupled Plasma Mass Spectrometry (2D HPLC-ICP MS).

Key words Selenomethionine, Selenocysteine, HPLC-ICP MS, Biological fluids and tissues

1 Introduction

Selenium (Se) is an element readily adsorbed by the intestinal tract [1] and distributed via the blood into different organs (offal tissues, edible tissues, etc.) or animal products when appropriate (milk, eggs, etc.). Most of the Se present in these samples is known to be proteins bound [2] and to occur in the form of two selenoamino-acids. One of them, the selenocysteine (SeCys), is considered biologically active and forms the active center of selenoproteins like glutathione-peroxidases, thioredoxin reductases, and selenoprotein P. The other one, the selenomethionine (SeMet), is supposed to be randomly incorporated into methionine-containing proteins. The difference in function of these amino-acids demands accurate methods for their routine determination.

The HPLC-ICP MS coupling has been predominantly used for the speciation analysis of low-molecular weight selenospecies [3].

Laurent Chavatte (ed.), *Selenoproteins: Methods and Protocols*, Methods in Molecular Biology, vol. 1661, DOI 10.1007/978-1-4939-7258-6_11, © Springer Science+Business Media LLC 2018

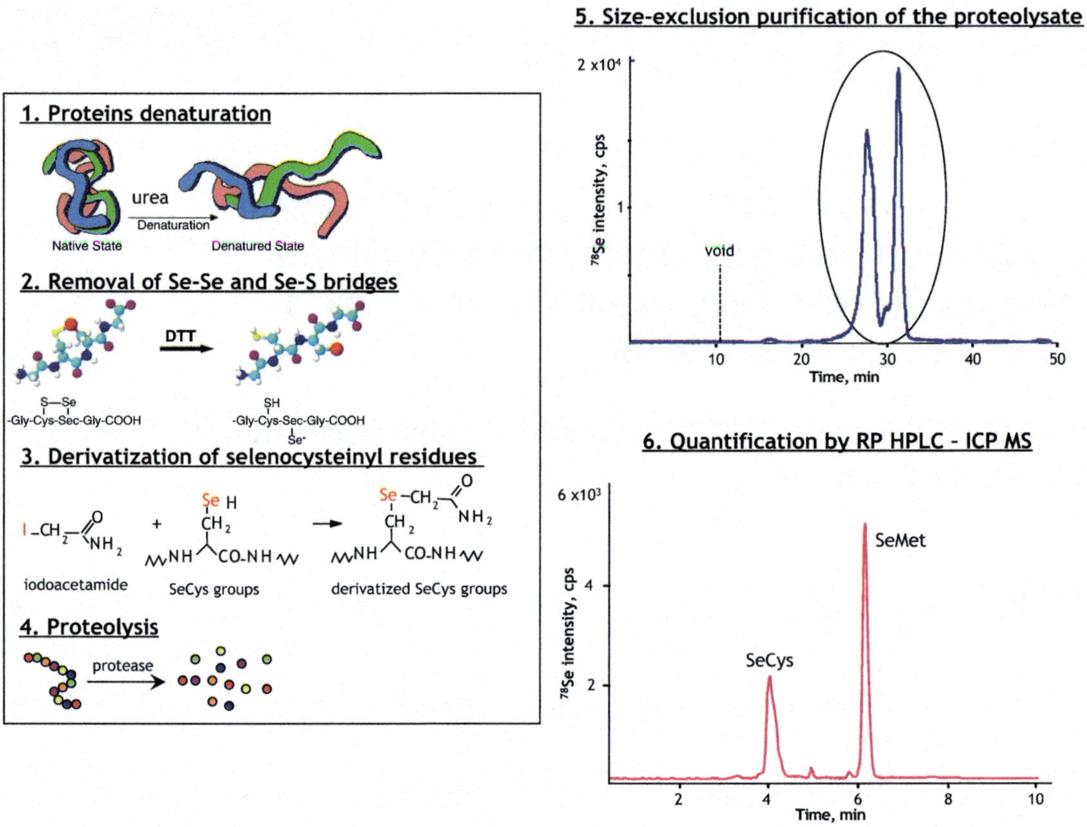

Fig. 1 Summary diagram of the analytical method allowing the determination of selenoamino-acids in biological fluids and tissues and animal products by 2D HPLC-ICP MS

In particular, the quantification of SeMet by HPLC-ICP MS following proteolytic digestion to release the free amino-acids has been addressed in different samples [4]. But the determination of SeCys is more complicate due to the instability of this species.

The aim of the paper is therefore to describe the analytical methods allowing the simultaneous determination of both SeMet and SeCys in biological fluids, tissues, and animal products by two-dimensional Liquid Chromatography coupled to Inductively Coupled Plasma Mass Spectrometry (2D HPLC-ICP MS). The methods are summarized in Fig. 1 and are described separately elsewhere [5–8].

2 Materials

Prepare all the solutions using ultrapure water (18.2 MΩ.cm) and analytical grade reagents. Prepare and store all the reagents at room temperature (unless stated otherwise). Diligently follow all waste disposal regulations when disposing waste materials.

2.1 Reagents

1. 100 mM Tris–HCl solution: dissolve 12 g of Tris (99.9%) in 1 L of ultrapure water and adjust the pH to 7.5 by dropwise addition of HCl.
2. 6 M urea solution: dissolve 3.6 g of urea (99%) in 10 mL of 100 mM Tris–HCl buffer.
3. 7 M urea solution: dissolve 4.2 g of urea (99%) in 10 mL of 100 mM Tris–HCl buffer.
4. 0.2 M dithiothreitol (DTT) solution: dissolve 0.31 g of DTT (≥99%) in 10 mL of 100 mM Tris–HCl buffer.
5. 0.5 M DTT solution: dissolve 0.78 g of DTT (≥99%) in 10 mL of 100 mM Tris–HCl buffer.
6. 0.5 M iodoacetamide (IAM) solution: dissolve 0.9 g of IAM (≥99%) in 10 mL of ultrapure water.
7. 2% protease solution: dissolve 0.2 g of protease (type XIV) in 10 mL of 100 mM Tris–HCl buffer.
8. 1% lipase solution: dissolve 0.1 g of lipase (type VII) in 10 mL of 100 mM Tris–HCl buffer.
9. 100 mM ammonium acetate solution: dissolve 7.7 g of ammonium acetate (>98%) in 1 L of ultrapure water and adjust the pH to 7.5 by dropwise addition of NH_4OH.
10. Methanol, HPLC grade.
11. 30% ammonium hydroxyde (NH_4OH) (Instra).
12. 0.1% heptafluorobutyric acid (HFBA) solution: add 100 μL of HFBA (HPLC grade) in 1 L of ultrapure water or methanol.
13. 37% hydrochloric acid (HCl) (Instra).
14. 1000 μg Se/mL selenomethionine (SeMet): dissolve 25 mg of SeMet (≥99%) in 10 mL of ultrapure water.
15. 10 μg Se/mL SeMet: dilute 50 μL of the 1000 μg Se/mL SeMet solution in 4.95 mL of ultrapure water.
16. Acetone, HPLC grade (only for milk samples).
17. Cyclohexane, HPLC grade (only for milk samples).

2.2 Instrumentation

1. Freezer.
2. Grinder (only for tissue samples).
3. Sieve (only for tissue samples).
4. Analytical grade balance.
5. Adjustable volume pipettes.
6. Plastic tubes.
7. Bottles fitting with the HPLC system.
8. Elliptic table.
9. Sonication probe (only for tissue samples).
10. Shaking water bath.

11. Centrifuge.
12. Ultracentrifuge (only for tissue samples).
13. Freeze-drier.
14. Fraction collector.
15. Proofer (only for milk samples).
16. Superdex HPLC column (300 × 10 mm) (Amersham Biosciences).
17. Altima C8 HPLC column (150 × 4,6 mm × 5 μm) (Alltech).
18. Binary HPLC pump coupled to an ICP MS equipped with a collision/reaction cell filled with H_2 and with an organic kit (*see* **Notes 1–3**).

3 Methods

Carry out all the procedures at room temperature unless stated otherwise.

3.1 Samples Storage and Pretreatment

1. For blood samples, a Se-free anti-coagulant has to be added at the time of the sampling in the sampling tube.
2. Fresh tissues and fluids samples have to be stored at −20 °C.
3. All the samples have to be freeze-dried prior to analysis. For tissues, an additional step of grinding and sieving has to be performed.

3.2 Protocol for Defatting

This is only applicable for milk and egg samples.

1. Take around 2.5 g of sample in a plastic tube.
2. Add 10 mL of cyclohexane.
3. Shake the tube on an elliptic table during 10 min.
4. Centrifuge (1000 × *g*, 3 min).
5. Remove the supernatant.
6. Repeat **steps 2–4** two times on the residue.
7. Dry the residue in a proofer at 37 °C during 17 h.

3.3 Protocol for Proteins Precipitation

This is only applicable for milk samples.

1. Add 10 mL of ultrapure water on the preliminary residue.
2. Add 30 mL acetone.
3. Shake 10 min on an elliptic table.
4. Centrifuge (1000 × *g*, 5 min).
5. Remove the supernatant.
6. Dry the residue in a proofer at 37 °C during 17 h.

Table 1
Concentrations and volumes of the reactants as a function the matrix (expressed as concentration/volume)

	Blood/plasma	Tissues	Milk	Eggs
Urea	6 M/300 μL	7 M/2 mL	7 M/2 mL	7 M/4 mL
DTT-1	0.2 M/30 μL	0.2 M/60 μL	0.5 M/80 μL	0.5 M/80 μL
IAM	0.5 M/40 μL	0.5 M/100 μL	0.5 M/320 μL	0.5 M/320 μL
DTT-2	0.2 M/375 μL	0.2 M/300 μL	0.5 M/480 μL	0.5 M/450 μL
Tris	N/A	100 mM/11 mL	100 mM/11 mL	100 mM/20 mL
SeMet	10 mg/L/5 μL	10 mg/L/5 μL	10 mg/L/5 μL	10 mg/L/10 μL
protease + lipase	2% + 1%/900 μL	2% + N/A/1 mL	2% + N/A/1 mL	2% + N/A/1 mL

3.4 Protocol for Extraction

This is applicable for all the matrices. The mass/volume uptakes and buffer/salts concentrations are given in Table 1.

1. Precisely weigh 100 mg (80 mg in the case of tissue samples) of each sample in a plastic tube. Include at least one tube devoted for a blank. One sample should be weighed in four different tubes (*see* **Note 4**).
2. Add 850 μL of ultrapure water (this is only applicable for blood and plasma samples).
3. Add the urea solution (*see* **Note 5**).
4. For tissue samples, sonicate the solution (amplitude 20%, pulse on 1, pulse off 1) keeping the tube in ice. Each sample should be sonicated two times not consecutive (*see* **Note 6**). Ultracentrifuge (20 min, 7500 × *g*, 4 °C) and transfer the supernatant into another plastic tube. For the other matrices than tissues, shake the tube 1 h on an elliptic table.
5. Add the DTT solution (DTT-1 in Table 1) (*see* **Note 7**).
6. Shake for 1 h on an elliptic table.
7. Add the IAM solution (*see* **Note 8**).
8. Shake for 1 h on an elliptic table away from light.
9. Add the DTT solution (DTT-2 in Table 1) (*see* **Note 9**).
10. Shake for 1 h on an elliptic table.
11. Add the Tris buffer (*see* **Note 10**).
12. Proceed to the addition of the SeMet standards in the tubes devoted for that. The volume given in Table 1 corresponds to the lowest spike. The intermediate one should be twice this volume and the highest one should be three times this volume.
13. Add the proteolytic solution (*see* **Note 11**).

14. Shake for 17 h at 37 °C.
15. Freeze-dry the solution.
16. Dissolve the dried material in 1 mL of ultrapure water.
17. Centrifuge the solution (8500 × *g*, 10 min) and put the supernatant in an HPLC vial.

3.5 Purification of the Extract by Size-Exclusion Chromatography (See Note 12)

1. Check the chromatographic profile on one sample (*see* **Note 13**). The chromatographic conditions are given Table 2.
2. For all the samples, collect the low molecular weight fraction starting around 1 min before the first peak and ending 1 min after the last peak (*see* **Note 15**).
3. Each maximum eight samples, inject the 0.2 M DTT solution without collecting fractions to clean the column.
4. Freeze-dry the fraction collected.
5. Dissolve the freeze-dried material in 500 μL of ultrapure water and put the solution in an HPLC vial.
6. Add 10 μL of the 0.2 M DTT solution.

Table 2
HPLC and ICP conditions for the coupling with the size-exclusion column

HPLC	
Column	Superdex
Mobile phase	100 mM ammonium acetate (pH 7.5)
Gradient	Isocratic
Volume injected	100 μL
Flow rate	0.7 mL/min
Duration	1 h
ICP MS	
Parameters	Value
Rf power	1500 W
Isotopes monitored	^{76}Se, ^{77}Se, ^{78}Se (*see* **Note 14**)
Integration time	0.15 s
Collision/reaction cell	Filled with H_2
Carrier gas flow, lens voltage, etc.	Optimized daily
Organic kit	No

Table 3
HPLC and ICP conditions for the coupling with the reversed-phase column

HPLC	
Column	Altima C_8
Mobile phase	A: 0.1% of HFBA in water B: 0.1% of HFBA in methanol
Gradient	0–5 min: 5% of B 5–30 min: up to 40% of B 30–40 min: 40% of B 40–42 min: down 5% of B 42–48 min: 5% of B
Volume injected	100 μL
Flow rate	0.9 mL/min
ICP MS	
Parameters	Value
Rf power	1500 W
Isotopes monitored	^{76}Se, ^{77}Se, ^{78}Se (*see* **Note 14**)
Integration time	0.15 s
Collision/reaction cell	Filled with H_2
Carrier gas flow, lens voltage, etc.	Optimized daily
Organic kit	Yes

3.6 Analysis by RP HPLC-ICP MS

1. The chromatographic conditions are detailed in Table 3.
2. The sequence of analysis should be as follows: the blanks, the standard addition curve in the rising order of the concentrations, and the samples.

3.7 Calculations

The validation of the standard addition curve should be made according to the rules in force in the laboratory.

1. The concentrations in the raw sample are calculated according the following equation:

$$C_i = \frac{A_i}{m \times S}$$

with C_i: concentration for the species i

A_i: area of the peak of the species i
S: slope of the standard addition curve (*see* **Notes 16** and **17**)
D: dilution factor when appropriate
m: sample uptake.

2. In the absence of Certified Reference Material for selenoamino-acids quantification, validation of the analysis by checking the sum SeMet + SeCys measured with the total Se content in the same sample.

4 Notes

1. The online coupling between HPLC and ICP MS allows performing the speciation of nonvolatile species without preliminary chemical treatment except extraction. As most of the mobile phases and flow rates used in HPLC are convenient with the ICP MS, the exit of the HPLC column is directly connected to the ICP MS. The species of interest are separated on the HPLC column and reach step by step the ICP MS where they are detected. The identification of the species is made on the basis of their retention time.
2. When the ICP MS is coupled to a reversed-phase column, the interface should be adapted because of the high content of organic solvent in the mobile phase that would, otherwise, switch off the plasma. This amendment consists in cooling the spray chamber, decreasing the inner diameter of the injector, adding O_2, and replacing the Ni cones by Pt ones. This adapted interface is called "organic kit."
3. To break the polyatomic interferences generated in the plasma, the use of an ICP MS equipped with a collision/reaction cell is now generalized. In the case of Se, even if different collision/reaction gas may be used, the most frequently used one remains H_2 [9].
4. In these three tubes, rising concentrations of SeMet will be spiked later to build a calibration curve.
5. Urea is a commonly used chaotropic agent added in order to help the solubilization of the proteins.
6. The two sonication steps must not be consecutive in order to avoid too high increase of the temperature in the sample.
7. The Se-H group of the SeCys can easily form bridges with other Se-H or S-H group present in the sample. DTT is a commonly used reducing agent added to break the Se-Se and Se-S bridges.
8. When SeCys is not protected by Se-Se or Se-S bridges, it undergoes an oxidation into selenoxide followed by syn-β-elimination of selenic acid [10]. A derivatization reaction, e.g., with IAM, can prevent this process [11].
9. This fresh DTT is added to destroy the excess of IAM.

10. Tris buffer is added in order to decrease concentration of urea to the appropriate level for proteolytic digestion.
11. Proteases are enzymes able to break unspecifically the bonds between amino-acids and therefore release the free amino-acids.
12. The raw extracts cannot be directly analyzed by RP HPLC-ICP MS. Indeed, because of a strong matrix effect, the chromatograms obtained in this way are not usable. Prior to the selenoamino-acids quantification, the extracts need to be purified. This can be achieved by size-exclusion chromatography. This purification consists of collecting the low molecular weight fraction (that contains the selenoamino-acids) at the exit of the size-exclusion column and therefore simplifying the matrix. Beforehand, the chromatographic profile should have been checked to evaluate the fraction time range to collect. This cleaning approach by size-exclusion presents two main advantages: (1) even if the resolution on this kind of column is low, strong matrices can be injected on them; (2) the separation is made mainly as a function of the size, the big compounds, such as proteins, eluting in or close to the void of the column, and the low molecular weight compounds, such as amino-acids, eluting later. The size-exclusion profile allows, besides determining the fraction time range to collect, checking the proper completeness proteins digestion and the release of the amino-acids reflected by the absence of peaks in or close to the void and the presence of peaks in the low molecular weight fraction of the chromatograms.
13. It is more reliable to check the profile on one of the spiked samples.
14. In case of doubt, ^{78}Se remains the most reliable isotope.
15. For long series of samples, the retention times may move a little bit. It is better to maintain a safety margin of 1 min at the beginning and the end of the fraction collected.
16. In the absence of standard of SeCys, the SeCys quantification is performed with the slope obtained for the SeMet.
17. To calculate the slope, consider the amount of SeMet in ng in the solution injected on the reversed-phase column.

References

1. Reilly C (2006) The biology of selenium. In: Selenium in food and health. Springer, New York, pp 1–206
2. Schrauzer GN (2000) Selenomethionine: a review of its nutritional significance, metabolism and toxicity. J Nutr 130:1653–1656
3. Bierla K, Szpunar J, Lobinski R (2008) Analytical techniques for speciation of selenium in food and food supplements: state of the art. In: Current advances in selenium research and applications. Wageningen Academic Publishers, Wageningen, The Netherlands, pp 107–132
4. Jagtap R, Maher W (2016) Determination of selenium species in biota with an emphasis on animal tissues by HPLC-ICP MS. Microchem J 124:422–529

5. Bierla K, Vacchina V, Szpunar J et al (2008) Simultaneous derivatization of selenocysteine and selenomethionine in animal blood prior to their determination by 2D size-exclusion ion-pairing reversed-phase HPLC-ICP MS. J Anal At Spectrom 23:508–513
6. Bierla K, Dernovics M, Vacchina V et al (2008) Determination of selenocysteine and selenomethionine in edible animal tissues by 2D size-exclusion reversed-phase HPLC-ICP MS following carbamidomethylation and proteolytic extraction. Anal Bioanal Chem 390:1789–1798
7. Bierla K, Szpunar J, Lobinski R (2008) Specific determination of selenoamino-acids in whole milk by 2D size-exclusion ion-pairing reversed-phase high performance liquid chromatography – inductively coupled plasma mass spectrometry (HPLC-ICP MS). Anal Chim Acta 634:195–202
8. Lipiec E, Siara G, Bierla K et al (2010) Determination of selenomethionine, selenocysteine and inorganic selenium in eggs by HPLC inductively coupled plasma mass spectrometry. Anal Bioanal Chem 397:731–741
9. Darrouzes J, Bueno M, Lespès G et al (2005) Operational optimisation of ICP MS–octopole collision/reaction cell–MS for applications to ultratrace selenium total and speciation determination. J Anal At Spectrom 20:88–94
10. Ma S, Caprioli RM, Hill KE et al (2003) Loss of selenium from selenoproteins: conversion of selenocysteine to dehydroalanine in vitro. J Am Soc Mass Spectrom 14:593–600
11. Encinar JR, Schaumlöffel D, Ogra Y et al (2004) Determination of selenomethionine and selenocysteine in human serum using speciated isotope dilution-capillary HPLC inductively coupled plasma collision cell mass spectrometry. Anal Chem 76:6635–6642

Chapter 12

Simultaneous Speciation of Selenoproteins and Selenometabolites in Plasma and Serum

Belén Callejón-Leblic, Gema Rodríguez-Moro, Tamara García-Barrera, and José Luis Gómez-Ariza

Abstract

Selenium is an essential element incorporated to different proteins with important biological functions in connection to antioxidant activity, cancer-protective properties, neurodegenerative pathologies, and prevention of effects of diabetes, among others. In addition, selenoamino acids play a basic role in the global equilibrium of key selenium-biomolecules synthesis, including selenoprotein P, selenoalbumin, and glutathione peroxidase. Homeostasis of these selenium-containing biomolecules involves different organs in living organisms including human, and bloodstream is the connection fluid in this process. Therefore, it is very important to have an analytical methodology suitable for selenium proteins and metabolites speciation in serum and plasma samples. For this purpose, a simultaneous speciation method for Se-containing biomolecules in serum/plasma is described on the basis of in series three-dimensional chromatography: size exclusion, affinity, and anion exchange high performance liquid chromatography (3D/SE-AF-AEC-HPLC), using different columns of each type and hyphenation to inductively coupled plasma-(quadrupole) mass spectrometry (ICP-MS). The method allows the quantitative simultaneous analysis of selenoprotein P (SeP), extracellular glutathione peroxidase (eGPx), selenoalbumin (SeAlb), selenite, and selenate in serum (from human and mouse) using species-unspecific isotope dilution (SUID). In addition, a simplified two-dimensional approach (2D/SE-AF-HPLC-SUID-ICP-MS) is described when selenium metabolites are globally analyzed. The method provides detection limits in the range 0.2–1.3 ng of Se g^{-1} and avoids typical interferences in this matrix from chloride and bromide with a chromatographic runtime less than 35 min.

Key words Selenoproteins, Selenometabolites, Column-switching, Isotopic dilution analysis, Inductively coupled plasma mass spectrometry, Serum, Plasma

1 Introduction

Selenium (Se) is an essential element for humans with an important role in biological functions mediated by different proteins as extracellular glutathione peroxidase (eGPx), selenoprotein P (SeP) and selenoalbumin (SeAlb). In some cases selenoamino acids as selenocysteine (SeCys) are specifically incorporated into the active sites of the proteins, such as selenoprotein P [1, 2] and GPx, while

Laurent Chavatte (ed.), *Selenoproteins: Methods and Protocols*, Methods in Molecular Biology, vol. 1661, DOI 10.1007/978-1-4939-7258-6_12, © Springer Science+Business Media LLC 2018

selenoalbumin (SeAlb) does not join specifically this element into the albumin moiety as selenomethionine (SeMet) [3, 4], being randomly replaced methionine (which is a constituent of albumin, HSA) by selenomethionine (SeMet). Selenoalbumin is not a selenoprotein properly because its synthesis is not genetically encoded (as the case of GPx and SelP), it can play the role of a potential bioindicator of selenium status, particularly correlated with nutrition [5]. In healthy persons SeAlb level tends to reduce with the age [6], unlike GPx and SelP, hence SeAlb might be the first depleted seleno-containing protein in case of selenium drop [7]. In addition, the presence of both SeP and SeAlb is interrelated, because Se bound to albumin is assumed to be transported to the liver for new synthesis of SeP that is then released into the bloodstream [8]. Nevertheless, despite the important role of Se-containing proteins and metabolites in serum an plasma in connection to the antioxidant activity of these species [9], cancer-protective properties [10], neurodegenerative pathologies [11], prevention of diabetes adverse effects [12], and others, no conclusive analytical methodologies for simultaneous determination of SeP, eGPx, SeAlb, and low molecular selenium species in plasma/serum, free of interferences and with high resolution, sensitivity, and precision have been proposed until now.

Different efforts have been made for analytical selenium species quantification in serum [13, 14] mainly based on the coupling of inductively coupled plasma mass spectrometry (ICP-MS) with previous chromatographic separation. ICP-MS is very suitable for selenium analysis due to its low detection limits, isotope ratios measurements, tolerance to matrix, and large linearity range [15–17]. In addition, this detector allows species unspecific isotope dilution mode (SUID), which is very useful either when the structure and composition of analyzed species is not exactly known or when the corresponding isotopically labeled compound is not commercially available [17]. On the other hand, a number of chromatographic methods have been proposed for complete selenium-species separation from mammalian plasma or serum, as anion exchange chromatography (AEC) [7, 13, 18], size exclusion chromatography (SEC) [19], and affinity chromatography (AFC) [13]. However, SEC cannot provide suitable species separation because of low chromatographic resolution, which causes overlapping between peaks. Alternatively, other approaches introduced affinity chromatographic separation (with heparine-sepharose columns) prior to SEC in order to improve the resolution by retention of SeP, but the time of analysis increases considerably [20]. Besides, AEC provides good recoveries of analytes but chromatographic resolution is not satisfactory either [13]. Finally, when only AFC is used, the weakly retained eGPx together with other nontarget matrix Se-components are not resolved and they are quantified together [13, 14, 21].

Moreover, it has to be considered the high concentration of Cl^- and Br^- in plasma and serum [22], which coelute with GPx and Se metabolites in the previously cited procedures [14, 21] representing a serious interference by the formation of polyatomic species $^{40}Ar_2^+$ and $^{79}Br^1H^+$ isobaric with ^{80}Se isotope in the ICP-MS. Most recently, multidimensional approaches based on AEC online coupled to double column affinity-high performance liquid chromatography (AE-AF-HPLC) hyphenated to ICP-MS have been developed for the speciation analysis of GPx, SeP, and SeAlb [14, 21]. Finally, a method for the simultaneous speciation of selenoproteins and selenometabolites in human serum has been developed on the basis of in series three-dimensional chromatography: size exclusion, affinity, and anion exchange high performance liquid chromatography (3D/SE-AF-AEC-HPLC), using different columns of each type and detection by inductively coupled plasma-(quadrupole) mass spectrometry (ICP-qMS). The method allows the quantitative simultaneous analysis of selenoprotein P (SeP), extracellular glutathione peroxidase (eGPx), selenoalbumin (SeAlb), selenite, and selenate in human serum using species-unspecific isotope dilution (SUID) [9, 23]. The 3D chromatographic separation is proposed to remove typical spectral interferences in this matrix from chloride and bromide. In addition, a simplified method based on 2D/SE-AF-HPLC has been proposed as a simpler alternative when low molecular mass selenium species are absent in the samples [23]. These later procedures will be detailed in the present chapter.

2 Materials

All reagents used for sample preparation were of the highest available purity. Phenylmethanesulfonyl fluoride (PMSF) and tris(2-carboxyethyl)phosphine hydrochloride (TCEP) (BioUl-tra grade, >98%) were obtained from Sigma–Aldrich (Steinheim, Germany). Hydrogen used as reaction gas in SEC-AF-AEC-ICP-ORC-MS system was of high-purity grade (>99.999%).

1. The mobile phase solutions used in SEC and AFC were 0.05 M ammonium acetate (mobile phase A) and 1.5 M ammonium acetate (mobile phase B) (Bio-ultra grade), which were daily prepared with ultrapure water (18 MΩ-cm) and adjusted to pH 7.4 with ammonia solution, this later prepared by dilution of 20% (v/v) ammonia solution (Suprapur) with ultrapure water.
2. The mobile phase solution used in AEC was 0.1 M sodium phosphate (mobile phase C) for analysis grade, which was daily prepared with ultrapure water and adjusted to pH 8 with a 50% (v/v) aqueous phosphoric acid solution.

3. Human serum certified reference material (CRM) BCR-637 from the Institute for Reference Materials and Measurements (IRMM, Geel, Belgium).
4. Lyophilized plasma (Clinchek of plasma control, Level II) and serum (Level I) (Recipe Chemicals).
5. Standard solutions containing 1000 mg L^{-1} of Se and 1000 mg L^{-1} of Br, both stabilized in 5% (v/v) Suprapur nitric acid.
6. Enriched ^{74}Se was obtained from Cambridge Isotope Laboratories (Andover, MA, USA) as elemental powder and it was dissolved in the minimum volume of nitric acid (Suprapur grade) and diluted to the appropriate volume with ultra-pure water. The concentration of this solution was established by reverse isotope dilution analysis as described elsewhere [13].
7. Inductively coupled plasma mass spectrometer equipped with collision/reaction cell for selenium-linked molecules analysis, ICP-MS measurement conditions are collected in Table 1.
8. HPLC equipment for chromatographic separations.

Table 1
Operating conditions of ICP-RS-MS detection

Parameter	ICP-MS conditions
Forward power	1500 W
Plasma gas flow rate	15 L min^{-1}
Auxiliary gas flow rate	1 L min^{-1}
Carrier gas flow rate	0.15 L min^{-1}
Sampling depth	7 mm
Sampling and skimmer cones	Ni
Nebulizer	Micromist (plass expancion)
Torch	Shield (with long life platinum shield plate)
Qoct	−18 V
Qp	−16 V
Points per peak	1
Integration time	0.3 per isotope
Replicates	1
Isotopes monitored	^{74}Se, ^{76}Se, ^{77}Se, ^{78}Se, ^{80}Se, ^{82}Se, ^{79}Br and ^{81}Br
Dead time detector	47 ns

9. Two 5 mL HiTrap® Desalting Columns (GE Healthcare, Uppsala, Sweden), in series connected to two affinity columns, with stationary phases of heparin-sepharose (HEP-HP) and blue-sepharose (BLU-HP).
10. Anionic exchange chromatography (AEC) with a Hamilton PRP-X100 column (25 cm × 4.6 mm, 5 μm) for selenite and selenate separation.

3 Methods

3.1 Analytical Procedure for Speciation of Se-Proteins in Serum/Plasma

1. Filter plasma and serum samples through Iso-Disc poly-(vinylidene difluoride) (PVDF) filters (20 mm diameter, 0.2 μm pore size).
2. Perform the speciation of Se-proteins (eGPx SeP, SeAlb) with the instrumental arrangement 2D/SE-AF-HPLC-SUID–ICP-ORS-MS, using a column switching valve with six ports (Fig. 1) (*see* **Notes 1–3**).
3. Separation of the analytes is carried out by in series stacking of two 5 mL HiTrap® Desalting columns in series connected with a dual affinity column arrangement comprising a 1 mL heparin-sepharose column (HEP-HP) and a 1 mL blue-sepharose

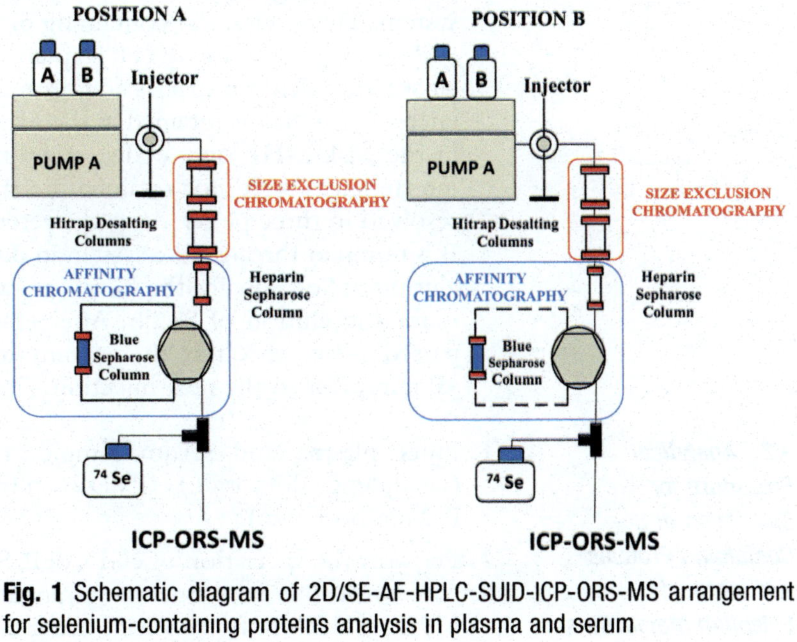

Fig. 1 Schematic diagram of 2D/SE-AF-HPLC-SUID-ICP-ORS-MS arrangement for selenium-containing proteins analysis in plasma and serum

Table 2
Chromatographic conditions of 2D/SE-AF-HPLC-SUID-ICP-ORS-MS

Sample loop	100 μL
Flow rate	1.3 mL min^{-1}
Mobile phase A	0.05 M ammonium acetate pH 7.4
Mobile phase B	1.5 M ammonium acetate pH 7.4
Gradient	0–7 min 100% A 6–18 min 100% B 18–20 min 100% A
Valve position	1–10 min inject 10–17 min load 17–20 min inject

column(BLU-HP) interconnected by the six-way switching column valve. The combination of two size exclusion (HiTrap) columns increases the resolution of the chromatographic separation. On the other hand, HEP-HP column retains selectively SeP whereas BLUE-HP column retains both SeP and SeAlb [7, 24, 25].

4. Operational details about optimum conditions for Se-proteins speciation are summarized in Tables 1 and 2.
5. Inject aliquots of plasma and serum samples with the system in position A (Fig. 2) with mobile phase A passing through the system. Then, selenium-containing proteins (eGPx, SeAlb, and SeP) are separated by SE (HiTrap columns) from low molecular weight selenium species and bromide and chloride interferences. Then, SeP is retained in the HEP-HPcolumn and SeAlb in the BLUE-HP one. eGPx, selenometabolites and Br + Cl interferences were not retained in affinity columns and elute resolved in three peaks. After the retention step, mobile phase B is pumped through the system in position B (Fig. 1) for the elution of SeP. Finally, the valve was switched again to position A for the elution of SeAlb. After elution, mobile phase A is passed again through the system for equilibration during 10 min prior to the next injection.

3.2 Analytical Procedure for Speciation of Se-Containing Proteins and Inorganic Selenium Metabolites in Serum/Plasma

1. Filter plasma and serum samples through Iso-Disc poly-(vinylidene difluoride) (PVDF) filters (20 mm diameter, 0.2 μm pore size).
2. Perform the speciation of eGPx SeP, SeAlb, selenite, and selenate with the instrumental arrangement 3D/SE-AF-AEC-HPLC-SUID–ICP-ORS-MS, using a column switching valve with eight ports (Fig. 2) (*see* **Notes 1** and **2**).

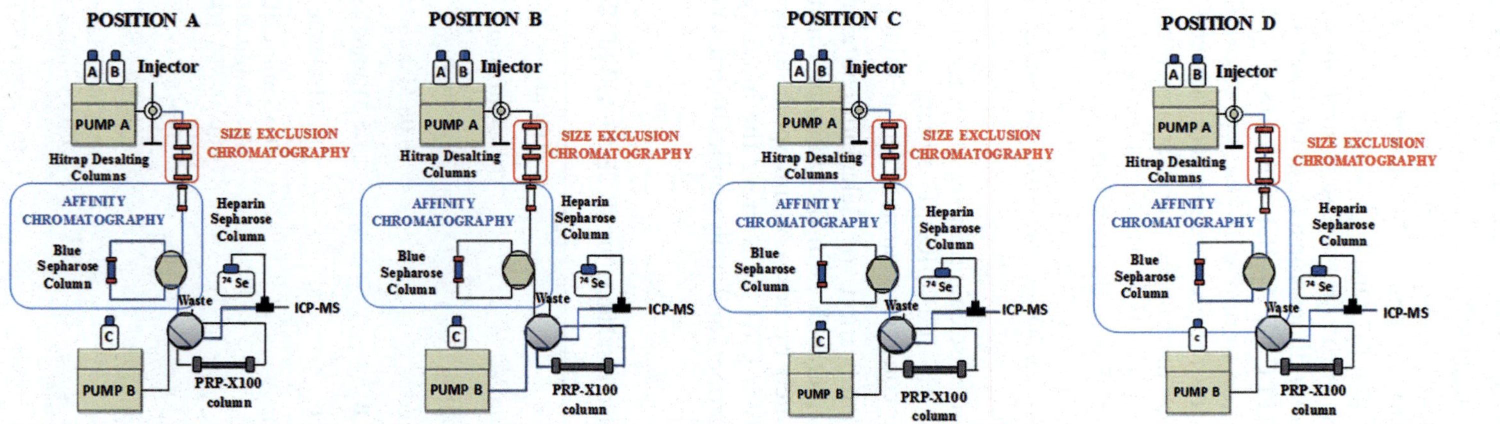

Fig. 2 Schematic diagram of 2D/SE-AF-AEC-HPLC-SUID-ICP-ORS-MS arrangement for selenium-containing proteins and inorganic selenium analysis in plasma and serum

3. Fractionation of selenium-containing proteins and inorganic selenium metabolites is carried out by a three-dimensional chromatographic arrangement based on double size exclusion prior to double affinity chromatography and anionic exchange chromatography (Fig. 2). The approach consists of two successive steps, in the first one separation of analytes is performed with the two-dimensional chromatographic arrangement (2D/SE-AF-HPLC-SUID–ICP-ORS-MS, Fig. 1), using the 6-port valve. In the second step, the 8-port column switching valve compensates the pressure gradient between low and high pressure pumps used in the global arrangement, Fig. 2. When the fraction containing low molecular weight selenium species fills the loop of 3 mL, the position of the 8-port valve is changed and these selenium species are driven by the pump B through the anion exchange column for separation and ICP-MS detection.
4. Speciation of selenium in serum using the 3D/SE-AF-AEC-HPLC-SUID–ICP-ORS-MS system is carried out using the operating conditions summarized in Tables 1 and 3.
5. The quantification of selenium-containing proteins and selenium-metabolites in the different chromatographic peaks is performed by post-column specie-unspecific isotopic dilution analysis [26], converting the intensity of different selenium isotopes and polyatomic interferences to mass flow

Table 3
Chromatographic conditions for 3D/SE-AF-AEC-HPLC-SUID–ICP-ORS-MS

Sample loop	100 μL
Flow rate Pump A	1.3 mL min^{-1}
Flow rate Pump B	1.3 mL min^{-1}
Mobile phase A	0.05 M ammonium acetate pH 7.4
Mobile phase B	1.5 M ammonium acetate pH 7.4
Mobile phase C	0.1 M sodium phosphate pH 8
Gradient Pump A	0–15 min 100% A 15–30 min 100% B 30–35 min 100% A
Isocratic Pump B	100% C
Valve position 6-port	1–10 min inject 10–26 min load 26–35 min inject
Valve position 8-port	0–6.5 min load 6.5–17 min inject 17–35 min load

chromatogram. Dead time correction using the procedure described by Vanhaecke et al. [27]. Mathematical treatments can be applied to correct BrH^+ and SeH^+ polyatomic interferences. Mass bias corrections can be applied using the $^{78}Se/^{74}Se$ and $^{80}Se/^{74}Se$ isotope ratios calculated (exponential mode) as previously described by Ruiz-Encinar et al. [28]. Finally, an online dilution equation is applied to each point of the chromatogram and the amount of selenium in each chromatographic peak calculated using suitable software.

3.3 Practical Applications (Case-Studies)

1. Quantification of selenium-containing proteins and low molecular weight selenium species in mouse plasma using 2D/SE-AF-HPLC-SUID-ICP-ORS-MS.

This analytical approach allows the effective separation of selenium species including eGPx, SeP, SeAlb, and selenium metabolites in serum/plasma samples and cytosolic tissues extracts. The combination of two in series size exclusion mini-columns (SE) with two AF columns enables the separation of before mentioned selenium species in less than 20 min avoiding chloride and bromide polyatomic interferences. A typical chromatogram is shown in Fig. 3a, the peak about 9 min (BrH^+) is not observed when ^{78}Se is used for the quantification of selenium species. In Fig. 3b the chromatogram is depicted corresponding to selenium speciation in plasma mouse in which ^{78}Se isotope was used for quantification. The absence of selenium metabolites was observed in the sample.

The method was checked in a human serum reference material (BCR-637) certified for total Se content ($Se_{Total} = 81 \pm 7$ ng mL^{-1}) that was additionally spiked with 50 ng $g^{-1}$of inorganic selenium (sodium selenate) [29] and results compared with the previous ones from Jitaru et al. [30], which use anion exchange solid phase extraction (AE-SPE) of the sample prior to the analysis by affinity chromatography (AF-HPLC–ICP-MS). The improvement

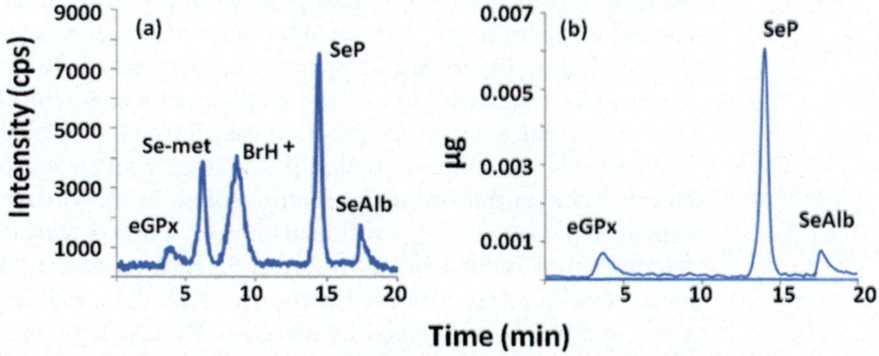

Fig. 3 (**a**) Conventional ^{80}Se chromatogram from mouse serum using 2D/SE-AF-HPLC-SUID-ICP-ORS-MS coupling (the interference of $^{79}Br^1H^+$ at 9 min is observed); (**b**) Chromatogram of mouse serum based on $^{78}Se/^{74}Se$ isotope ratios using the same instrumental arrangement

Table 4
Selenoproteins concentration (ng g−1) in hepatic cytosolic extract and serum of mice after inorganic mercury and selenium exposure for 10 days

Samples	Group	GPx	Se-metabolites	SeP	SeAlb
Liver cytosolic extracts	Control group	46.5 ± 5.	47.7 ± 6	195 ± 15	20.2 ± 3
	Hg-exposed group	102 ± 8	111 ± 10	297 ± 18	34.5 ± 4
	Hg/Low-Se-exposed group	117 + 11	90.3 ± 12	422 ± 31	48.6 ± 7
	Hg/High-Se-exposed group	129 ± 9	118 ± 8	434 ± 18	59.4 ± 10
Serum	Control group	8.72 ± 2	52.1 ± 2	251 ± 16.2	14.3 ± 2
	Hg-exposed group	14.4 ± 3	5.41 ± 1	266 ± 12	9.66 ± 1
	Hg/Low-Se-exposed group	13.0 ± 2	12.4 ± 2	463 ± 24	10.7 ± 2
	Hg/High-Se-exposed group	14.9 ± 2	26.3 ± 4	514 ± 31	13.1 ± 2

provided by the double SEC unit in the chromatographic resolution is clearly demonstrated by the results. The elimination of conventional interferences from Cl and Br in plasma/serum samples is particularly remarkable.

2. Quantification of selenium-containing proteins and selenium metabolites in serum and cytosolic extracts of tissues from mouse exposed to mercury and selenium by 2D/SE-AF-HPLC-SUID-ICP-ORS-MS [31].

Application of the methodology reveals that SeP levels increase in serum and hepatic cytosolic extracts after Hg exposure and this effect is more pronounced with increasing selenite supplementation (Table 4). This fact reflects the synthesis of SeP from selenite and the antagonistic interaction between Hg and Se in mice serum and liver, since the Hg-SelP complex plays an important role as a vehicle for Hg detoxification [32]. In relation to this, decreased levels of Se-metabolites in serum and correlatively increased in liver cytosolic extracts are concordant with the requirement of selenite for SelP synthesis in the liver, which are then transferred to the bloodstream [33]. SelP levels are upregulated in the liver, which can explain the elevated concentration of this selenoprotein in this organ after Hg exposure period, and this fact is also more marked with increasing selenite supplementation, as in serum. In relation to this, decreased levels of SeAlb were observed in mouse serum when Hg is supplied alone and was accompanied by increased liver SeAlb concentration (Table 4), since this transport selenoprotein is transferred to the liver for the synthesis of SelP [33]. More pronounced effects were observed when Se was administered together with Hg at different increasing doses.

3. Quantification of selenium species in maternal and umbilical cord serum by /SE-AF-HPLC-SUID-ICP-ORS-MS.

Selenium is related to the metabolism of thyroid hormones, so that the intake of this element is a critical parameter during pregnancy and lactation, and supplementation has beneficial effects on thyroid function, besides being related to the oxidative equilibrium of the metabolism. Moreover, transfer processes of essential elements from mother to fetus provide the minerals necessary for fetal growth and development. Therefore, the methodology proposed is very useful to know the selenium status and transference in the mother/fetus interface using samples of maternal and umbilical cord serum.

The analytical methodology proposed allows a deeper knowledge about the possible deficiency of selenium in mothers and infants, as well as possible causes, since different chemical forms of selenium can be independently quantified. The results obtained reveal that SeAlb can be considered a possible route of transmission of selenium from mother to fetus [34].

4 Notes

1. Connect directly the outlet of the chromatographic column to the T connector direct to the Micromist nebulizer inlet of the ICP-MS by means of a 30 cm PEEK tubing (0.6 mm i.d.).
2. Post column isotope dilution analysis can be performed by the introduction of ^{74}Se via the T connector. The quality control of the HPLC-SUID–ICP-ORS-MS system to overcome problems related to contamination, loss, and stability of species can be performed with the procedure described elsewhere [35].
3. Selenium isotopes intensity is converted in mass flow chromatogram for the quantification of selenium species using mathematical equation corrections. The peak of the interference of BrH^+ (at 9 min) is eliminated when ^{78}Se is used in the quantification process.

Acknowledgments

The authors thanks to Spanish Ministry of Economy and Competitiveness (Grant CTM2015-67902-C2-1-P and Regional Ministry of Economy, Innovation, Science and Employment (Andalusian Government, Spain), Grant P12-FQM-0442. Gema Rodríguez-Moro thanks Spanish Ministry of Economy and Competitiveness for a predoctoral scholarship BES-2013-064501, and Belén Callejón-Leblic thanks the Ministerio de Educación for a predoctoral scholarship FPU13/03615. Finally, the authors are grateful to FEDER (European Community) for financial support, grants number UNHU13-1E-1611 and UNHU15-CE-3140.

References

1. Ursini F, Heim S, Kiess M, Maiorino M, Roveri A, Wissing J, Flohé L (1999) Dual function of the selenoprotein PHGPx during sperm maturation. Science 285:1393–1396. doi:10.1126/science.285.5432.1393
2. Yoneda S, Suzuki KT (1997) Equimolar Hg-Se complex binds to selenoprotein P. Biochem Biophys Res Commun 231:7–11. doi:10.1006/bbrc.1996.6036
3. Mostert V (2000) Selenoprotein P: properties, functions, and regulation. Arch Biochem Biophys 376:433–438. doi:10.1006/abbi.2000.1735
4. Thomson CD (1998) Selenium speciation in human body fluids. Analyst 123:827–831. doi:10.1039/a707292i
5. Rayman MP, Infante HG, Sargent M (2008) Food-chain selenium and human health: spotlight on speciation. Br J Nutr 100:238–253. doi:10.1017/S0007114508922522
6. Roman M, Jitaru P, Agostini M, Cozzi G, Pucciarelli S, Nitti D, Bedin C, Barbante C (2012) Serum seleno-proteins status for colorectal cancer screening explored by data mining techniques – a multidisciplinary pilot study. Microchem J 105:124–132. doi:10.1016/j.microc.2012.02.004
7. Jitaru P, Goenaga-Infante H, Vaslin-Reimann S, Fisicaro P (2010) A systematic approach to the accurate quantification of selenium in serum selenoalbumin by HPLC-ICP-MS. Anal Chim Acta 657:100–107. doi:10.1016/j.aca.2009.10.037
8. Suzuki KT, Ogra Y (2002) Metabolic pathway for selenium in the body: speciation by HPLC-ICP MS with enriched Se. Food Addit Contam 19:974–983. doi:10.1080/02652030210153578
9. García-Sevillano MA, García-Barrera T, Navarro F, Gómez-Ariza JL (2014) Absolute quantification of superoxide dismutase in cytosol and mitochondria of mice hepatic cells exposed to mercury by a novel metallomic approach. Anal Chim Acta 842:42–50. doi:10.1016/j.aca.2014.07.014
10. Lunøe K, Martínez-Sierra JG, Gammelgaard B, Alonso JIG (2012) Internal correction of spectral interferences and mass bias for selenium metabolism studies using enriched stable isotopes in combination with multiple linear regression. Anal Bioanal Chem 402:2749–2763. doi:10.1007/s00216-012-5747-7
11. Humann-Ziehank E, Ganter M, Michalke B (2016) Selenium speciation in paired serum and cerebrospinal fluid samples of sheep. J Trace Elem Med Biol 33:14–20. doi:10.1016/j.jtemb.2015.07.010
12. Roman M, Lapolla A, Jitaru P, Sechi A, Cosma C, Cozzi G, Cescon P, Barbante C (2010) Plasma selenoproteins concentrations in type 2 diabetes mellitus – a pilot study. Transl Res 156:242–250. doi:10.1016/j.trsl.2010.07.006
13. Hinojosa Reyes L, Marchante-Gayón JM, García Alonso JI, Sanz-Medel A (2003) Quantitative speciation of selenium in human serum by affinity chromatography coupled to post-column isotope dilution analysis ICP-MS. J Anal At Spectrom 18:1210–1216. doi:10.1039/b305455a
14. Jitaru P, Cozzi G, Gambaro A, Cescon P, Barbante C (2008) Simultaneous speciation analysis of glutathione peroxidase, selenoprotein P and selenoalbumin in human serum by tandem anion exchange-affinity HPLC and on-line isotope dilution ICP-quadrupole MS. Anal Bioanal Chem 391:661–669. doi:10.1007/s00216-008-2043-7
15. Sariego Muñiz C, Marchante-Gayón JM, García Alonso JI, Sanz-Medel A (1999) Multi-elemental trace analysis of human serum by double-focusing ICP-MS. J Anal At Spectrom 14:193–198
16. García-Barrera T, Gómez-Ariza JL, González-Fernández M, Moreno F, García-Sevillano MA, Gómez-Jacinto V (2012) Biological responses related to agonistic, antagonistic and synergistic interactions of chemical species. Anal Bioanal Chem 403:2237–2253. doi:10.1007/s00216-012-5776-2
17. Rodríguez-González P, Marchante-Gayón JM, García Alonso JI, Sanz-Medel A (2005) Isotope dilution analysis for elemental speciation: a tutorial review. Spectrochim Acta Part B 60:151–207. doi:10.1016/j.sab.2005.01.005
18. Palacios Ò, Ruiz Encinar J, Schaumlöffel D, Lobinski R (2006) Fractionation of selenium-

containing proteins in serum by multiaffinity liquid chromatography before size-exclusion chromatography-ICPMS. Anal Bioanal Chem 384:1276–1283. doi:10.1007/s00216-005-0286-0

19. Xu M, Yang L, Wang Q (2008) Quantification of selenium-tagged proteins in human plasma using species-unspecific isotope dilution ICP-DRC-qMS coupled on-line with anion exchange chromatography. J Anal At Spectrom 23:1545–1549. doi:10.1039/b804935a
20. Koyama H, Omura K, Ejima A, Kasanuma Y, Watanabe C, Satoh H (1999) Separation of selenium-containing proteins in human and mouse plasma using tandem high-performance liquid chromatography columns coupled with inductively coupled plasma-mass spectrometry. Anal Biochem 267:84–91. doi:10.1006/abio.1998.2949
21. Jitaru P, Prete M, Cozzi G, Turetta C, Cairns W, Seraglia R, Traldi P, Cescon P, Barbante C (2008) Speciation analysis of selenoproteins in human serum by solid-phase extraction and affinity HPLC hyphenated to ICP-quadrupole MS. J Anal At Spectrom 23:402–406. doi:10.1039/B712693J
22. Jitaru P, Cozzi G, Seraglia R, Traldi P, Cescon P, Barbante C (2010) Towards an improved qualitative and quantitative determination of glutathione peroxidase, selenoprotein P and selenoalbumin in human serum by HPLC coupled to ICP-MS. Anal Methods 2:1382–1387. doi:10.1039/c0ay00173b
23. García-Sevillano MA, García-Barrera T, Gómez-Ariza JL (2014) Simultaneous speciation of selenoproteins and selenometabolites in plasma and serum by dual size exclusion-affinity chromatography with online isotope dilution inductively coupled plasma mass spectrometry. Anal Bioanal Chem 406:2719–2725. doi:10.1007/s00216-014-7629-7
24. Deagen JT, Butler JA, Zachara BA, Whanger PD (1993) Determination of the distribution of selenium between glutathione peroxidase, selenoprotein P, and albumin in plasma. Anal Biochem 208:176–181. doi:10.1006/abio.1993.1025
25. Harrison I, Littlejohn D, Fell GS (1996) Distribution of selenium in human blood plasma and serum. Analyst 121:189–194. doi:10.1039/an9962100189
26. Sariego Muñiz C, Marchante Gayón JM, García Alonso JI, Sanz-Medel A (2001) Speciation of essential elements in human serum using anion-exchange chromatography coupled to post-column isotope dilution analysis with double focusing ICP-MS. J Anal At Spectrom 16:587–592. doi:10.1039/b100947h
27. Vanhaecke F, De Wannemacker G, Moens L, Dams R, Latkoczy C, Prohaska T, Stingeder G (1998) Dependence of detector dead time on analyte mass number in inductively coupled plasma mass spectrometry. J Anal At Spectrom 13:567–571
28. Encinar JR, Schaumlöffel D, Ogra Y, Lobinski R (2004) Determination of Selenomethionine and Selenocysteine in Human Serum Using Speciated Coupled Plasma Collision Cell Mass Spectrometry. Anal Chem 76:6635–6642
29. García-Sevillano MA, García-Barrera T, Gómez-Ariza JL (2013) Development of a new column switching method for simultaneous speciation of selenometabolites and selenoproteins in human serum. J Chromatogr A 1318:171–179. doi:10.1016/j.chroma.2013.10.012
30. Jitaru P, Roman M, Barbante C, Vaslin-Reimann S, Fisicaro P (2010) Challenges in the accurate speciation analysis of selenium in humans: first report on indicative levels of selenoproteins in a serum certified reference material for total selenium (BCR-637). Accredit Qual Assur 15:343–350
31. García-Sevillano MA, Rodríguez-Moro G, García-Barrera T, Navarro F, Gómez-Ariza JL (2015) Biological interactions between mercury and selenium in distribution and detoxification processes in mice under controlled exposure. Effects on selenoprotein. Chem Biol Interact 229:82–90. doi:10.1016/j.cbi.2015.02.001
32. Chen C, Yu H, Zhao J, Li B, Qu L, Liu S, Zhang P, Chai Z (2006) The roles of serum selenium and selenoproteins on mercury toxicity in environmental and occupational exposure. Environ Health Perspect 114:297–301. doi:10.1289/ehp.7861
33. Shiobara Y, Suzuki KT (1998) Binding of selenium (administered as selenite) to albumin after efflux from red blood cells. J Chromatogr B Biomed Sci Appl 710:49–56
34. Santos C, García-Fuentes E, Callejón-Leblic B, García-Barrera T, Gómez-Ariza JL, Rayman MP, Velasco I (2017) Selenium, selenoproteins and selenometabolites in mothers and babies at the time of birth. Br J Nutr 117:1304–1311. doi:10.1017/S0007114517001155
35. Gonzalez-Fernández M, García-Sevillano MA, Jara-Biedma R, García-Barrera T, Vioque A, López-Barea J, Pueyo C, Gómez-Ariza JL (2011) Size characterization of metal species in liver and brain from free-living (Mus spretus) and laboratory (Mus Musculus) mice by SEC-ICP-MS: application to environmental contamination assessment. J Anal At Spectrom 26:141–149. doi:10.1039/C0JA00127A

Chapter 13

Radioactive ^{75}Se Labeling and Detection of Selenoproteins

Sun Hee Yim, Ryuta Tobe, Anton A. Turanov, and Bradley A. Carlson

Abstract

The trace element selenium (Se) is incorporated into proteins as the amino acid selenocysteine (Sec), which is cotranslationally inserted into specific proteins in response to a UGA codon. Proteins containing Sec at these specific positions are called selenoproteins. Most selenoproteins function as oxidoreductases, while some serve other important functions. There are 25 known selenoprotein genes in humans and 24 in mice. The use of Sec allows selenoproteins to be detected by a convenient method involving metabolic labeling with ^{75}Se. Labeling of cells and whole animals are used for the examination of selenoprotein expression profiles and the investigation of selenoprotein functions. In mammals, nonspecific ^{75}Se insertion is very low, and sensitivity and specificity of selenoprotein detection approaches that of Western blotting. This method allows for the examination of selenoprotein expression and Se metabolism in model and non-model organisms. Herein, we describe experimental protocols for analyzing selenoproteins by metabolic labeling with ^{75}Se both in vitro and in vivo. As an example, the procedure for metabolic labeling of HEK293T human embryonic kidney cells is described in detail. This approach remains a method of choice for the detection of selenoproteins in diverse settings.

Key words ^{75}Se metabolic labeling, Selenium, Selenocysteine, Selenoproteins, Cell culture, Animal models

1 Introduction

^{75}Se is a very convenient tool for the detection of selenium (Se)-containing proteins (selenoproteins), which co-translationally incorporate Se in the form of selenocysteine (Sec), the 21st proteinogenic amino acid [1]. For this, a solution containing this isotope is added to a cell culture medium, is injected into an animal, or provided with a food source. The radioisotope then enters Se metabolic pathways. ^{75}Se is a gamma emitter, requiring proper handling and protection methods. It does not occur naturally and is normally made in a synchrotron. ^{75}Se contains 34 protons and 41 neutrons with the isotopic mass of 74.9225234. Its half-life is ~120 days with an electron capture decay mode to produce daughter isotope, ^{75}As. ^{75}Se has been used to label biological samples since the early 1950s [2–11]. As a selenious acid, ^{75}Se was first used

Laurent Chavatte (ed.), *Selenoproteins: Methods and Protocols*, Methods in Molecular Biology, vol. 1661,
DOI 10.1007/978-1-4939-7258-6_13, © Springer Science+Business Media LLC 2018

in humans to localize tumors [7, 12] and assess nutritional status of selenium [13]. It was also used in various animal studies, including in monkeys, sheep, swine, dogs, rabbits, rats, mice, and chickens [6, 9, 14–20] and was extended for use in primary mammalian cells [21], viruses [22], and transformed cell culture platforms [23, 24]. Studies using cells in culture and animal models have shown that expression of selenoproteins is differentially regulated by Se availability. Some selenoproteins are more sensitive to Se deficiency, while others are less regulated. ^{75}Se labeling is also useful for examining selenoproteins expression in normal and cancer cells, and different cancer cell types. For example, ^{75}Se labeling of human nonsmall cell lung carcinoma (NSCLC), small cell lung carcinoma (SCLC), and hepatocellular carcinoma (HCC) cell lines allows for the examination of selenoproteins expression profiles and demonstrates that the selenoprotein thioredoxin reductase 1 (TXNRD1) is elevated in NSCLC cell lines when compared to SCLC and HCC cell lines (*see* Fig. 1). ^{75}Se labeling can be also used to study Se metabolism and Sec incorporation machinery as ^{75}Se is incorporated into Sec-tRNA$^{[Ser]Sec}$ [25]. In addition, metabolic labeling with ^{75}Se has proven to be a valuable tool for investigating selenoprotein expression in non-conventional species [26–28] such as *Euplotes crassus* [24], *Chlamydomonas reinhardtii* [29], *Strongyloides stercoralis* [30], *Emiliania huxleyi* [29], *Aureococcus anophagefferens* [31], as well as in prokaryotes [32–34] (e.g., *Escherichia coli and Enterococcus faecalis* [35]).

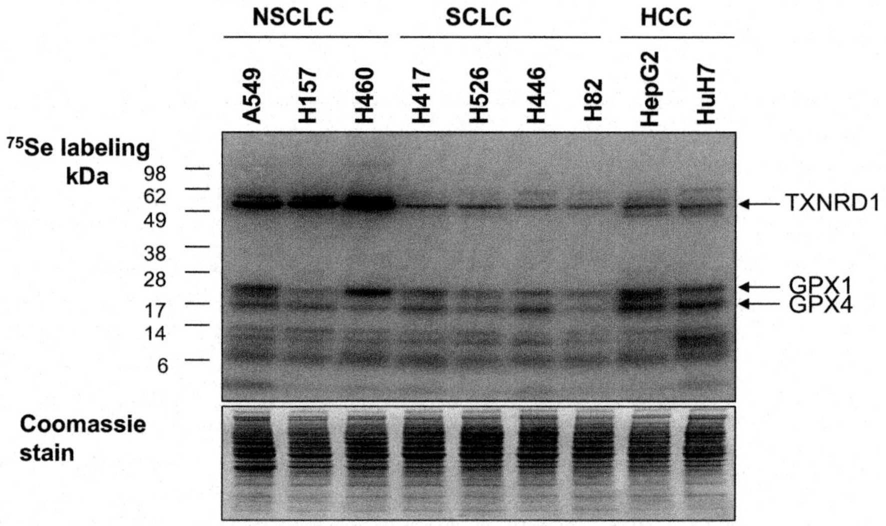

Fig. 1 Representative ^{75}Se-labeling patterns of various human lung and liver cancer cells. Human nonsmall cell (NSCLC) and small cell (SCLC) lung carcinoma and hepatocellular carcinoma (HCC) cell lines were metabolically labeled with ^{75}Se for 16 h, protein extracted, quantified, separated by SDS-PAGE and dried gels were exposed to a PhosporImager (*Upper panel*). Molecular weight markers are indicated on the *left*. The positions of representative selenoproteins are indicated on the *right*. Coomassie staining is shown in the *lower panel*. (R. Tobe, B. A. Carlson and D. L. Hatfield, unpublished data)

2 Materials

All solutions are prepared with molecular biology grade H_2O and all reagents are of the highest grade commercially available. Routine reagents with no listed supplier should be purchased from your supplier of choice. Cell culture media are stored at 4 °C. Buffers are stored at room temperature unless otherwise noted. Purchased reagents are stored as recommended by the supplier.

2.1 Obtaining and Quantifying Radiolabeled Selenium (^{75}Se)

2.1.1 Obtaining ^{75}Se

In the United States, ^{75}Se is available from University of Missouri Research Reactor Center, Columbia, MO, (MURR, http://www.murr.missouri.edu/ps_radio_isotopes.php). The isotope is produced biannually, typically in March and September, and is provided in the form of selenious acid ($H_2{}^{75}SeO_3$) in nitric acid (HNO_3) solution. The quantity of ^{75}Se can be purchased from 1 mCi to 5 mCi per vial.

2.1.2 Calculation of Radioactivity

As stated above, the half-life of ^{75}Se is ~120 days and radioactivity can be calculated by the following equation:

$$A = A_0 \cdot e^{-(0.693t/T_{½})}$$

where, A-Final activity of radioactive material
A_0-Initial activity
t-Radiation decay time
$T_{1/2}$-Isotope half-life.

2.2 Materials for Cell Culture, ^{75}Se Labeling, SDS-PAGE Analysis, and Autoradiography

2.2.1 Cell Culture and ^{75}Se Labeling

1. Appropriate cell culture medium (*see* **Note 1**).
2. Antibiotic-antimycotic solution.
3. Cell culture dishes.
4. Phosphate-buffered saline (PBS): 10 mM $PO_4{}^{3-}$, 137 mM NaCl, 2.7 mM KCl, pH 7.4.
5. Fetal bovine serum (FBS).
6. Trypsin-EDTA.
7. 1 M NaOH.
8. pH indicator paper.
9. ATCC® Medium 1120 (American Type Culture Collection (ATCC), Manassas, VA).
10. M9 buffer: 22 mM KH_2PO_4, 42 mM Na_2HPO_4, 85 mM NaCl, 1 mM $MgSO_4$.
11. Luria Broth (LB) medium.
12. Nematode growth medium-containing agar plates.
13. Suspension buffer: 25 mM Tris–HCl, 1 mM PMSF, 1 mM EDTA, 1 mM DTT, pH 7.5.

Table 1
Artificial seawater recipe[a]

Reagent	Quantity (for 1 L)	Final concentration[b]
NaCl	26.29 g	450 mM
KCl	0.74 g	10 mM
$CaCl_2$	0.99 g	9 mM
$MgCl_2 \cdot 6H_2O$	6.09 g	30 mM
$MgSO_4 \cdot 7H_2O$	3.94 g	16 mM

[a]Cold Spring Harbor Protocols, http://cshprotocols.cshlp.org/content/2012/2/pdb.rec068270.full

[b]Adjust to pH 7.8, autoclave, and store at 4 °C

14. Sf-900 II SM serum-free insect medium.
15. Dextrose.
16. p-hydroxybenzoic acid methyl ester.
17. Low melting point agarose.
18. Sterile 25-gauge injection needle and 1 mL syringe.
19. Artificial seawater (*see* Table 1).

2.2.2 Extraction and Isolation of ^{75}Se-Labeled Proteins from Cells and Tissues

1. PBS, ice-cold.
2. Sterile 15 mL centrifuge tubes.
3. Tissue homogenizer.
4. Vacuum aspirator with trap for collecting liquid radioactive waste.
5. CelLytic protein extraction buffer and cOmplete, EDTA-free protease inhibitor cocktail (Sigma-Aldrich, St. Louis, MO).
6. Microcentrifuge tubes.
7. Sonicator.
8. Bicinchoninic acid (BCA) protein assay kit (Thermo Fisher Scientific, Waltham, MA). and microplate reader capable of measuring absorbance at 562 nm (*see* **Note 2**).

2.2.3 SDS-PAGE Gel Electrophoresis and Staining

1. SDS-PAGE polyacrylamide gel system (*see* **Note 3**).
2. Protein gel electrophoresis chamber and power supply.
3. Gel fixing solution: 50% methanol, 10% glacial acetic acid.
4. Gel staining solution: 0.1% Coomassie Brilliant Blue R-250, 50% methanol, 10% glacial acetic acid.
5. Gel destaining solution: 40% methanol, 10% glacial acetic acid.
6. Gel drying apparatus.
7. PVDF membrane.
8. Plastic wrap.

2.2.4 Autoradiography and Image Analysis

1. Intensifying screen was obtained from KODAK and Typhoon FLA 7000 laser scanner (GE Healthcare Life Sciences, Pittsburgh, PA).
2. Image processing and analysis are performed with ImageJ, (https://imagej.nih.gov/ij/), ImageQuant (GE Healthcare Life Sciences, Pittsburgh, PA) or other similar software.

3 Methods

Caution! When working with radioactive materials, take appropriate precautions to avoid contamination. Carry out the experiment and dispose of waste in an appropriately designated area, following guidelines as provided by the local radiation safety department and animal care and use committee. Use recommended lead shielding to minimize exposure to ^{75}Se including benches, incubators, centrifuges, and other instruments designated for work with radioactive materials. Consult with the radiation safety office regarding dose limits, particularly for pregnant workers or workers with other health concerns. Some lab members may prefer not to use radioactivity. For this reason, we found that ^{75}Se labeling experiments work best if carried out on weekends, when fewer people are in the lab.

3.1 *In Vitro* Labeling

Metabolic labeling of mammalian cells with ^{75}Se has been used extensively in selenium research [36, 37]. As growth rate, doubling time, cell culture condition, and protein yield from various cell types vary greatly, the detailed method for metabolic labeling with ^{75}Se should be optimized accordingly. Cells with extremely low selenoprotein expression levels may require additional optimization to generate the resulting digital image that allows for quantification of subtle signal intensity differences within the dynamic range. In the case of labeling HEK293T cells with ^{75}Se, cells are labeled for 24 h. In cell lines with lower metabolic rates, slower doubling time, or lower selenoprotein expression levels, ^{75}Se labeling of up to 3 days may be required for achieving maximal signals. *See* Fig. 1 for representative ^{75}Se-labeling patterns of various human lung and liver cancer cells.

3.1.1 ^{75}Se Labeling of Adherent Mammalian Cells

1. Grow adherent cells in culture medium supplemented with 10% fetal bovine serum and antibiotic/antimycotic solution in a humidified incubator at 37 °C in 5% CO_2 [35] unless otherwise specified by medium or cell type (*see* **Note 4**).
2. Grow cells to 80–90% confluency in 100 mm cell culture plates and split cells 1:4 the day before metabolic labeling.
3. Incubate cells for 24 h.

4. Wash cells gently two-three times with 10 mL of pre-warmed (37 °C) PBS, and replace with 10 mL of fresh medium (*see* **Note 5**).
5. Prepare ^{75}Se solution by neutralization with NaOH (*see* **Note 6**) in cell culture medium.
6. Add 50 μCi (5 μCi/mL of medium) of freshly neutralized ^{75}Se to cells.
7. Incubate cells for 18–72 h (will vary by cell type).
8. Carefully aspirate medium containing ^{75}Se and wash cells gently twice with 10 mL of ice-cold PBS to remove unincorporated ^{75}Se, taking care not to detach the cells (*see* **Note 7**).
9. Harvest cells with trypsin-EDTA treatment, then add 10 mL of ice-cold PBS to suspend cells and transfer the cells to a 15 mL centrifuge tube.
10. Centrifuge cells at 300 × *g* for 5 min at 4 °C, aspirate PBS and keep cell pellet on ice (*see* **Note 8**).
11. Add freshly prepared ice-cold protein extraction buffer to each cell pellet or tissue (for recommended volumes, *see* **Note 9**).
12. Transfer the lysate to a clean microcentrifuge tube.
13. Sonicate cell lysate three times for 15 s/pulse, maintaining tubes in ice.
14. Centrifuge sonicated lysate at 16,000 × *g* for 20 min at 4 °C.
15. Transfer the supernatant to a clean microcentrifuge tube, being careful not to disturb pellet.
16. Determine protein concentration using a suitable method, such as BCA assay and proceed to Subheading 3.3.

3.1.2 ^{75}Se Labeling of Mammalian Cells in Suspension

1. Harvest ~0.5–2 × 10^7 cells (*see* **Note 10**) in suspension by centrifuging at 300 × *g* for 5 min at room temperature in sterile 15 mL centrifuge tube.
2. Wash cells gently with 10 mL of pre-warmed (37 °C) PBS by resuspending cells and centrifuging at 300 × *g* for 5 min at room temperature. Repeat wash step two to three times.
3. Resuspend cells in 10 mL of fresh medium (*see* **Note 5**) and transfer cells to culture dish.
4. Prepare ^{75}Se solution by neutralization with NaOH (*see* **Note 6**) in cell culture medium.
5. Add 50 μCi (5 μCi/mL of medium) of freshly neutralized ^{75}Se to cells.
6. Incubate cells for 18–72 h (will vary by cell type).
7. Centrifuge cells at 300 × *g* for 5 min at 4 °C and carefully aspirate the medium containing ^{75}Se.

8. Add ice-cold PBS and centrifuge cells at 300 × *g* for 5 min at 4 °C. Repeat wash step at least twice and keep cell pellet on ice (*see* **Note 8**).
9. Add freshly prepared ice-cold protein extraction buffer to each cell pellet or tissue (for recommended volumes, *see* **Note 9**).
10. Transfer the lysate to a clean microcentrifuge tube.
11. Sonicate cell lysate three times for 15 s/pulse, maintaining tubes in ice.
12. Centrifuge sonicated lysate at 16,000 × *g* for 20 min at 4 °C.
13. Transfer supernatant to a clean microcentrifuge tube, being careful not to disturb pellet.
14. Determine protein concentration using a suitable method, such as BCA assay and proceed to Subheading 3.3.

3.2 In Vivo Labeling

3.2.1 Labeling of Bacterial Selenoproteins

Bacterial selenoproteins may be labeled with ^{75}Se by supplementing the culture with the isotope [32, 33]. *Clostridium kluyveri* (*C. kluyveri*; ATCC 8527) is rod-shaped and gram-positive bacteria, grows anaerobically on ethanol and acetate as sole energy sources. Selenium-containing proteins can be purified from *C. kluyveri* after being labeled with ^{75}Se [33, 38].

1. *C. kluyveri* is grown on ATCC® Medium 1120 at 37 °C and anaerobic gas mixture (80% N_2, 20% CO_2).
2. To label *C. kluyveri* with ^{75}Se, culture *C. kluyveri* to attain maximum turbidity in a selenium-deficient complex medium. Dilute the culture into equal volumes of fresh medium (ATCC® Medium 1120).
3. Add 50–500 μCi of freshly neutralized ^{75}Se mixture per liter of culture (*see* **Note 6**).
4. Incubate the cultures anaerobically at 37 °C for the times specified in the various experiments (frequently, 150–240 min).
5. Harvest cells, resuspend in 50 mM potassium phosphate, pH 7.1, with 1 mM 1,4-dithiothreitol and rupture by sonication of cell lysate three times for 15 s/pulse, maintaining tubes in ice.
6. Centrifuge sonicated lysate at 16,000 × *g* for 20 min at 4 °C and assay protein concentration using a suitable method, such as BCA assay.
7. To purify selenium-labeled proteins from *C. kluyveri*, apply the crude extract to a DEAE-cellulose column equilibrated with 50 mM potassium phosphate, pH 7.1. Elute the adsorbed labeled protein depending on protein of interest [33, 38].

8. Monitor ^{75}Se radioactivity by a gamma counter in the eluted fractions to identify those containing the labeled selenoprotein.
9. Alternatively, the labeled proteins in the crude extract (from **Step 6**) can be analyzed by SDS-PAGE followed by a PhosphorImager [30, 39, 40] as described in Subheading 3.3.

3.2.2 Labeling of Euplotes crassus (E. crassus)

The free-crawling marine ciliate, *E. crassus*, can be metabolically labeled with ^{75}Se to examine the occurrence of selenoproteins [26].

1. Culture *E. crassus* (strain CT5) in artificial sea water.
2. Supplement the culture with 50 μCi of freshly neutralized ^{75}Se (*see* **Note 6**) per gram of *E. crassus* and continuously culture for 48–72 h.
3. After labeling, collect cells by centrifugation at 1000 × *g* for 15 min at 4 °C and carefully aspirate the medium containing ^{75}Se.
4. Wash cells twice with ice-cold PBS to remove unincorporated ^{75}Se (*see* **Note 7**).
5. Resuspend cells in five volumes of ice-cold PBS and sonicate cell lysate three times for 15 s/pulse, maintaining tubes in ice.
6. Centrifuge sonicated lysate at 16,000 × *g* for 20 min at 4 °C.
7. Transfer the supernatant to a clean microcentrifuge tube being careful not to disturb the pellet.
8. Determine protein concentration using a suitable method, such as BCA assay and proceed to Subheading 3.3.

3.2.3 Labeling of Caenorhabditis elegans (C. elegans)

C. elegans is an important model organism that has only one selenoprotein, thioredoxin reductase (TXNRD1). For ^{75}Se labeling of *C. elegans,* worms can be grown in either ^{75}Se-containing media [41] or fed with a ^{75}Se-labeled food source [42, 43].

1. To label *C. elegans* with ^{75}Se-labeled food source (*E. coli* strain OP50), grow *E. coli* strain OP50 overnight in 1 mL of LB medium in the presence of 20–100 μCi of freshly neutralized ^{75}Se [42, 43].
2. Harvest the ^{75}Se-labeled bacteria by centrifugation at 6500 × *g* for 15 min at 4 °C, wash twice with ice-cold PBS, and resuspend in M9 buffer.
3. Dispense the labeled bacteria in a 60 mm plate containing 3 mL of nematode growth medium-containing agar plate. Take care not to spread the lawn all the way to the edges.
4. Transfer the worms on lawns of ^{75}Se-labeled *E. coli* and incubate at 22 °C for 24 h or until bacteria are depleted.

5. Collect worms in M9 buffer, wash three times with ice-cold PBS by centrifugation at 800 × *g* for 5 min (*see* **Note 7**), and pellet.
6. Suspend pellets in four volumes of suspension buffer and sonicate on ice.
7. Centrifuge the resulting crude extract at 16,000 × *g* for 20 min at 4 °C.
8. Transfer supernatant to a clean microcentrifuge tube and maintain on ice.
9. Determine protein concentration using a suitable method, such as BCA assay and proceed to Subheading 3.3.

3.2.4 *Labeling of* Drosophila melanogaster (D. melanogaster)

The number of selenoproteins in insects varies significantly. *Drosophila willistoni* does not encode selenoproteins or the selenoprotein insertion machinery, while *D. melanogaster* encodes three selenoproteins: selenophosphate synthetase 2 (Sephs2), selenoprotein H (Selenoh, also known as BthD), and selenoprotein K (Selenok, also known as G-rich), and the respective selenoprotein machinery [44, 45].

1. Maintain adult *D. melanogaster* on a Sf-900 II SM serum-free insect medium supplemented with 3% dextrose, 0.01% p-hydroxybenzoic acid methyl ester, and 2% low melting point agarose.
2. For labeling, add 35 μCi of freshly neutralized ^{75}Se to the above media and maintain flies at 25 °C in a 12 h light/dark cycle for 3 days [45, 46].
3. Homogenize 35 flies per group in 350 μL of ice-cold PBS buffer containing protease inhibitors.
4. Remove tissue debris by centrifugation at 16,000 × *g* for 20 min at 4 °C.
5. Transfer the supernatant to clean microcentrifuge tube and maintain on ice.
6. Determine protein concentration using a suitable method, such as BCA assay and proceed to Subheading 3.3.

3.2.5 *Labeling of* Mus musculus

Caution! An approved animal protocol is necessary before proceeding with the below protocol. Mice, bedding, and cages will be radioactive. Therefore, a designated area for radioisotope use will be needed. Please consult with the proper institutional animal care and use committee for guidance on these issues and for protocol approval. For proper care and handling of radiolabeled mice, please consult with a radiation safety officer to determine how to handle and dispose of animals, tissues, carcasses and bedding contaminated by these materials.

1. Prepare fresh ^{75}Se solution by neutralization with NaOH in PBS (*see* **Note 6**).
2. Inject mice intraperitoneally with ^{75}Se solution (25 μCi per gram of body weight) using a 25-gauge × 5/8″ needle [33, 43].
3. Provide food and water to mice according to standard and approved practices and maintain in a cage for 24–48 h.
4. Euthanize mice according to established and approved procedures for animal and radioisotope handling.
5. Excise mouse tissues and immediately freeze tissues in liquid N_2.
6. Store tissues at −80 °C behind recommended and approved shielding until ready to use.
7. Add freshly prepared ice-cold protein extraction buffer to each tissue (for recommended volumes, *see* **Note 11**).
8. Homogenize tissue using homogenizer until the tissue is completely disrupted and a uniform suspension is obtained, keeping tissue on ice.
9. Transfer the lysate to a clean microcentrifuge tube.
10. Sonicate cell or tissue lysate three times for 15 s/pulse, maintaining tubes in ice.
11. Centrifuge sonicated lysate at 16,000 × *g* for 20 min at 4 °C.
12. Transfer the supernatant to a clean microcentrifuge tube, being careful not to disturb pellet.
13. Determine protein concentration using a suitable method, such as BCA assay and proceed to Subheading 3.3.

3.3 SDS-PAGE Gel Electrophoresis and Gel Staining

1. Prepare SDS-PAGE gel and apparatus as described by gel manufacturer or by standard lab protocol (*see* **Note 12**).
2. Mix the cell lysate with gel loading buffer and heat sample at 70 °C for 10 min.
3. Load 30 μg of total cell lysate per well on a 12-well gel, along with molecular weight markers (*see* **Note 13**).
4. Run the electrophoresis at 150 V for ~1 h or until the lowest molecular weight marker reaches the bottom of the gel.
5. Transfer gel to a container containing 100 mL of gel fixing solution and gently shake for at least 30 min, but can be shaken overnight.
6. Stain gel in 100 mL of a gel staining solution for 30 min to 1 h or until the gel is a uniform blue color.
7. Destain gel for ~4–24 h in gel destaining solution, changing solution as needed until the desired background is reached.

8. Dry stained gel using vacuum dryer system or other preferred gel drying techniques.
9. Alternatively, following electrophoresis (**step 4**), proteins can be transferred onto a PVDF membrane (*see* **Note 14**).
10. Following transfer, wrap the PVDF membrane with plastic wrap.

3.4 Detection and Quantitation of ^{75}Se-Labeled Selenoproteins

1. Expose dried gel or PVDF membrane in plastic wrap to a PhosphorImager screen for 24 h or other desired length of time (*see* **Note 15**).
2. Quantitate lanes or bands using ImageJ, ImageQuant or similar software (*see* **Note 16**).

3.5 Example: Labeling of HEK293T Mammalian Cells with ^{75}Se (See Fig. 2)

Day 1. Plate cells in a 100 mm plate (6×10^3 viable cells/cm^2) in DMEM culture media supplemented with 10% FBS and antibiotics.

Day 2. Start ^{75}Se labeling. Hypothetically, ^{75}Se was purchased 50 days ago from MURR. At that time, 3.218 mCi of ^{75}Se were in the volume of 203 μL. The current radioactivity of ^{75}Se is calculated as given above (*see* Subheading 2.1.2).

At day 50, ^{75}Se radioactivity is

$$= A_0\, e^{-(0.693 \times 50/120)}$$
$$= A_0\, e^{-(0.28875)}$$
$$= A_0 \times 0.74920$$

Thus, the current radioactivity of ^{75}Se solution is 2.41 mCi/203 μL, or 11.87 μCi/μL. To prepare a 5 μCi/mL DMEM media for labeling HEK293T cells seeded in a 100 mm plate with 10 mL media, 4.21 μL (equal to 50 μCi) of the selenious acid solution is required. To neutralize the ^{75}Se solution, first add 12.63 μL of 1 M NaOH solution and titrate with 252.6 μL of 1 M Tris–HCl, pH 8.0 buffer. Alternatively, 1 M HEPES buffer can be used. Now, the

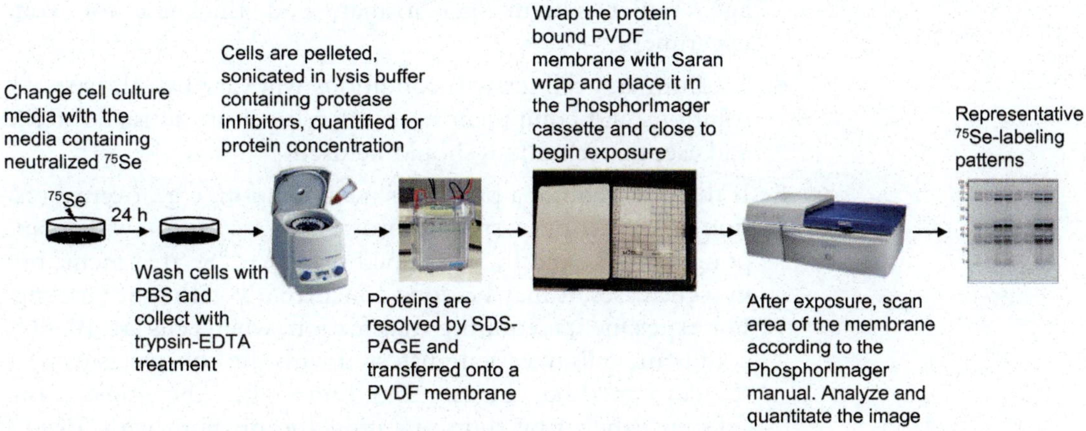

Fig. 2 Schematic diagram showing procedure for ^{75}Se labeling of cells and detection of selenoproteins

total volume of the neutralized ^{75}Se solution is 269.44 μL. Prepare 9.730 mL of fresh DMEM media, add 269.44 μL of neutralized ^{75}Se solution, mix well, and apply the medium to HEK293 cells. Cells are cultured at 37 °C in the 5% CO_2 humidified mammalian cell culture incubator for 24 h.

Day 3. After 24 h, cells are washed twice with PBS, trypsinized, collected in a 15 mL centrifuge tube, and pelleted. Cell pellets are washed twice with ice-cold PBS, lysed in protein extraction buffer containing protease inhibitor, and sonicated for 2 min on ice. The soluble fraction of the cell lysate is collected by centrifugation at 16,000 × *g* for 20 min at 4 °C. The protein concentration of the soluble fraction is determined using a BCA protein assay or similar standard technique. A portion of 30 μg of total soluble protein per well is resolved by SDS–PAGE and transferred onto a PVDF membrane. The PVDF membrane is wrapped with plastic wrap, and placed on an intensifying screen for 24 h.

Day 4. The intensifying screen is scanned with a PhosphorImager and the proteins that are labeled with ^{75}Se are identified against the protein molecular weight markers and known positive and negative controls.

4 Notes

1. Depending on the selenoprotein of interest and experimental design, this procedure can be adapted for any mammalian cell line, required medium, and growth conditions.
2. Any established technique may also be used for measuring total protein levels.
3. Commercially available precast gel or handcast gel systems (including gel, sample loading buffer, and running buffer) may be used, depending on your preference. To obtain the best results, it is important to choose the correct gel percentage, buffer system, gel format, and thickness for your experiment.
4. Medium and cell growth conditions will vary by cell type. All solutions and equipment used for cell culture must be sterile and aseptic techniques should be used.
5. If the cells require a particular manipulation, e.g., overexpression or targeting a specific protein or treatment with compounds, it is typically performed prior to ^{75}Se treatment, but in some cases, it may be done concurrently with ^{75}Se labeling. For experiments requiring transfection, when cells are 40–60% confluent, cells may be transfected with the constructs using a standard method. Twenty-four hours after the transfection, cells are labeled by supplementing the medium with 50 μCi of freshly prepared and neutralized ^{75}Se.

6. Neutralize ^{75}Se with NaOH and monitor pH of ^{75}Se solution using pH indicator paper and pH indicator in medium.
7. It is important to carefully wash the labeled cells or animal tissues with PBS at least twice to remove excess, unincorporated ^{75}Se. Mammalian cells incorporate only about 2% of ^{75}Se from the medium, and insufficient washing would result in a high background.
8. Cell pellets can be stored at −80 °C, if not immediately used.
9. Use ~500 μL of protein extraction buffer per 1×10^7 cells, but please note that this value may require optimization for each cell type.
10. Check guidelines for a specific cell line for recommended cell culturing densities.
11. For tissues, use ~500 μL of protein extraction buffer per 100 mg tissue.
12. Gel percentage, size and thickness as well as the buffer system used will depend on factors such as the selenoprotein of interest being studied, the overall experimental design (e.g., reducing vs. nonreducing conditions), and the amount and concentration of the sample.
13. Amount of protein loaded will depend on the number of wells and thickness of the gel.
14. The PVDF membrane can be further used for protein normalization and/or Western blotting. In mammals, nonspecific ^{75}Se insertion is very low and sensitivity of selenoprotein detection is similar to that of Western blotting. Combining ^{75}Se labeling with Western blotting (using the same membrane) is an added benefit of the method.
15. Length of exposure will vary. In general, a 24 h exposure of radiolabeled gel or membrane to a PhosphorImager screen should be adequate. When the ^{75}Se signal is detected in a PhosphorImager representing a selenoprotein pattern, the molecular weight is not readily discernible, as protein standards are not radioactive. To visualize molecular weight standards, after transfer of the proteins to the PVDF membrane, small aliquots of a ^{75}Se solution may be spotted on the bands of the protein standards in the PVDF membrane, followed by drying and exposure to a PhosphorImager screen. As these spots are radioactive, they will be visualized in a PhosphorImager analysis, indicating the migration of standards.
16. Background is typically very low when ^{75}Se is applied to mammalian cells. This is because selenium is not inserted nonspecifically into sulfur-containing amino acids (mammals do not synthesize methionine de novo and cysteine is made from methionine). In contrast, in prokaryotes, ^{75}Se is inserted into

proteins nonspecifically in place of cysteine and methionine [47, 48]. Therefore, background is common and corresponds approximately to the intensity of protein staining. Selenoproteins are then detected against this background.

Acknowledgments

This work was supported by the Intramural Research Program of the National Institutes of Health, NCI, Center for Cancer Research to BAC.

References

1. Gladyshev VN, Arner ES, Berry MJ, Brigelius-Flohe R, Bruford EA, Burk RF et al (2016) Selenoprotein gene nomenclature. J Biol Chem 291:24036–24040
2. Bostrom H, Jorpes E, Odeblad E (1953) An autoradiographic study on the uptake of Se75-labelled sodium selenate by different organs of the adult rat. Acta Physiol Scand 28:255–260
3. Claycomb CK, Gatewood DC, Sorenson FM, Jump EB (1960) Presence of Se75 in rat saliva after intracardiac injection of radioactive sodium selenite. J Dent Res 39:1264
4. Colonna G, Burch GE, Walsh JJ, Darby TD (1963) In vivo transfer of radioselenium, Se75, across the pericardium of dogs. Tex Rep Biol Med 21:331–336
5. Hansson E, Blau M (1963) Incorporation of Se75-selenomethionine into pancreatic juice proteins in vivo. Biochem Biophys Res Commun 13:71–74
6. Hansson E, Jacobsson SO (1966) Uptake of [75Se] selenomethionine in the tissues of the mouse studied by whole-body autoradiography. Biochim Biophys Acta 115:285–293
7. Cavalieri RR, Scott KG, Sairenji E (1966) Selenite (75Se) as a tumor-localizing agent in man. J Nucl Med 7:197–208
8. Campo RD, Wengert PA Jr, Tourtellotte CD, Kirsch MA (1966) A comparative study of the fixation of 75Se and 35S onto protein-polysaccharides of bovine costal cartilage. Biochim Biophys Acta 124:101–108
9. Breccia A, Trenta A, Badiello RR, Moretti S, Mattii M (1966) Rate of uptake of 75Se, as selenourea-75Se, in rats in vivo. Experientia 22:475
10. Anghileri LJ (1966) Effects of cystine and methionine on the 75Se selenocystine mtabolism in mice. Naturwissenschaften 53:256
11. Mc CK, Van Loon EJ (1955) Distribution of Se75 in serum proteins as determined by paper electrophoresis. J Biol Chem 212:747–750
12. Haynie TP, Otte WK, Wright JC (1964) Visualization of a hyperfunctioning parathyroid adenoma using Se75 selenomethionine and the photoscanner. J Nucl Med 5:710–714
13. Lee M, Dong A, Yano J (1969) Metabolism of 75Se-selenite by human whole blood in vitro. Can J Biochem 47:791–797
14. Muth OH, Pendell HW, Watson CR, Oldfield JE, Weswig PH (1967) Uptake and retention of parenterally administered 75Se in ewes on different selenium regimens. Am J Vet Res 28:1397–1406
15. Withrow G, Bell MC (1969) Erythrocytic life span estimations in growing sheep and swine using 75Se. J Anim Sci 28:240–245
16. Vigorito T, Celander DR, Celander E (1968) Relative effictiveness of 75Se and 125-I as lebels for rabbit anticanine urokinase. Nucl Med (Stuttg) 7:136–147
17. Kupic EA, Kasenter AG (1969) Experimental pancreatic scanning: preliminary results using intra-arterial 75Se-selenomethionine and hormone stimulation. Radiology 93:1376–1379
18. Ochoa-Solano A, Gitler C (1968) Incorporation of 75Se-selenomethionine and 35S-methionine into chicken egg white proteins. J Nutr 94:243–248
19. Zuidema GD, Kirsh M, Turcotte JG, Gaisford WD, Powers W, Kowalczyk RS (1963) Pancreatic uptake of Se75–selenomethionine. Ann Surg 158:894–897
20. Jensen LS, Walter ED, Dunlap JS (1963) Influence of dietary vitamin E and selenium on distribution of Se75 in the chick. Proc Soc Exp Biol Med 112:899–901
21. Cheng HF, Peterson RE, Evans TC (1969) Incorporation of 75Se-selenomethionine in

human peripheral lymphocytes as an in vitro test of globulin-synthesizing capacity. J Nucl Med 10:63–67
22. Jensik SC, Northrop RL (1971) Incorporation of radioactive seleno-(75Se)-methionine into mumps virus. Appl Microbiol 21:451–455
23. Leibold W, Bridge S (1979) 75Se-release: a short and long term assay system for cellular cytoxicity. Z Immunitatsforsch Immunobiol 155:287–311
24. Gutman GA, Warner NL, Harris AW, Bowles A (1978) Use of [75Se]selenomethionine in immunoglobulin biosynthetic studies. J Immunol Methods 21:101–109
25. Xu XM, Turanov AA, Carlson BA, Yoo MH, Everley RA, Nandakumar R et al (2010) Targeted insertion of cysteine by decoding UGA codons with mammalian selenocysteine machinery. Proc Natl Acad Sci U S A 107:21430–21434
26. Turanov AA, Lobanov AV, Fomenko DE, Morrison HG, Sogin ML, Klobutcher LA et al (2009) Genetic code supports targeted insertion of two amino acids by one codon. Science 323:259–261
27. Lobanov AV, Gromer S, Salinas G, Gladyshev VN (2006) Selenium metabolism in Trypanosoma: characterization of selenoproteomes and identification of a Kinetoplastida-specific selenoprotein. Nucleic Acids Res 34:4012–4024
28. Lobanov AV, Fomenko DE, Zhang Y, Sengupta A, Hatfield DL, Gladyshev VN (2007) Evolutionary dynamics of eukaryotic selenoproteomes: large selenoproteomes may associate with aquatic life and small with terrestrial life. Genome Biol 8:R198
29. Novoselov SV, Rao M, Onoshko NV, Zhi H, Kryukov GV, Xiang Y et al (2002) Selenoproteins and selenocysteine insertion system in the model plant cell system, Chlamydomonas reinhardtii. EMBO J 21:3681–3693
30. Aikens LM, Schad GA (1989) Radiolabeling of infective third-stage larvae of Strongyloides stercoralis by feeding [75 Se]selenomethionine-labeled Escherichia coli to first- and second-stage larvae. J Parasitol 75:735–739
31. Gobler CJ, Lobanov AV, Tang YZ, Turanov AA, Zhang Y, Doblin M et al (2013) The central role of selenium in the biochemistry and ecology of the harmful pelagophyte, aureococcus anophagefferens. ISME J 7:1333–1343
32. Sliwkowski MX, Stadtman TC (1985) Incorporation and distribution of selenium into thiolase from clostridium kluyveri. J Biol Chem 260:3140–3144
33. Hartmanis MG, Stadtman TC (1982) Isolation of a selenium-containing thiolase from clostridium kluyveri: identification of the selenium moiety as selenomethionine. Proc Natl Acad Sci U S A 79:4912–4916
34. Huber RE, Segel IH, Criddle RS (1967) Growth of Escherichia coli on selenate. Biochim Biophys Acta 141:573–586
35. Zhang Y, Turanov AA, Hatfield DL, Gladyshev VN (2008) In silico identification of genes involved in selenium metabolism: evidence for a third selenium utilization trait. BMC Genomics 9:251
36. Turanov AA, Lobanov AV, Hatfield DL, Gladyshev VN (2013) UGA codon position-dependent incorporation of selenocysteine into mammalian selenoproteins. Nucleic Acids Res 41:6952–6959
37. Kryukov GV, Kryukov VM, Gladyshev VN (1999) New mammalian selenocysteine-containing proteins identified with an algorithm that searches for selenocysteine insertion sequence elements. J Biol Chem 274:33888–33897
38. Seedorf H, Fricke WF, Veith B, Bruggemann H, Liesegang H, Strittmatter A et al (2008) The genome of Clostridium kluyveri, a strict anaerobe with unique metabolic features. Proc Natl Acad Sci U S A 105:2128–2133
39. Ching WM, Stadtman TC (1982) Selenium-containing tRNAGlu from Clostridium sticklandii: correlation of aminoacylation with selenium content. Proc Natl Acad Sci U S A 79:374–377
40. Ching WM, Wittwer AJ, Tsai L, Stadtman TC (1984) Distribution of two selenonucleosides among the selenium-containing tRNAs from Methanococcus vannielii. Proc Natl Acad Sci U S A 81:57–60
41. Boehler CJ, Raines AM, Sunde RA (2013) Deletion of thioredoxin reductase and effects of selenite and selenate toxicity in Caenorhabditis elegans. PLoS One 8:e71525
42. Buettner C, Harney JW, Berry MJ (1999) The Caenorhabditis elegans homologue of thioredoxin reductase contains a selenocysteine insertion sequence (SECIS) element that differs from mammalian SECIS elements but directs selenocysteine incorporation. J Biol Chem 274:21598–21602
43. Gladyshev VN, Krause M, Xu XM, Korotkov KV, Kryukov GV, Sun QA et al (1999) Selenocysteine-containing thioredoxin reductase in C. elegans. Biochem Biophys Res Commun 259:244–249
44. Chapple CE, Guigo R (2008) Relaxation of selective constraints causes independent

selenoprotein extinction in insect genomes. PLoS One 3:e2968

45. Martin-Romero FJ, Kryukov GV, Lobanov AV, Carlson BA, Lee BJ, Gladyshev VN et al (2001) Selenium metabolism in Drosophila: selenoproteins, selenoprotein mRNA expression, fertility, and mortality. J Biol Chem 276:29798–29804
46. Shchedrina VA, Kabil H, Vorbruggen G, Lee BC, Turanov AA, Hirosawa-Takamori M et al (2011) Analyses of fruit flies that do not express selenoproteins or express the mouse selenoprotein, methionine sulfoxide reductase B1, reveal a role of selenoproteins in stress resistance. J Biol Chem 286:29449–29461
47. Behne D, Kyriakopoulos A, Scheid S, Gessner H (1991) Effects of chemical form and dosage on the incorporation of selenium into tissue proteins in rats. J Nutr 121:806–814
48. Burk RF, Hill KE, Motley AK (2001) Plasma selenium in specific and non-specific forms. Biofactors 14:107–114

Chapter 14

Nonradioactive Isotopic Labeling and Tracing of Selenoproteins in Cultured Cell Lines

Jordan Sonet, Sandra Mounicou, and Laurent Chavatte

Abstract

Selenium (Se) is an essential component of genetically encoded selenoproteins, in the form of a rare amino acid, namely the selenocysteine (Sec). Radioactive ^{75}Se has been widely used to trace selenoproteins in vitro and in vivo (cell models and animals). Alternatively, its unique isotopic pattern can be used to detect and characterize nonradioactive Se-compounds in cellular extracts using molecular or elemental mass spectrometry at ppm levels. However, when studying trace levels of Se-compounds, such as selenoproteins (ppt levels), the distribution of the signal between its six naturally abundant isotopes reduces its sensitivity. Here, we describe the use of isotopically enriched forms of Se as an alternative strategy to radioactive ^{75}Se, for the labeling and tracing of selenoproteins in cultured cell lines.

Key words Selenium, Selenoproteins, Nonradioactive isotopes, ICP MS, Cultured cells

1 Introduction

Human selenoproteins are encoded by 25 genes, where Se is present as selenocysteine, and altogether compose the selenoproteome [1]. Selenoproteins are involved in a wide variety of physiological functions that include redox signaling, redox homeostasis, and antioxidant defense [2–6]. The selenoproteome is primarily controlled by Se bioavailability, which causes the prioritization of selenoprotein biosynthesis, when this trace element is deficient. This regulation of selenoprotein is often referred to as the "selenoprotein hierarchy" [7–10]. The hierarchical regulation of the selenoproteome can also be controlled by other exogenous stimuli, cellular stressors, or pathophysiological conditions [11, 12]. This aspect of selenoprotein expression is poorly understood, mainly due to the lack of routine analytical methods to study the selenoproteome.

In mammals, Sec is almost exclusively present in a single copy per selenoprotein, with the noticeable exception of SELENOP (also referred to as SELP or SEPP1 [13]) where Sec is present between 10 and 17 times, depending on the species. The relative

Laurent Chavatte (ed.), *Selenoproteins: Methods and Protocols*, Methods in Molecular Biology, vol. 1661,
DOI 10.1007/978-1-4939-7258-6_14, © Springer Science+Business Media LLC 2018

quantification of selenoproteins from selenium detection becomes straightforward since one molecule of Se equals one molecule of selenoprotein (except for SELENOP), no matter if selenium is radioactive or not. Different highly sensitive methods are available nowadays for the detection and quantification of selenium, and therefore selenoproteins. First, radioactive ^{75}Se has been widely used to label selenoproteins in vivo and in vitro (*see* Chapter 13). Its use is similar to the one of radioactive ^{35}S, with the significant difference that ^{75}Se can be incorporated into proteins using selenite while ^{35}S should be in the form of [^{35}S]-methionine or [^{35}S]-cysteine. Another difference is that ^{35}S is a weak radioactive emitter and therefore its manipulation does not need important infrastructure while ^{75}Se is a gamma-emitter with a rather long half-life (120 days). In both cases, radiolabeled-proteins are separated by different biochemical methods, including chromatography or SDS-PAGE, and revealed by gamma-counter or autoradiography. Thus, the use of ^{75}Se is restricted to several laboratories worldwide. Alternatively, selenium signal from selenoprotein can be detected by elemental (inductively coupled plasma-mass spectrometry, ICP MS) and molecular (electrospray ionization-mass spectrometry, ESI-MS) mass spectrometry. While the detection of selenium by ICP MS is now reaching its limits at the ppt level (1 ng/L for liquid detection, [3]), the detection limits for proteins by ESI-MS/MS have been particularly improved in the past decade with the last generation of high-resolution mass spectrometer. In large-scale proteomic analyses performed with ESI-MS/MS, more than 10,000 proteins could be detected and quantified, almost routinely, with several orders of magnitude from relatively low amount of cellular/tissue extracts (www.proteomicsdb.org). The presence of selenium in a molecule, such as a peptide, can be inferred from the high resolution isotopic pattern [14]. Indeed, Se has six stable isotopes (Table 1) with a rather constant natural relative

Table 1
Selenium has six naturally occurring isotopes

Isotope	Z(p)	N(n)	Atomic mass	Natural abundance (%)	Half-life
^{74}Se	34	40	73.922474	0.89	Stable
^{76}Se	34	42	75.919214	9.36	Stable
^{77}Se	34	43	76.919912	7.63	Stable
^{78}Se	34	44	77.917309	23.78	Stable
^{80}Se	34	46	79.916522	49.61	Stable
^{82}Se	34	48	81.916700	8.73	~10^{20} Years

The atomic number (Z) is the number of protons of an atom of a chemical element. The neutron number (N) is number of neutrons of an atom of a chemical element. The molecular weight of natural Se is 78.96 g/mol

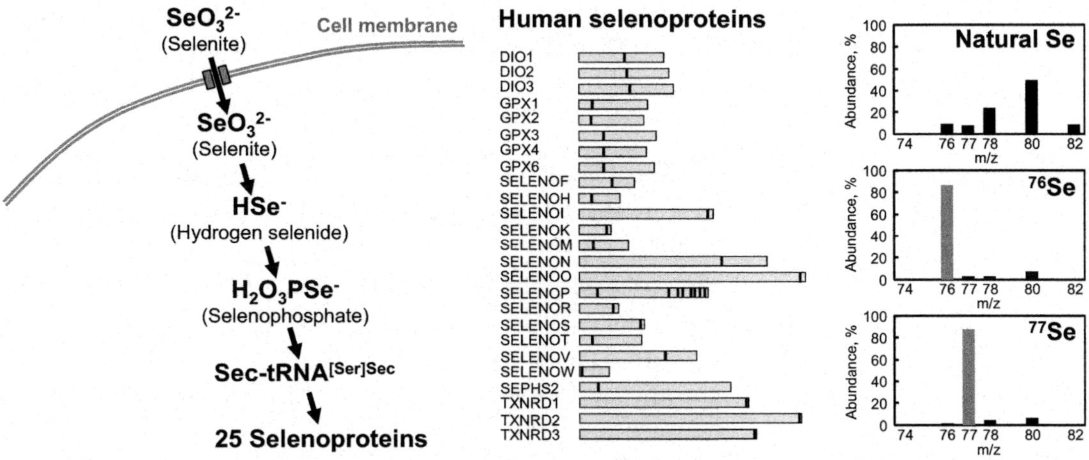

Fig. 1 Schematic of selenium metabolism from selenite to selenoprotein within mammalian cells. Therefore, selenite (either natural or isotopically enriched) is readily bioavailable at low dose for an efficient incorporation in selenoprotein as part of selenocysteine amino acid. We consider that a concentration of 30–100 nM selenite is optimal for optimal expression of most selenoproteins

abundance: ^{74}Se (0.89%), ^{76}Se (9.37%), ^{77}Se (7.63%), ^{78}Se (23.77%), ^{80}Se (49.61%), and ^{82}Se (8.73%). The mass difference between the different isotopes is a signature for each element of the periodic table, especially for metals and heteroelements.

Interestingly, each of these Se-isotopes can be enriched to >99% and isotopically labeled Se compounds such as selenite, selenomethionine or selenocysteine can readily be synthesized. When selenium is added as pure isotope, its incorporation into selenoproteins can be followed by ICP MS or ESI-MS/MS. Here, we describe a simple procedure to label and trace selenoprotein in culture cells that can be applied to many other cell models. We used human cell lines using LNCaP as an example. This protocol can be applied to other mammalian cell lines, either adherent or not, either cancerous or not. The efficiency of Se exchange in proteins is analyzed by size exclusion chromatography (SEC) coupled with ICP MS (SEC ICP MS). As illustrated in Fig. 1, selenite is an interesting form of selenium for rapid and efficient incorporation in selenoproteins. Due to its chemical stability over time and its action at very low dose, selenite is often found for in vitro experiments in the literature.

2 Materials

Prepare all solutions using ultrapure water (18 MΩ cm at 25 °C) and analytical grade reagents. Concerning the solutions in contact with cells in culture, they should be sterilized with an autoclave or filtrated using a 0.22 μm syringe filter. Prepare and store all reagents at room temperature (unless stated otherwise). Diligently follow all waste disposal regulations when disposing waste materials. Routine reagents should be purchased from your supplier of choice. Suppliers are indicated only when important for the method.

2.1 Cell Culture

1. LNCaP cell lines. These cells originate from a human prostate biopsy in the left supraclavicular lymph node. They derived from metastatic site of a 50 years adult (*see* **Note 1**).
2. Dulbecco's Modified Eagle Medium (DMEM) containing L-Glutamine (2 mM) and Pyruvate (1 mM).
3. Optionally, the media can be supplemented with penicillin and streptomycin antibiotics (100 U/mL) to prevent cell culture contamination by bacteria.
4. Fetal Calf Serum (FCS) is commercially available from numerous companies. Keep at −20 °C.
5. Maintenance medium: DMEM supplemented with 10% FCS.
6. Isotopically enriched selenite solution (1000×). Isotopically enriched sodium selenite is available in very few companies (*see* **Note 2**). Prepare a solution at 0.1 mM in water (*see* **Note 3**). Keep at −20 °C.
7. Natural selenite solution (1000×). Sodium selenite with natural abundance is prepared at 0.1 mM in water (1000×). Keep at −20 °C.
8. Human insulin solution at 10 mg/mL (1000×). Cell culture tested recombinant human insulin is commercially available. Keep at +4 °C.
9. Human transferrin solution at 10 mg/mL (2000×). Dilute 100 mg of powder, cell culture tested in 10 mL of PBS. Keep at −20 °C.
10. Hydrocortisone solution at 0.1 mM (2000×). Prepare the dilutions in absolute ethanol. Keep at −20 °C.
11. T3 hormone solution at 0.1 μM (10,000×). Prepare the dilutions of L-3,3′,5-triiodothyronine (T3 hormone), sodium salt, in NaOH 0.1 M. Keep at −20 °C.
12. Se-depleted medium. To 98 mL of DMEM (containing glutamine, pyruvate, and optionally antibiotics), add 2 mL of FCS (*see* **Note 4**), 100 μL of human insulin solution (1000×), 50 μL of human transferrin solution (2000×), 50 μL of hydrocortisone solution (2000×), and 10 μL of T3 hormone solution.
13. Se-supplemented medium. To 100 mL of Se-depleted medium, add 100 μL of isotopically enriched selenite solution (1000×) or natural selenite solution (1000×) to reach a final concentration of 100 nM selenite.

14. 0.05% Trypsin/EDTA (1×).
15. Dulbecco's Phosphate Buffered Saline (DPBS) with no magnesium, no calcium: 137 mM NaCl, 26.8 mM KCl, 40.2 mM Na_2HPO_4, 17.6 mM KH_2PO_4, pH 7.4.
16. Passive lysis buffer: 25 mM Tris-phosphate, 2 mM DTT, 2 mM EDTA, 1% Triton X100, 10% glycerol, pH 7.8.
17. Sterile 75 and 150 cm^2 culture dish for adherent cells.
18. Incubator CO_2 at 37 °C, humidified.
19. Cell culture hood.
20. Centrifuge for cells for 15 mL tubes.
21. Centrifuge for 1.5 mL tubes.
22. Sterile pipettes.
23. Microscope.
24. Autoclave.

2.2 Size Exclusion Chromatography Coupled with ICP MS

1. Protein quantitation kit (such as Lowry, BCA, or Bradford).
2. Mobile phase for SEC: 100 mM ammonium acetate buffered with NH_4^+ at pH 7.4.
3. Long-term conservation solution: 20% Ethanol in water.
4. SEC Protein calibration kit.
5. SEC chromatography Pump with UV detector.
6. HPLC pump with UV detector.
7. Size exclusion chromatography column for protein (Superdex 200), range: 15–300 kDa.
8. ICP MS with collision/reaction cell.
9. ICP MS tuning solution containing 1 μg/L of Lithium, Thallium, Cerium, Magnesium, Yttrium.

3 Methods

3.1 Cellular Growth

1. LNCaP are grown and amplified in 75 or 150 cm^2 plates in the maintenance medium. Cells were cultivated in 5% CO_2 at 37 °C in humidified atmosphere. They are plated at a density of 10^4 cells/cm^2 and replated when reaching 90% confluency. Confluency is usually reached in 3 or 4 days (approximately 10^5 cells/cm^2). Although adherent LNCaP cells can easily be detached from the culture dish and therefore need to be manipulated with care.
2. Cells are grown for one passage in Se-depleted medium. Then, LNCaP cells are plated at 10^4 cells/cm^2 cell density (1.5×10^6 cells) in Se-supplemented medium in 150 cm^2 dish containing

either ^{76}Se, ^{77}Se, or natural selenium (100 nM), or in Se-depleted medium. After 2 days, the medium is changed with the respective fresh one.

3.2 Cell Harvest and Extract

1. After 72 h of growth, cells are washed with ice-cold DPBS and harvested with trypsin solution.
2. Cells suspensions are transferred in 15 mL centrifugation tubes and pelleted by centrifugation at 1000 × *g* for 5 min.
3. Cell pellets are washed with cold DPBS, then DPBS is discarded. At this stage, cell pellets can be flash frozen and stored at −80 °C for later analysis.
4. Cells extracts are performed by the addition of 300 μL of passive lysis buffer, and incubated for 15 min on ice.
5. Cell extracts are transferred in 1.5 mL centrifugation tubes and pelleted at 10,000 × *g* for 30 min at +4 °C.
6. Supernatants are transferred in new 1.5 mL centrifugation tubes and analyzed for protein concentration using your commercially available protein concentration kit (*see* **Note 5**).

3.3 Cell Analysis by SEC-ICPMS

1. Equilibrate the SEC chromatography system with mobile phase at 0.2 mL/min if the column was in long-term conservation solution. Otherwise start at 0.7 mL/min.
2. Calibrate your SEC with the protein calibration kit. Note the elution time of each protein using the UV signal. Verify that a linear regression is observed between the log (MW) and the elution time.
3. Switch on the ICP MS and optimize its detection with the ICP MS tuning solution.
4. Select the detection of all selenium isotope masses (74, 76, 77, 78, 80, and 82), with 0.1 ms integration time and with time resolved analysis (TRA) as acquisition mode.
5. Connect the SEC chromatography system to the inlet of the introduction system of ICP MS (*see* refs. 15, 16 for details).
6. Dilute your cell extracts with mobile phase to have 500 μg of proteins in 100 μL (*see* **Note 6**).
7. Inject 100 μL of each cell extract, and acquire the UV^{280nm} and ICP MS signal. A typical run time for a superdex 200 at 0.7 mL/min is 80 min (*see* **Note 7**).
8. For long time storage of the SEC column (>2 weeks), the mobile phase should be exchanged with long-term conservation solution (Fig. 2).

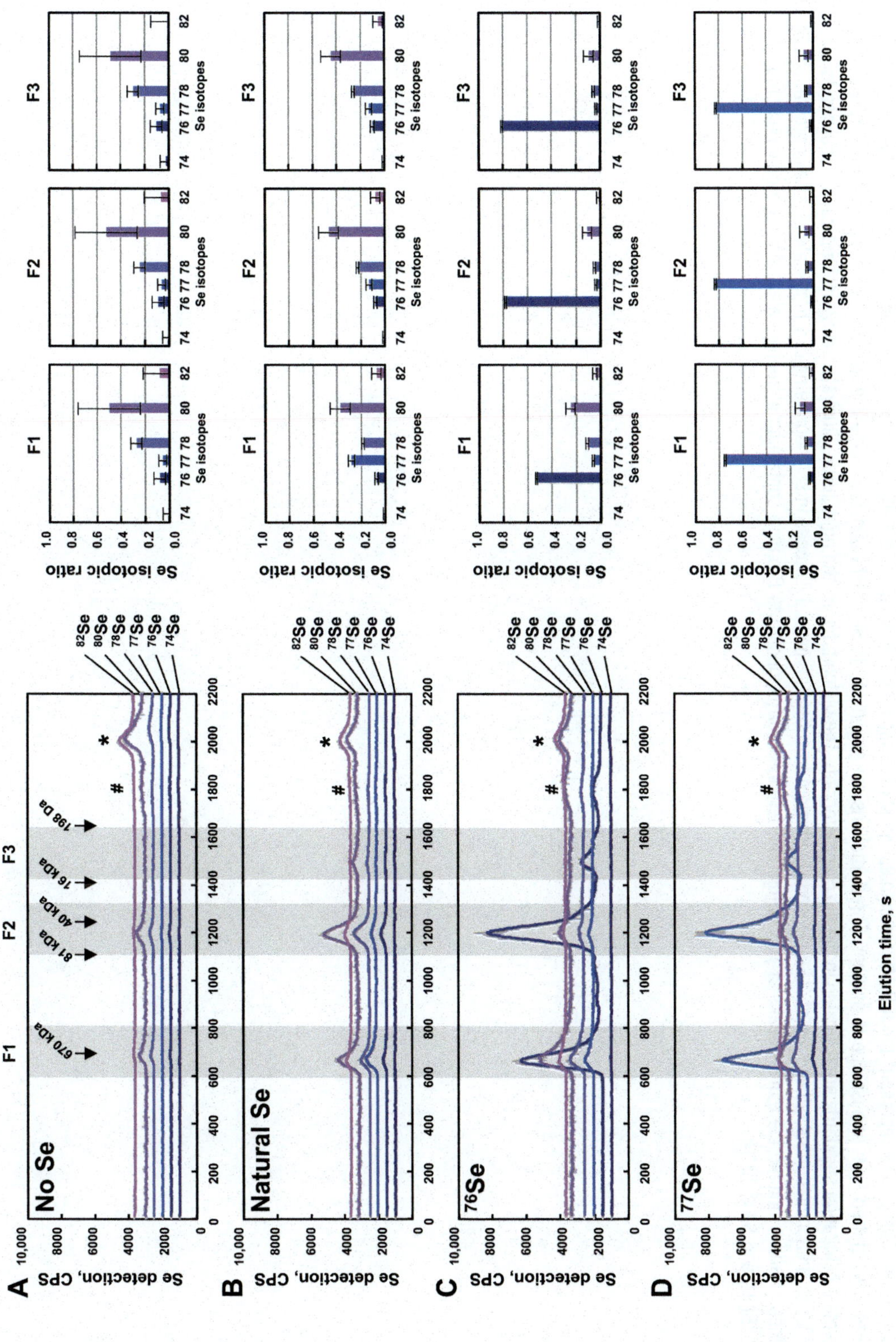
A
No Se
B
Natural Se
C
^{76}Se
D
^{77}Se
Se detection, CPS
Elution time, s
F1
F2
F3
670 kDa
81 kDa
40 kDa
16 kDa
198 Da
#
*
^{82}Se
^{80}Se
^{78}Se
^{77}Se
^{76}Se
^{74}Se
Se isotopic ratio
Se isotopes
74 76 77 78 80 82

Fig. 2 Illustration of a typical Se-chromatogram obtained with LNCaP grown with 100 nM of either natural (**b**), ^{76}Se (**c**), or ^{77}Se selenite (**d**) as compared with no Se added (**a**). *Left panels* represent the ICP MS Se-signals (all six isotopes) as a function of the elution time. To increase the clarity of the figure Se-signals is artificially shifted from one isotope to another. Three fractions corresponding to Se most abundant signal are defined as F1, F2, and F3. Elution times of SEC protein calibration standards are indicated in the top chromatogram. The histograms, on the right inside, illustrate the Se-isotopic ratio calculated for F1, F2, and F3. Our data indicate that either ^{76}Se or ^{77}Se are equally efficient to substitute natural isotopes in selenoproteins in a timescale of 3 days growth of LNCaP cell lines. # and * indicate the elution peaks of polyatomic interferences corresponding to Bromide containing molecules since they are only detected at mass 80 ($^{79}Br^{1}H$) and 82 ($^{81}Br^{1}H$)

3.4 Data Treatment and Analysis

1. ICP MS data are collected as .csv format.
2. Represent Se-signal for individual isotope as a function of time.
3. The background value for individual Se-isotope is evaluated in the acquisition between the injection and void volume (V_0, also referred to as dead volume). This value can be subtracted to the raw data in order to calculate a more accurate isotopic ratio.
4. To evaluate the Se-isotopic ratio, the sum of background-subtracted Se-signal from an elution time window is calculated. Then the isotopic ratio for each isotope is calculated relative to overall counts from a defined time window.

4 Notes

1. Other cell lines have been successfully tested in the laboratory and include HEK293, HaCaT, and HepG2 cells.
2. The price of Se isotopes usually depends on their natural abundance, which is inversely proportional to the amount and effort needed for its purification. For example, ^{74}Se-enriched isotope, which naturally presents at 0.89%, is much more expensive than ^{78}Se-enriched isotope, present at a relative abundance of 23.77%.
3. Be careful when diluting your isotope dilution that the molecular weight of selenite depends on its isotope.
4. 10% FCS is commonly used. Since selenium is originating from the serum, a five-time reduction of serum results in a five-time reduction of selenium in the final Se-depleted medium. The concentration of selenium in FCS varies from 100 to 400 nM between lot numbers. This should be evaluated by ICP MS. In the Se-depleted medium (2% FCS), the selenium concentration should be lower than 4 nM.
5. Verify that your protein concentration kit is not interfered by 1% Triton X100 present in the passive lysis buffer.
6. Lower protein amount could be used, but it will limit the definition of the different elution peaks for selenium. Alternatively, higher protein amount could be loaded to the column, but increasing the injection volume will increase the width of elution peaks.
7. Verify that the selenium signal is back to baseline between each run.

Acknowledgments

This work was supported by the CNRS (ATIP program to LC), the Fondation pour la Recherche Médicale (LC), the Ligue Contre le Cancer (Comité de l'Essonne, LC), the programme interdisciplinaire de recherche du CNRS longévité et vieillissement (LC), the Association pour la recherche sur le cancer [grants numbers 4849, LC] and the Agence Nationale de la Recherche [grant number ANR-09-BLAN-0048 to LC]. JS is a recipient of a Ph.D fellowship from the French Ministry for Research.

References

1. Kryukov GV, Castellano S, Novoselov SV, Lobanov AV, Zehtab O, Guigo R, Gladyshev VN (2003) Characterization of mammalian selenoproteomes. Science 300 (5624):1439–1443
2. Touat-Hamici Z, Legrain Y, Sonet J, Bulteau A-L, Chavatte L (2016) Alteration of selenoprotein expression during stress and in aging. In: Hatfield DL, Su D, Tsuji PA, Gladyshev VN (eds) Selenium: its molecular biology and role in human health, 4th edn. Springer Science +Business Media, LLC, New York, pp 539–551
3. Sonet J, Bulteau A-L, Chavatte L (2016) Selenium and selenoproteins in human health and diseases. In: Michalke B (ed) Metallomics: analytical techniques and speciation methods. Wiley-VCH Verlag GmbH & Co KGaA, Weinheim, Germany, pp 364–381. doi:10.1002/9783527694907.ch13
4. Labunskyy VM, Hatfield DL, Gladyshev VN (2014) Selenoproteins: molecular pathways and physiological roles. Physiol Rev 94 (3):739–777. doi: 94/3/739 [pii] 10.1152/physrev.00039.2013
5. Latrèche L, Chavatte L (2008) Selenium incorporation into selenoproteins, implications in human health. Met Ions Biol Med X:731–737
6. Papp LV, Lu J, Holmgren A, Khanna KK (2007) From selenium to selenoproteins: synthesis, identity, and their role in human health. Antioxid Redox Signal 9(7):775–806
7. Bulteau A-L, Chavatte L (2015) Update on selenoprotein biosynthesis. Antioxid Redox Signal 23(10):775–794
8. Latreche L, Duhieu S, Touat-Hamici Z, Jean-Jean O, Chavatte L (2012) The differential expression of glutathione peroxidase 1 and 4 depends on the nature of the SECIS element. RNA Biol 9(5):681–690
9. Latreche L, Jean-Jean O, Driscoll DM, Chavatte L (2009) Novel structural determinants in human SECIS elements modulate the translational recoding of UGA as selenocysteine. Nucleic Acids Res 37(17):5868–5880
10. Driscoll DM, Copeland PR (2003) Mechanism and regulation of selenoprotein synthesis. Annu Rev Nutr 23:17–40
11. Touat-Hamici Z, Legrain Y, Bulteau AL, Chavatte L (2014) Selective up-regulation of human selenoproteins in response to oxidative stress. J Biol Chem 289(21):14750–14761. doi: M114.551994 [pii] 10.1074/jbc.M114.551994
12. Legrain Y, Touat-Hamici Z, Chavatte L (2014) Interplay between selenium levels, selenoprotein expression, and replicative senescence in WI-38 human fibroblasts. J Biol Chem 289 (9):6299–6310
13. Gladyshev VN, Arner ES, Berry MJ, Brigelius-Flohe R, Bruford EA, Burk RF, Carlson BA, Castellano S, Chavatte L, Conrad M, Copeland PR, Diamond AM, Driscoll DM, Ferreiro A, Flohe L, Green FR, Guigo R, Handy DE, Hatfield DL, Hesketh J, Hoffmann PR, Holmgren A, Hondal RJ, Howard MT, Huang K, Kim H-Y, Kim IY, Kohrle J, Krol A, Kryukov GV, Lee BJ, Lee BC, Lei XG, Liu Q, Lescure A, Lobanov AV, Loscalzo J, Maiorino M, Mariotti M, Sandeep Prabhu K, Rayman MP, Rozovsky S, Salinas G, Schmidt EE, Schomburg L, Schweizer U, Simonovic M, Sunde RA, Tsuji PA, Tweedie S, Ursini F, Whanger PD, Zhang Y (2016) Selenoprotein gene nomenclature. J Biol Chem 291(46):24036–24040
14. Bianga J, Touat-Hamici Z, Bierla K, Mounicou S, Szpunar J, Chavatte L, Lobinski R (2014) Speciation analysis for trace levels of selenoproteins in cultured human cells. J Proteome 108:316–324. doi:10.1016/j.jprot.2014.05.025

15. Mounicou S, Bierla K, Szpunar J Detection and determination of heteroatom-containing molecules by HPLC: inductively coupled plasma mass spectrometry. In: Shalliker RA (ed) Hyphenated and alternative methods of detection in chromatography, vol 2011. CRC Press, Boca Raton, FL, pp 161–185
16. Mounicou S, Szpunar J, Lobinski R (2009) Metallomics: the concept and methodology. Chem Soc Rev 38(4):1119–1138. doi:10.1039/b713633c

Chapter 15

Detection of Selenoproteins by Laser Ablation Inductively Coupled Plasma Mass Spectrometry (LA-ICP MS) in Immobilized pH Gradient (IPG) Strips

Jordan Sonet, Sandra Mounicou, and Laurent Chavatte

Abstract

In contrast to other trace elements that are cofactors of enzymes and removed from proteins under denaturing conditions, Se is covalently bound to proteins when incorporated into selenoproteins, since it is a component of selenocysteine aminoacid. It implies that selenoproteins can undergo several biochemical separation methods in stringent and chaotropic conditions and still maintain the presence of selenium in the primary sequence. This feature has been used to develop a method for the detection of trace levels of human selenoproteins in cell extracts without the use of radioactive isotopes. The selenoproteins are separated as a function of their isoelectric point (pI) using iso-electrofocusing (IEF) electrophoretic strips and detected by laser ablation-inductively coupled plasma mass spectrometry (LA-ICP MS). This method, therefore referred to as IEF-LA-ICP MS, allowed the detection of several selenoproteins in human cell lines, including Gpx1, Gpx4, TXNRD1, TXNRD2, and SELENOF.

Key words Selenium, IEF, Laser ablation, ICP MS, ESI MS/MS, Glutathione peroxidase, Thioredoxin reductase, SELENOF, Cell culture

1 Introduction

Selenium (Se) is an essential trace element implicated in many facets of human health and disease. Most of its beneficial effects are attributed to its presence as selenocysteine (Sec) in a small, but vital group of proteins, namely the selenoproteins [1–6]. Selenocysteine is a structural and functional analog of cysteine, where a selenium atom replaces sulfur. For selenoproteins with a characterized function, selenocysteine is often found in the catalytic site of enzymes where its presence is vital for redox reactions. Twenty-five selenoprotein genes have been identified in human genome (24 in rodents), and their expression constitutes the selenoproteome [6, 7].

The expression of the selenoproteome is primarily controlled by selenium bioavailability that induces prioritization of protein biosynthesis, when this trace element is deficient at the scale of

Laurent Chavatte (ed.), *Selenoproteins: Methods and Protocols*, Methods in Molecular Biology, vol. 1661,
DOI 10.1007/978-1-4939-7258-6_15, © Springer Science+Business Media LLC 2018

organism, tissue, or cell lines. This regulation that maintains the expression of essential selenoenzymes at the expense of others is often referred to as the hierarchy of selenoproteins [8–13]. In order to evidence a link between selenium, selenoprotein, and pathophysiological conditions and eventually identify biomarkers for selenium status, scientists are dependent on reliable and sensitive methods to detect and quantify selenoproteins and selenium levels in various body fluids or in cellular/tissue extracts. As illustrated in Table 1, human selenoproteins could potentially be well separated

Table 1
List of human selenoprotein genes

Gene name	Protein name	Synonyms	Acc number	pI	MW, Da
SELENOF	Selenoprotein F	Selenoprotein 15, Sel15, Sep15	O60613	4.96	17,744
SELENON	Selenoprotein N	SelN	Q9NZV5	5.38	65,793
SELENOM	Selenoprotein M	SelM	Q8WWX9	5.39	16,185
SEPHS2	Selenophosphate 2	Selenide, water dikinase 2, SPS2	Q99611	5.63	47,305
SELENOO	Selenoprotein O	SelO	Q9BVL4	5.69	73,442
TXNRD1	Thioredoxin reductase 1	TR1, TRXR1	Q16881	6.07	54,707
SELENOI	Selenoprotein I	SelI	Q9C0D9	6.12	45,182
GPX1	Glutathione peroxidase 1	Glutathione peroxidase 1	P07203	6.15	21,899
GPX6	Glutathione peroxidase 6	Glutathione peroxidase 6	P59796	6.21	24,924
DIO3	Iodothyronine deiodinase 3	D3	P55073	6.24	31,404
TXNRD3	Thioredoxin reductase 3	TR2, TRXR3, TGR	Q86VQ6	6.36	63,540
DIO2	Iodothyronine deiodinase 2	D2	Q92813	6.60	29,600
TXNRD2	Thioredoxin reductase 3	TR3, TRXR2, mitochondrial thioredoxin reductase	Q9NNW7	7.23	56,460
GPX2	Glutathione peroxidase 2	Glutathione peroxidase 2	P18283	7.60	21,907
SELENOP	Selenoprotein N	SelP, SEPP1, SEPP, SeP	P49908	7.92	42,705

(continued)

Table 1
(continued)

Gene name	Protein name	Synonyms	Acc number	pI	MW, Da
GPX3	Glutathione peroxidase 3	Glutathione peroxidase 3	P22352	8.20	25,505
MSRB1	Methionine sulfoxide reductase B1	SelR, SelX, SEPX1	Q9NZV6	8.57	12,713
GPX4	Glutathione peroxidase 4	Glutathione peroxidase 4	P36969	8.64	22,128
SELENOT	Selenoprotein T	SelT	P62341	8.69	22,277
DIO1	Iodothyronine deiodinase 1	D1	P49895	8.81	28,878
SELENOW	Selenoprotein W	SelW	P63302	9.20	9270
SELENOV	Selenoprotein V	SelV	P59797	9.57	36,754
SELENOH	Selenoprotein H	SelH	Q8IZQ5	9.65	13,407
SELENOS	Selenoprotein S	SelS, SEPS1, VIMP	Q9BQE4	9.65	21,116
SELENOK	Selenoprotein K	SelK	Q9Y6D0	10.54	10,598

Only the reference selenoproteins (according to Uniprot) are shown here for each selenoprotein gene. All reported protein isoforms from alternative spice sites or alternative initiation sites can be found in various online databases. Selenoproteins are ranged according to their theoretical pI

as a function of their theoretical pIs, since selenoproteins range from 4.96 to 10.54. Therefore, the IPG strip with a nonlinear 3–10 immobilized gradient (3-10NL) is an interesting choice to cover the maximum members of the human selenoproteome. The nature of IPG strips, obviously, could be adapted to the set of selenoproteins studied. In this chapter, we describe an innovative strategy to analyze the selenoproteome in eukaryotic cellular extracts using laser ablation of IPG strips coupled to ICP MS detection (IEF-LA-ICP MS) [14, 15]. Detection of selenium and selenoproteins by ICP MS is still in constant evolution, and has been described in Chapters 10, 11, 14, and 16. In the current chapter, we will focus on the cellular extraction, precipitation, and IEF separation of proteins, prior to their analysis by LA-ICP MS. Typical examples obtained with natural Se and isotopically enriched Se-isotopes grown cell lines are shown.

2 Materials

Prepare all the solutions using ultrapure water (18 MΩ cm at 25 °C) and analytical grade reagents. Concerning the solutions in contact with cells in culture, they should be sterilized with an autoclave or filtrated using a 0.22 μm syringe filter. Routine reagents should be purchased from your supplier of choice. Suppliers are indicated only when important for the method. Prepare and store all the reagents at room temperature (unless stated otherwise). Diligently follow all waste disposal regulations when disposing waste materials. All the chemicals (except for cell culture) should be IEF grade.

2.1 Reagents

1. LNCaP cell lines. These cells originate from a human prostate biopsy in the left supraclavicular lymph node. They derived from metastatic site of a 50 years adult.
2. Human embryonic kidney 293 T-REx cells originally derived from epithelial kidney cells from a single and healthy fetus. T-REx Cell Lines are an engineered subclone that stably expresses the tetracycline repressor protein. This cell line will be referred to as HEK293 cells throughout this chapter.
3. Isotopically enriched selenite solution (1000×). Isotopically enriched sodium selenite is available in very few companies (*see* **Note 1**). Prepare a solution at 0.1 mM in water (*see* **Note 2**). Keep at −20 °C.
4. Natural selenite solution (1000×). Sodium selenite with natural abundance is prepared at 0.1 mM in water (1000×). Keep at −20 °C.
5. Se-depleted medium. To 98 mL of DMEM (containing glutamine, pyruvate, and optionally antibiotics), add 2 mL of FCS, 100 μL of human insulin solution (1000×), 50 μL of human transferrin solution (2000×), 50 μL of hydrocortisone solution (2000×), and 10 μL of T3 hormone solution.
6. Se-supplemented medium. To 100 mL of Se-depleted medium, add 100 μL of isotopically enriched selenite solution (1000×) or natural selenite solution (1000×) to reach a final concentration of 100 nM selenite.
7. Fetal Calf Serum (FCS) is commercially available from numerous companies. Keep at −20 °C.
8. Maintenance medium: DMEM supplemented with 10% FCS.
9. Optionally, penicillin and streptomycin antibiotics (100 U/mL) can be added to the media to prevent cell culture contamination by bacteria.
10. 0.05% Trypsin/EDTA (1×).
11. Dulbecco's Phosphate-Buffered Saline (DPBS) with no magnesium, no calcium: 137 mM NaCl, 26.8 mM KCl, 40.2 mM Na_2HPO_4, 17.6 mM KH_2PO_4, pH 7.4.
12. Passive lysis buffer: 25 mM Tris-phosphate, 2 mM DTT, 2 mM EDTA, 1% Triton X100, 10% glycerol, pH 7.8.
13. Protein quantitation kit (such as Lowry, BCA or Bradford).

14. Immobilized pH gradient (IPG) strips. Several companies provide IPG strips with a wide range of pH gradient. In this work, we used 3–10 NL gradient from GE Healthcare.
15. Methanol.
16. Chloroform.
17. Tris (IEF grade).
18. CHAPS (IEF grade).
19. Urea (IEF grade).
20. Thiourea (IEF grade).
21. Destreak reagent (GE Healthcare) (*see* **Note 3**).
22. Rehydration buffer: 8 M urea, 2 M thiourea, 2% (w/v) CHAPS, 12 μL/mL destreak reagent (GE Healthcare), 0. 5% (v/v) IPG Buffer (3–10 from GE Healthcare) and traces of bromophenol blue. Should be prepared fresh, do not store.
23. Mineral oil for IEF.
24. IPG fixing solution: 10% glycerol and 30% ethanol in water.
25. 2% HNO_3: dilute 14 mL of 69–70% HNO_3 in 486 mL of ultrapure water.
26. ICP MS tuning solution containing 1 ppb of Lithium, Thallium, Cerium, Magnesium, Yttrium.

2.2 Instrumentation

1. Incubator CO_2 at 37 °C, humidified.
2. Cell culture hood.
3. Centrifuge for cells for 15 mL tubes.
4. Centrifuge for 1.5 mL tubes.
5. Sterile pipettes.
6. Microscope.
7. Autoclave.
8. ICP MS with collision/reaction cell filled with H_2 and/or He.
9. Reswelling tray for IPG Strips.
10. System for the migration of IPG strips. Here, we used the IPGphor II from GE Healthcare.

3 Methods

We have developed and optimized a method to separate the human selenoproteome on IPG strips that is compatible for subsequent analyses with our LA-ICP MS system. Based on high-throughput proteomic data, one should be able to detect at least the most abundant selenoproteins expressed in your cell lines (*see* Fig. 1 for HEK293 and LNCaP cell lines used in this chapter). In order to

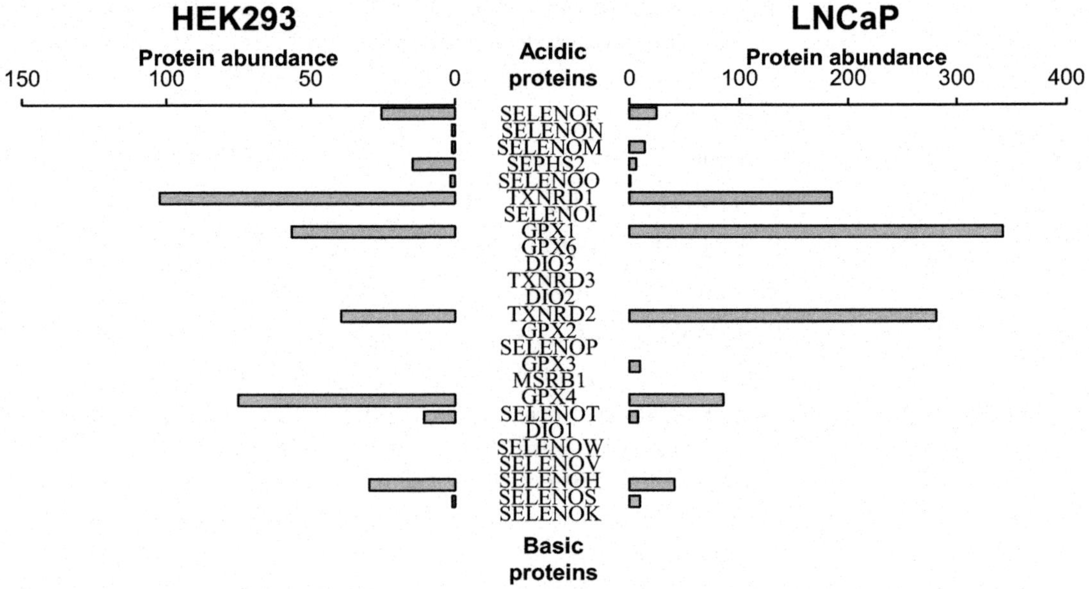

Fig. 1 Illustration of relative selenoprotein abundance in HEK293 and LNCaP cell lines, as obtained by high-throughput mass spectrometry analysis (www.proteomicsdb.org). For each cell line, the abundance is calculated relative to the lowest abundant, but detected, selenoprotein, set at one (SELENON and SELENOO, respectively)

have enough Se signal, we load the maximum capacity of protein extracts per strip (i.e., 500 μg for an 18 cm strip) (*see* **Note 4**). The use of 3–10 NL strips allowed a broad and reproducible picture of the detectable selenoproteome (as inferred from Table 1) (*see* **Note 5**).

3.1 Cell Culture

1. HEK293 and LNCaP cells are grown and amplified in 75 or 150 cm^2 plates in maintenance medium. Cells were cultivated in 5% CO_2 at 37 °C in humidified atmosphere. They are plated at a density of 10^4 cells/cm^2 and replated when reaching 90% confluency. Confluency is usually reached in 3 or 4 days (approximately 10^5 cells/cm^2).
2. Then the cells are plated in 150 cm^2 dish and grown for 3 days Se-supplemented media containing either 100 nM of natural selenite, or nonradioactive isotopically enriched ^{76}Se.
3. Cells are washed with ice-cold DPBS and harvest with trypsin solution. Se-depleted media is used to stop trypsin digestion and transfer the cells in a 15 mL tube.
4. Cells are pelleted by centrifugation at 1000 × *g* for 5 min.
5. Cells are resuspended with ice-cold DPBS and pelleted by centrifugation at 1000 × *g* for 5 min.
6. DPBS is removed and cell pellets can be flash frozen and stored at −80 °C. Otherwise move to Subheading 3.2.

3.2 Cell Lysis, Protein Extraction, Quantification, and Precipitation

1. Cell lysis is performed by adding 300 μL of passive lysis buffer to the cell pellet and 15 min incubation on ice (*see* **Note 6**).
2. Lysed cells are transferred in a 1.5 mL centrifugation tube and pelleted at 10,000 × *g* for 30 min at +4 °C.
3. Supernatants are transferred in a new 1.5 mL centrifugation tube and analyzed for protein concentration using your commercially available protein concentration kit.
4. Take the volume of cell extract corresponding to 500 μg of proteins and complete to 200 μL with water in a 2 mL centrifugation tube (*see* **Note 7**).
5. Add 800 μL of 100% methanol, vortex well and spin briefly.
6. Add 200 μL of chloroform, vortex well and spin briefly.
7. Add 600 μL of water, vortex well and centrifuge for 2 min at 10,000 × *g*. Protein should appear at interface.
8. Remove carefully the upper layer without discarding the upper layer.
9. Add 600 μL of methanol, vortex well and centrifuge for 2 min at 10,000 × *g*. Protein should appear as a pellet.
10. Remove the supernatant, air dry the pellet, and move directly to Subheading 3.3 (*see* **Note 8**).

3.3 Passive Rehydration of IPG Strips with Protein Sample

1. Protein pellets are resuspended in 340 μL of rehydration solution by several up and down pipetting; do not vortex to avoid any air bubble. The protein pellet should be completely solubilized; otherwise, mix again the sample by pipetting.
2. Load all 340 μL of the protein sample in a reswelling tray (*see* **Note 9**), taking care not to create air bubbles.
3. Take the 18 cm immobiline dry strip (3–10 NL) from the freezer. Remove the plastic protection which is on the gel side and place the strip gel-side-down on the top of the protein sample in the reswelling tray.
4. Add mineral oil to cover the strip according to the manufacturer's instruction and let the passive rehydration occur at room temperature for at least 8 h up to overnight.

3.4 Protein Migration of IPG Strips

1. Clean and dry the IPGphor bed before placing the manifold tray on the unit. Ensure that the manifold is level.
2. Place the rehydrated IPG strip gel-side-up in one channel of the manifold tray.
3. Prepare two paper wicks per IPG strip on a parafilm square. Moisten them with 150 μL of water. Add 1.2% destreak reagent (v/v) in the solution that will moisten the paper wick placed at the cationic electrode (−, minus side of the strip) (*see* **Note 10**).

Table 2
IEF-electrophoresis conditions for an 18 cm IPG strip

Step	Step voltage mode	Voltage (V)	Time (h:min)	Volt-hours (kV h)
1	Step and hold	150	1:00	150
2	Step and hold	200	1:00	200
3	Gradient	200–1000	8:00	3000–5000
4	Step and hold	1000	4:00	4000
5	Gradient	1000–8000	3:00	10,000–14,000
6	Step and hold	8000	2:00	16,000
	Total			33,000–39,000

4. Place the paper wicks at the edge of the strips. Half of the wicks should be in contact with the gel.
5. Place the electrodes in contact with the paper wicks.
6. Cover all the wells including empty ones with mineral oil.
7. Close the IPGPhor system and start the migration using the program shown in Table 2. Maximum intensity per strip is 50 μA.

3.5 IPG Strip Drying and Preparation for LA Analysis

1. After the migration, the strips are removed from the manifold and placed in a tray, which could be the reswelling tray, and fixed in IPG fixing solution for 30 min. IPG strips are then air dried under atmospheric condition (*see* **Notes 11** and **12**).
2. Dried strips can be kept at +4 °C between parafilm papers at least for several months, and not lose Se signal by LA-ICP MS.
3. To fit in our ablation cell, the sample should fit in a glass plate of 50 mm width and 26 mm height. A double-sided adhesive tape is placed on the glass slide (*see* **Note 13**).
4. The IPG strips (18-cm long) are divided into five pieces: four pieces of 4 cm long and one of 2 cm (*see* **Note 14**).
5. Then, carefully place the strip pieces gel-side-up on a glass slide, which has been previously prepared with a double-sided adhesive tape.

3.6 Setup and Tuning of LA-ICP MS System

1. In our setup, an UP-213 laser (NewWave Research, Fremont, CA) is connected to an Agilent 7500cs ICP MS (Agilent Technologies, Tokyo, Japan), as described in [14, 15] (*see* **Note 15**).
2. The laser is operated in a focused spot mode at the repetition rate of 20 Hz, spot size of 250 μm and 100 μm/s scan speed. Ablated material is carried with He gas flow (500 mL/min) and sent to the ICP MS used in collision/reaction mode (*see* **Note 16**).

3. For the ICP MS, a 1.5 mm i.d. injector torch and Pt cones are used. The ICP MS is tuned off-line using a commercially available tuning solution for the maximum Se signal-to-noise ratio. The integration time is set at 0.25 ms for each Se isotope.
4. When the LA-ICP MS in on-line. The sensitivity of the whole system is checked by ablating a polished glass standard (NIST glass standard).
5. Alternatively, a dried SDS-PAGE containing a selenoprotein standard, such as glutathione peroxidase 1, can be also used as an internal laboratory standard when the system is dedicated to detection of seleno-compounds.
6. Place the glass plate with the taped IPG strip samples in the ablation chamber.
7. The samples are then scanned with the laser. A typical run for an 18 cm strip is 40 min (*see* **Note 17**).

3.7 Data Treatment and Analysis

1. ICP MS data are collected as .csv format.
2. Se-signal for individual isotope is represented as a function of time. Depending on the ICP MS, one may have to manually remove the signal recorded between the end of one strip piece to the beginning of the next strip piece.
3. Raw data can be further treated to reduce the noise of ICP MS by applying a local average treatment, ranging from 5 to 10.
4. A peak is considered due to the detection of selenoprotein, and not an artifact, when a sufficient number of detection points are detected for the different Se isotopes (Figs. 2 and 3).

3.8 ESI MS/MS Identification of Selenoproteins

In the ablated IPG strip, which is 2 mm wide, only 250 μm has been ablated all along. Therefore, 1.8 mm of the gel is left for subsequent analyses. This remaining material can be used for another scan or for gel excision followed by trypsin digestion. Thus, protein identification by classical Mass Spectrometry strategy can be performed to annotate the electropherogram when performed on a new species or when new Se peaks are visible, as described in [14].

4 Notes

1. Other cell lines or tissue extracts have been successfully used.
2. Be careful when diluting your isotope dilution that the molecular weight of selenite depends on its isotope.
3. Destreak reagent can be replaced by DTT.
4. One needs to refer to the manufacturer's instruction for the maximum protein loading capacity of the strips. Alternatively to in gel rehydration of the strips, the protein loading can be

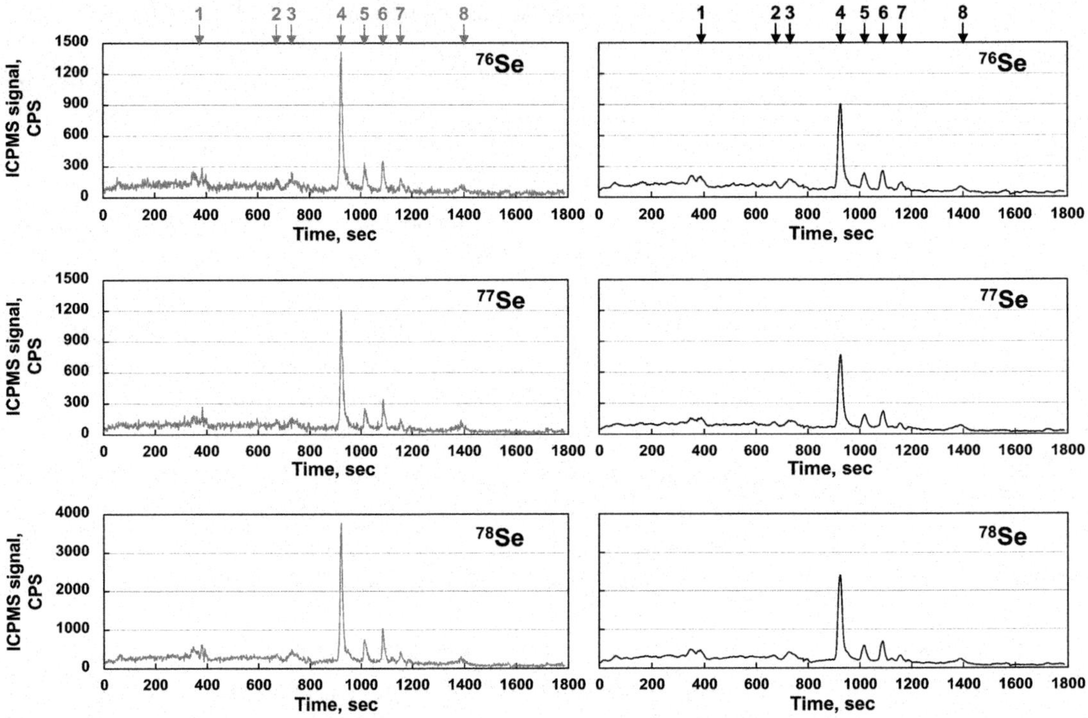

Fig. 2 Illustration of typical Se-electropherograms obtained with protein extracts from HEK293 cells grown with 100 nM of natural sodium selenite. Raw Se signals obtained for ^{76}Se, ^{77}Se, and ^{78}Se are shown in the *left panels* as a function of scan time. The *right panels* illustrate the treatment of the raw-data by local average calculation (set at 10). The signal is smoother but the intensity of the every peaks is reduced. Eight Se-peaks are visible, from which five selenoproteins could be detected by ESI-MS/MS as described in [14]: (1) SELENOF, (2) not identified, (3) not identified, (4) Gpx1, (5) Txnrd1, (6) Txnrd1, (7) Txnrd2, (8) Gpx4

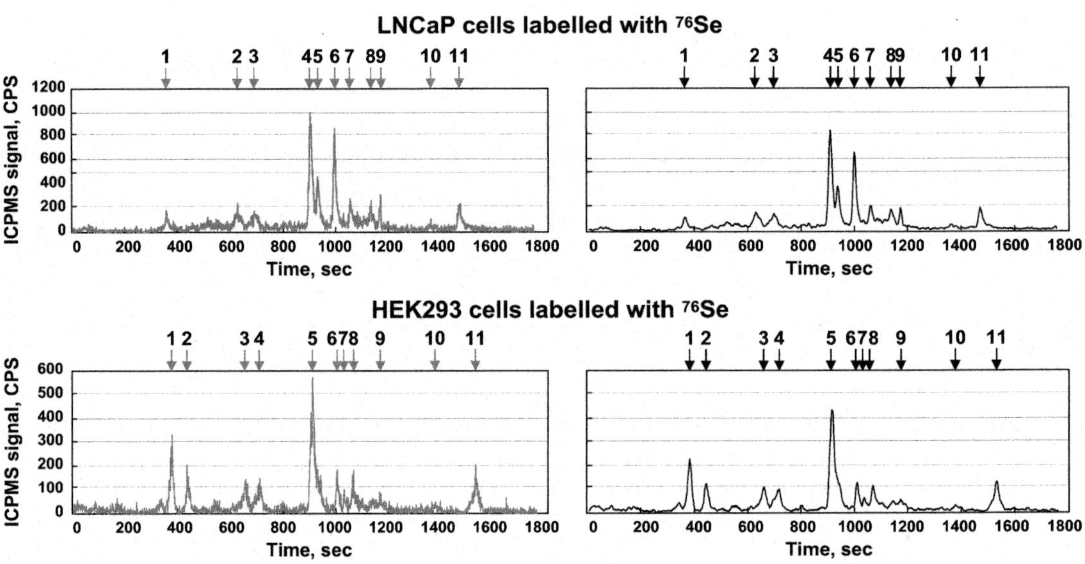

Fig. 3 Illustration of typical electropherograms obtained with protein extracts from HEK293 and LNCaP cells grown with 100 nM ^{76}Se isotopically enriched selenite. As compared with Fig. 2, more Se-peaks can be detected when using a Se-enriched isotope in the culture medium

improved in terms of quantity by cup-loading, although this strategy has not been tested during our work.

5. However, one potential downside of our strategy is the non-overlapping selenoprotein profile from one species to another, since the pI of selenoproteins is generally much more variable from one species to another than the size of the proteins. This implies that using our IEF-LA-ICP MS strategy for rodent selenoproteome analysis, for example, would necessitate to re-identify the selenoproteins from individual Se-peak in the electropherogram.
6. Using this soft lysis protocol will leave the nucleus in the pellet. If you expect your protein of interest to be in the nucleus, you can add a sonication step to your protocol.
7. If protein concentration is below 2.5 mg/mL, a higher volume of cell extract can be used. Then, all other volumes should be adapted accordingly. The ratio of protein:MeOH: chloroform: water is 1:4:1:3.
8. This precipitation method is efficient to remove SDS, Triton X-100, salts, and beta-mercaptoethanol [16].
9. The reswelling trays can be used only once to avoid cross contamination, or the presence of potential Se contaminants.
10. Addition of destreack reagent at this electrode is used to prevent precipitation of the basic proteins at this side of the strip, since the destreak (and the DTT) is migrating to the anionic electrode and therefore depleted at the cationic side.
11. To accelerate the drying process, one can tape the edges of the strips to a whatman 3MM paper, and then place under the chemical hood.
12. Due to the presence of glycerol in the fixing solution, the gels will never completely dry out, and will remain a little sticky.
13. Try to be consistent and precise in the way you cut the strips from one experiment to the other, especially in order to avoid cutting right in the middle of a Se peak.
14. From our laboratory convention, we always start the scan from the plus side toward the minus side of the strip.
15. Alternative setup with different lasers and/or ICP MS is commercially available in other laboratories worldwide. The sensitivity and robustness of the different apparatus are in constant evolution.
16. In collision/reaction mode, H_2 or He gas could be equally used for roughly similar sensitivity depending on ICP MS system.

17. A slower scan speed of 50 μm/s can be used to improve the precision of the electropherogram by increasing the number of detection points. However, the scan time will be doubled.

Acknowledgments

This work was supported by the CNRS (ATIP program to LC), the Fondation pour la Recherche Médicale (LC), the Ligue Contre le Cancer (Comité de l'Essonne, LC), the programme interdisciplinaire de recherche du CNRS longévité et vieillissement (LC), the Association pour la recherche sur le cancer [grants numbers 4849, LC] and the Agence Nationale de la Recherche [grant number ANR-09-BLAN-0048 to LC]. JS is a recipient of a Ph.D fellowship from the French Ministry for Research.

References

1. Touat-Hamici Z, Legrain Y, Sonet J, Bulteau A-L, Chavatte L (2016) Alteration of selenoprotein expression during stress and in aging. In: Hatfield DL, Su D, Tsuji PA, Gladyshev VN (eds) Selenium: its molecular biology and role in human health, 4th edn. Springer Science +Business Media, LLC, New York, pp 539–551
2. Sonet J, Bulteau A-L, Chavatte L (2016) Selenium and Selenoproteins in human health and diseases. In: Michalke B (ed) Metallomics: analytical techniques and speciation methods. Wiley-VCH Verlag GmbH & Co KGaA, Weinheim, pp 364–381. doi:10.1002/9783527694907.ch13
3. Labunskyy VM, Hatfield DL, Gladyshev VN (2014) Selenoproteins: molecular pathways and physiological roles. Physiol Rev 94 (3):739–777. doi: 94/3/739 [pii] 10.1152/physrev.00039.2013
4. Latrèche L, Chavatte L (2008) Selenium incorporation into selenoproteins, implications in human health. Met Ions Biol Med X:731–737
5. Papp LV, Lu J, Holmgren A, Khanna KK (2007) From selenium to selenoproteins: synthesis, identity, and their role in human health. Antioxid Redox Signal 9(7):775–806
6. Kryukov GV, Castellano S, Novoselov SV, Lobanov AV, Zehtab O, Guigo R, Gladyshev VN (2003) Characterization of mammalian selenoproteomes. Science 300 (5624):1439–1443
7. Gladyshev VN, Arner ES, Berry MJ, Brigelius-Flohe R, Bruford EA, Burk RF, Carlson BA, Castellano S, Chavatte L, Conrad M, Copeland PR, Diamond AM, Driscoll DM, Ferreiro A, Flohe L, Green FR, Guigo R, Handy DE, Hatfield DL, Hesketh J, Hoffmann PR, Holmgren A, Hondal RJ, Howard MT, Huang K, Kim H-Y, Kim IY, Kohrle J, Krol A, Kryukov GV, Lee BJ, Lee BC, Lei XG, Liu Q, Lescure A, Lobanov AV, Loscalzo J, Maiorino M, Mariotti M, Sandeep Prabhu K, Rayman MP, Rozovsky S, Salinas G, Schmidt EE, Schomburg L, Schweizer U, Simonovic M, Sunde RA, Tsuji PA, Tweedie S, Ursini F, Whanger PD, Zhang Y (2016) Selenoprotein gene nomenclature. J Biol Chem 291(46):24036–24040
8. Bulteau A-L, Chavatte L (2015) Update on selenoprotein biosynthesis. Antioxid Redox Signal 23(10):775–794
9. Touat-Hamici Z, Legrain Y, Bulteau AL, Chavatte L (2014) Selective up-regulation of human selenoproteins in response to oxidative stress. J Biol Chem 289(21):14750–14761. doi: M114.551994 [pii] 10.1074/jbc.M114.551994
10. Legrain Y, Touat-Hamici Z, Chavatte L (2014) Interplay between selenium levels, selenoprotein expression, and replicative senescence in WI-38 human fibroblasts. J Biol Chem 289 (9):6299–6310
11. Latreche L, Duhieu S, Touat-Hamici Z, Jean-Jean O, Chavatte L (2012) The differential expression of glutathione peroxidase 1 and 4 depends on the nature of the SECIS element. RNA Biol 9(5):681–690
12. Latreche L, Jean-Jean O, Driscoll DM, Chavatte L (2009) Novel structural determinants in human SECIS elements modulate the translational recoding of UGA as selenocysteine. Nucleic Acids Res 37(17):5868–5880

13. Driscoll DM, Copeland PR (2003) Mechanism and regulation of selenoprotein synthesis. Annu Rev Nutr 23:17–40
14. Bianga J, Touat-Hamici Z, Bierla K, Mounicou S, Szpunar J, Chavatte L, Lobinski R (2014) Speciation analysis for trace levels of selenoproteins in cultured human cells. J Proteome 108:316–324. doi:10.1016/j.jprot.2014.05.025
15. Bianga J, Ballihaut G, Pecheyran C, Touat Z, Preud'homme H, Mounicou S, Chavatte L, Lobinski R, Szpunar J (2012) Detection of selenoproteins in human cell extracts by laser ablation-ICP MS after separation by polyacrylamide gel electrophoresis and blotting. J Anal Atom Spectrom 27(1):25–32
16. Wessel D, Flugge UI (1984) A method for the quantitative recovery of protein in dilute-solution in the presence of detergents and lipids. Anal Biochem 138(1):141–143. doi:10.1016/0003-2697(84)90782-6

Chapter 16

Imaging of Selenium by Laser Ablation Inductively Coupled Plasma Mass Spectrometry (LA-ICP-MS) in 2-D Electrophoresis Gels and Biological Tissues

Elisa Castañeda Santa Cruz, J. Susanne Becker, J. Sabine Becker, and Alessandra Sussulini

Abstract

Selenium and selenoproteins are important components of living organisms that play a role in different biological processes. Laser ablation inductively coupled plasma mass spectrometry (LA-ICP-MS) is a powerful analytical technique that has been employed to obtain distribution maps of selenium in biological tissues in a direct manner, as well as in selenoproteins, previously separated by their molecular masses and isoelectric points using two-dimensional polyacrylamide gel electrophoresis (2-D PAGE). In this chapter, we present the protocols to perform LA-ICP-MS imaging experiments, allowing the distribution visualization and determination of selenium and/or selenoproteins in biological systems.

Key words Selenium, Selenoproteins, LA-ICP-MS, Mass spectrometry imaging, Bioimaging, Gel electrophoresis

1 Introduction

Recent studies revealed that the risk of different kinds of cancers, such as colorectal, liver, prostate, and lung cancer, could be reduced by selenium supplementation in diet [1–4]. This can be achieved by staple food, since selenium is associated with proteins in animal tissues, such as muscle, organ meats and seafood, and selenium is also present in grains and seeds [5–7].

The proteins that contain selenium are classified into three groups. The first group is named selenoproteins, which are in the form of genetically encoded selenocysteine (SeCys). In a second group, selenium is incorporated nonspecifically into proteins in the place of methionine by replacing sulfur and those proteins are termed selenium-containing proteins; therefore, this category is mainly found in methionine-rich proteins. The third group comprises the specific proteins in which selenium is only attached to the molecule

Laurent Chavatte (ed.), *Selenoproteins: Methods and Protocols*, Methods in Molecular Biology, vol. 1661,
DOI 10.1007/978-1-4939-7258-6_16, © Springer Science+Business Media LLC 2018

[7–9]. A total of 25 selenoprotein genes have been identified in humans, some of them produced proteins that have been characterized as antioxidant enzymes, mitigating the damage from reactive oxygen species (ROS) [10]. It has been suggested that some plants can utilize specific selenium incorporation to produce selenoproteins, being essential to plant's growth or survival [11, 12].

In proteomics research, different studies were developed in the last decade to characterize selenium-containing proteins using different experimental procedures, including chromatographic and mass spectrometric techniques [13]. Some examples include mass spectrometry imaging (MSI) for studying phospho- and metal-containing proteins in neurodegenerative diseases of different brain tissue regions using laser ablation inductively coupled plasma mass spectrometry (LA-ICP-MS) in combination with matrix-assisted laser desorption/ionization time-of-flight mass spectrometry (MALDI-TOF-MS) [14]. Another application example involves the determination of selenium distribution using LA-ICP-MS in proteins of African catfish fillets studying the soluble proteins separated by one-dimensional sodium dodecyl sulfate polyacrylamide gel electrophoresis (SDS-PAGE) and nano-liquid chromatography (nanoLC) coupled to electrospray-tandem mass spectrometry (ESI-MS/MS) [15]. Selenium storage proteins in wheat were also determined by using 2-D gel electrophoresis and LA-ICP-MS followed by capillary high performance liquid chromatography (HPLC) coupled to ICP-MS and electrospray linear trap quadrupole (LTQ)-Orbitrap MS [16].

Selenium can also be directly measured on thin sections of biological tissues. One example is the analysis of slugs (genus Arion) performed to identify regions containing high amount of selenium after treatment of the slugs with selenium-enriched water and/or lettuce leaves [17]. Samples from selenium-treated slugs were compared with samples from slugs not treated with selenium by LA-ICP-MS using a big ablation chamber for the measurement. The highest amount of selenium was detected in the gut and in the skin of the selenium-treated slugs, as well as in the digestive gland. The developed analytical technique [18] allowed the quantitative imaging of selenium together with selected metals in thin sections of biological tissue with limits of detection at the sub-microgram per gram range. From the selenium-enriched slug tissues, sample extraction was performed for following 2-D gel electrophoresis separation. LA-ICP-MS measurement of the dried gels did not show any proteins containing selenium from these samples. Future studies were planned to be performed with longer selenium treatment of slugs.

Other recent applications for selenium/selenoproteins determination by LA-ICP-MS imaging can be found in Table 1.

For mass spectrometry imaging (MSI), the main ion sources employed are MALDI and DESI (desorption electrospray

Table 1
Recent applications for selenium/selenoprotein determination by LA-ICP-MS imaging

Analyte	Sample	References
Selenium	Animal tissue	[44, 45]
	Plant tissue	[16]
Selenoproteins	Plant tissue	[16, 36, 46]
	Yeast	[47]
	Animal tissue	[48]

ionization) for biomolecular imaging, and SIMS (secondary ion mass spectrometry) and ICP for elemental bioimaging. LA-ICP-MS is being considered an important tool for different applications in life sciences [19–22], since it has become routine for the analysis of different metals, semimetals, and non-metals related to biomolecules in various diseases [23–25]. This technique is also widely used because it provides the spatial distribution in situ of a variety of analytes at ultratrace level directly on the surface of tissue samples without any labeling or staining agents and with high analytical throughput, minimal sample preparation, and reduced mass spectral interferences [19, 26].

According to the literature, the mostly used laser for bioimaging is the Nd:YAG, working at wavelengths of 266 or 213 nm [27]. The two mass spectrometers that can be used for elemental bioimaging are quadrupole-based ICP-MS (ICP-QMS) equipped with a collision cell, and double-focusing sector field ICP-MS (ICP-SFMS) equipped with single ion collector [27, 28]. ICP-SFMS offers a variable slit system so that the mass resolution ($m/\Delta m$ from 300 to ~10,000) can be optimized for specific applications. Moreover, it has the advantages of improving the precision of isotopic analysis in low resolution ($m/\Delta m$ ~300) and showing higher element sensitivity than ICP-QMS, with very low detection limits (down to the pg/L concentration range) [17, 29, 30]. As the cost of this equipment is very high, ICP-QMS equipped with a collision cell using He as a nonreactive gas is a very useful option. The collision cell provides effective removal of the interferences such as isobaric overlap from a second element with an isotope at the same nominal mass or from polyatomic species at the same nominal mass as the target analyte [31].

In the following sections, we present general protocols for selenium or selenoproteins determination by LA-ICP-MSI directly in biological tissues or in proteins previously separated by 2-D gel electrophoresis. Figure 1 shows a scheme of the experimental analytical workflow for LA-ICP-MSI.

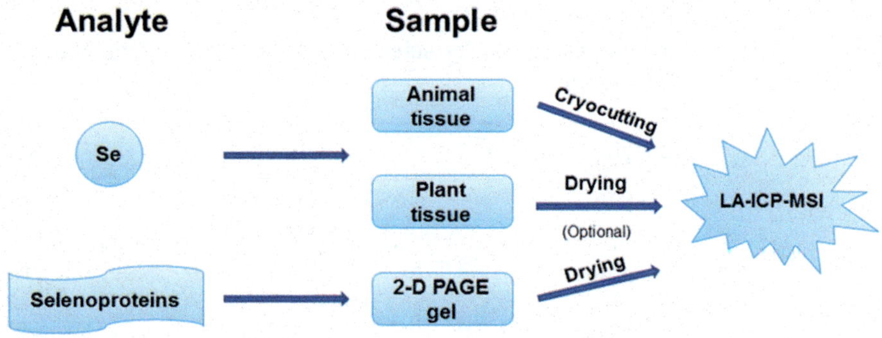

Fig. 1 Experimental analytical workflow for bioimaging of selenium and selenoproteins

2 Materials

1. Quadrupole (Q) or sector field (SF) inductively coupled plasma mass spectrometer coupled with a laser ablation system (Nd:YAG laser).
2. Certified reference material (CRM), laboratory made matrix-matched standards or standard solutions for calibration.
3. Milli-Q water purification system.
4. Cryostat microtome.
5. Microwave oven.
6. Acetate double-sided adhesive tape.
7. Gel electrophoresis apparatus.
8. 3MM Whatman chromatography paper.
9. Gel drying apparatus.
10. Imaging software.

3 Methods

3.1 Imaging of Selenoproteins in Two-Dimensional Electrophoresis Gels

1. The proteins extracted from tissue or biofluid samples are separated using native 2-D gel electrophoresis. The first dimension, isoelectric focusing (IEF), is carried out using immobilized pH gradient (IPG) strips of pI 3–10NL (nonlinear) or pI 4–7 (*see* **Note 1**).
2. The IPG strips are rehydrated overnight by passive rehydration using a solution containing the extracted proteins and a buffer composed of 8 M urea, 2% (m/v) CHAPS, 40% (v/v) carrier ampholytes, and 1% (m/v) bromophenol blue.
3. After rehydration, IPG strips are placed in an IEF chamber used together with the power supply. The system is cooled to 10 °C and the temperature is kept constant during the run.

4. After IEF, equilibration for reduction and alkylation of the proteins is omitted.
5. The second dimension is performed in a cooled chamber. 10% acrylamide: bis-acrylamide (30%: 0.8% m/v) gels are generally used for protein separation by molecular mass [21].
6. The gel is dried on 3MM Whatman chromatography paper under vacuum at 80 °C in a gel dryer apparatus.
7. LA-ICP-MS imaging is performed by moving the dried gel mounted on the target holder in the laser ablation chamber and applying the focused laser beam line by line in a way that the selected gel fragment is completely ablated.
8. Laser ablation conditions should be optimized for the maximum sensitivity using a laboratory-made standard gel. The most common conditions are the following: spot size 250 μm, repetition rate 20 Hz, output energy 100%, fluence 7 J/cm^2. Scan speed is usually adjusted between 100 and 50 μm/s [16] (*see* **Note 2**).
9. Images of selenium distribution in proteins are constructed using an appropriate software (*see* **Note 3**).

3.2 Selenium Imaging in Biological Tissues

1. Samples commonly are frozen in low temperature conditions, either inside an ultra-low temperature freezer, or in powdered dry ice, liquid nitrogen, or liquid nitrogen-chilled isopentane [32] (*see* **Note 4**).
2. For plants, the thickness in current LA-ICP-MS imaging studies is about 50 μm, providing a good compromise between optimum imaging performance and practicality [33].
3. For mammalian tissues: section thickness between 5 and 20 μm is recommended for analysis of low molecular mass molecules and <5 μm thickness for high molecular mass proteins (m/z >9000) [34, 35] (*see* **Note 5**).
4. Samples can be fixed onto acetate double-sided adhesive tape or on glass slides, using glutaraldehyde and placed into the ablation chamber for analysis [36] (*see* **Note 6**).
5. For the blank correction, tissue pieces of the same sample are spiked with the selected standard solution [37].
6. The determination of selenium free of interferences is based on the reaction of Se^+ species formed in the plasma with oxygen in the reaction cell of the instrument. Therefore, the monitored species consist of the product ion SeO^+ at m/z 96. To compensate for the possible variations in the ablation process, $^{12\text{-}13}C^+$ is chosen as the internal standard to normalize the analytical signals [36].
7. Tissue sections are ablated line by line (the whole cross section is ablated) by approximately 300 μm between dots, and it is set the total time necessary for one section ablation.

8. To obtain two-dimensional (2-D) images, all sample section fixed onto the glass slide is scanned. ICP-MS data files for each sample section are converted into "txt"-files and transformed in 2-D images of element distribution. The images are plotted using an appropriate imaging software [38] (*see* **Note 3**).

4 Notes

1. Denaturating detergents, such as DTT (dithiothreitol) and SDS (sodium dodecyl sulfate), are preferentially avoided in protein extraction.
2. The measurement time for LA-ICP-MS imaging (up to several hours) depends on the size of the gel or tissue section analyzed, the number of lines set, and the scan speed applied [21].
3. LA-ICP-MS images can be constructed using MATLAB [39], Microsoft Excel [40], or freely available softwares, like IMAGENA [41] and LA-iMageS [42].
4. This kind of storage can help the sectioning procedure using a cryostat. For plant samples, drying can be performed in a microwave oven and sectioning is made at room temperature using a microtome [43].
5. Frozen mammalian tissues can be partially embedded in OCT (optimal cutting temperature) compound and further sectioned in a cryostat at approximately −20 °C. This compound helps to hold the sample in the cryostat during the cutting process, but it should not be in contact with the sectioned sample to be analyzed, in order to avoid any polymer contamination.
6. For tissues sectioned in a cryostat, this step is not necessary, since the sliced frozen sample adheres to the glass slide after a soft heating (placing heated fingers under the slide and incubating at room temperature). Mounted sections can be further stored at −80 °C until analysis.

Acknowledgments

A.S. thanks FAPESP (grant number 2015/13229-1) for financial support. E.C.S.C. thanks CNPq for her Ph.D. scholarship.

References

1. Rayman MP (2005) Selenium in cancer prevention: a review of the evidence and mechanism of action. Proc Nutr Soc 64:527–542
2. Hawk ET, Levin B (2005) Colorectal cancer prevention. J Clin Oncol 23:378–391
3. Irons R, Tsuji PA, Carlson BA, Ouyang P, Yoo MH, XM X, Hatfield DL, Gladyshev VN, Davis CD (2010) Deficiency in the 15-kDa selenoprotein inhibits tumorigenicity and metastasis of colon cancer cells. Cancer Prev Res 3:630–639

4. Yagublu V, Arthur JR, Babayeva SN, Nicol F, Post S, Keese M (2011) Expression of selenium-containing proteins in human colon carcinoma tissue. Anticancer Res 31:2693–2698
5. Holben DH, Smith AM (1999) The diverse role of selenium within selenoproteins: a review. J Am Diet Assoc 99:836–843
6. Palace VP, Halden NM, Yang P, Evans RE, Sterling G (2007) Determining residence patterns of rainbow trout using laser ablation inductively coupled plasma mass spectrometry (LA-ICP-MS) analysis of selenium in otoliths. Environ Sci Technol 41:3679–3683
7. Jayasinghe SB, Caruso JA (2011) Investigation of Se-containing proteins in Bertholletia excelsa H.B.K. (Brazil nuts) by ICPMS, MALDI-MS and LC–ESI-MS methods. Int J Mass Spectrom 307:16–27
8. Bansal MP, Cook RG, Danielson KG, Medina D (1989) A 14-kilodalton selenium-binding protein in mouse-liver is fatty acid-binding protein. J Biol Chem 264:13780–13784
9. Gladyshev VN, Khangulov SV, Stadtman TC (1994) Nicotinic acid hydroxylase from Clostridium barkeri: electron paramagnetic resonance studies show that selenium is coordinated with molybdenum in the catalytically active selenium-dependent enzyme. Proc Natl Acad Sci U S A 91:232–236
10. Reeves MA, Hoffmann PR (2009) The human selenoproteome: recent insights into functions and regulation. Cell Mol Life Sci 66:2457–2478
11. Fu L-H, Wang X-F, Eyal Y, She Y-M, Donald LJ, Standing KG, Ben-Hayyim G (2002) A Selenoprotein in the plant kingdom. Mass spectrometry confirms that an opal codon (UGA) encodes selenocysteine in Chlamydomonas reinhardtii gluththione peroxidase. J Biol Chem 277:25983–25991
12. Novoselov SV, Rao M, Onoshko NV, Zhi HJ, Kryukov GV, Xiang YB, Weeks DP, Hatfield DL, Gladyshev VN (2002) Selenoproteins and selenocysteine insertion system in the model plant cell system, Chlamydomonas reinhardtii. EMBO J 21:3681–3693
13. Mangiapane E, Pessione A, Pessione E (2014) Selenium and Selenoproteins: an overview on different biological systems. Curr Protein Pept Sci 15:598–607
14. Becker JS, Becker JS, Zoriy MV, Dobrowolska J, Matusch A (2007) Imaging mass spectrometry in biological tissues by laser ablation inductively coupled plasma mass spectrometry. Eur J Mass Spectrom 13:1–6
15. Pedrero Z, Madrid Y, Camara C, Schram E, Luten JB, Feldmann I, Waentig L, Hayen H, Jakubowski N (2009) Screening of selenium containing proteins in the tris-buffer soluble fraction of African catfish (Clarias gariepinus) fillets by laser ablation-ICP-MS after SDS-PAGE and electroblotting onto membranes. J Anal Atom Spectrom 24:775–784
16. Bianga J, Govasmark E, Szpunar J (2013) Characterization of selenium incorporation into wheat proteins by two-dimensional gel electrophoresis-laser ablation ICP MS followed by capillary HPLC-ICP MS and electrospray linear trap Quadrupole Orbitrap MS. Anal Chem 85:2037–2043
17. Becker JS, Matusch A, Depboylu C, Dobrowolska J, Zoriy MV (2007) Quantitative imaging of selenium, copper, and zinc in thin sections of biological tissues (slugs–genus Arion) measured by laser ablation inductively coupled plasma mass spectrometry. Anal Chem 79:6074–6080
18. Becker JS, Zoriy M, Becker JS, Dobrowolska J, Matusch A (2007) Laser ablation inductively coupled plasma mass spectrometry (LA-ICP-MS) in elemental imaging of biological tissues and in proteomics. J Anal Atom Spectrom 22:736–744
19. Hanc A, Malecka A, Kutrowska A, Bagniewska-Zadworna A, Tomaszewska B, Baralkiewicz D (2016) Direct analysis of elemental biodistribution in pea seedlings by LA-ICP-MS, EDX and confocal microscopy: imaging and quantification. Microchem J 128:305–311
20. Hanć A, Piechalak A, Tomaszewska B, Barałkiewicz D (2014) Laser ablation inductively coupled plasma mass spectrometry in quantitative analysis and imaging of plant's thin sections. Int J Mass Spectrom 363:16–22
21. Becker JS, Lobinski R, Becker JS (2009) Metal imaging in non-denaturating 2D electrophoresis gels by laser ablation inductively coupled plasma mass spectrometry (LA-ICP-MS) for the detection of metalloproteins. Metallomics 1:312–316
22. Konz I, Fernandez B, Fernandez ML, Pereiro R, Sanz-Medel A (2012) Laser ablation ICP-MS for quantitative biomedical applications. Anal Bioanal Chem 403:2113–2125
23. Clark LC, Combs GF Jr, Turnbull BW, Slate EH, Chalker DK, Chow J, Davis LS, Glover RA, Graham GF, Gross EG, Krongrad A, Lesher JL Jr, Park HK, Sanders BB Jr, Smith CL, Taylor JR (1996) Effects of selenium supplementation for cancer prevention in patients with carcinoma of the skin: a randomized controlled trial. JAMA 276:1957–1963
24. M-M P, Merle U, Weiskirchen R, Becker JS (2013) Bioimaging of copper deposition in Wilson's diseases mouse liver by laser ablation

inductively coupled plasma mass spectrometry imaging (LA-ICP-MSI). Int J Mass Spectrom 354–355:281–287

25. Zoriy MV, Dehnhardt M, Matusch A, Becker JS (2008) Comparative imaging of P, S, Fe, Cu, Zn and C in thin sections of rat brain tumor as well as control tissues by laser ablation inductively coupled plasma mass spectrometry. Spectroc Acta Part B 63:375–382
26. Matusch A, Depboylu C, Palm C, Wu B, Höglinger GU, Schäfer MKH, Becker JS (2010) Cerebral Bioimaging of Cu, Fe, Zn, and Mn in the MPTP mouse model of Parkinson's disease using laser ablation inductively coupled plasma mass spectrometry (LA-ICP-MS). J Am Soc Mass Spectrom 21:161–171
27. Becker JS, Matusch A, Wu B (2014) Bioimaging mass spectrometry of trace elements—recent advance and applications of LA-ICP-MS: a review. Anal Chim Acta 835:1–18
28. Rubakhin SS, Sweedler JV (eds) (2010) Mass spectrometry imaging: principles and protocols. Humana Press, New York, NY
29. Becker JS, Dietze H-J (1998) Ultratrace and precise isotope analysis by double-focusing sector field inductively coupled plasma mass spectrometry. J Anal Atom Spectrom 13:1057–1063
30. Becker JS (2007) Inorganic mass spectrometry: principles and applications. Wiley, New York, NY
31. McCurdy E, Woods G (2004) The application of collision/reaction cell inductively coupled plasma mass spectrometry to multi-element analysis in variable sample matrices, using He as a non-reactive cell gas. J Anal Atom Spectrom 19:607–615
32. Bong Y, Li B, Malitsky S, Rogachev L, Aharoni A, Kaftan F, Svatos A, Franceschi P (2016) Sample preparation for mass spectrometry imaging of plant tissues: a review. Front Plant Sci 7:16
33. Peukert M, Matros A, Lattanzio G, Kaspar S, Abadia J, Mock HP (2012) Spatially resolved analysis of small molecules by matrix-assisted laser desorption/ionization mass spectrometric imaging (MALDI-MSI). New Phytol 193:806–815
34. Sugiura Y, Shimma S, Setou M (2006) Thin sectrioning improves the peak intensity and signal-to-noise ratio in direct tissue mass spectrometry. J Mass Spectrom Soc Jpn 54:45–48
35. Becker JS (2010) Bioimaging of metals in brain tissue from micrometre to nanometre scale by laser ablation inductively coupled plasma mass spectrometry: state of the art and perspectives. Int J Mass Spectrom 289:65–75
36. da Silva MAO, Arruda MAZ (2013) Laser ablation (imaging) for mapping and determining Se and S in sunflower leaves. Metallomics 5:62–67
37. Dobrowolska J, Dehnhardt M, Matusch A, Zoriy M, Palomero-Gallagher N, Koscielniak P, Zilles K, Becker JS (2008) Quantitative imaging of zinc, copper and lead in three distinct regions of the human brain by laser ablation inductively coupled plasma mass spectrometry. Talanta 74:717–723
38. Becker JS, Dietrich RC, Matusch A, Pozebon D, Dressler VL (2008) Quantitative images of metals in plant tissues measured by laser ablation inductively coupled plasma mass spectrometry. Spectroc Acta Part B 63:1248–1252
39. Zoriy MV, Dehnhardt M, Reifenberger G, Zilles K, Becker JS (2006) Imaging of Cu, Zn, Pb and U in human brain tumor resections by laser ablation inductively coupled plasma mass spectrometry. Int J Mass Spectrom 257:27–33
40. Uerlings R, Matusch A, Weiskirchen R (2016) Reconstruction of laser ablation inductively coupled plasma mass spectrometry (LA-ICP-MS) spatial distribution images in Microsoft Excel 2007. Int J Mass Spectrom 395:27–35
41. Osterholt T, Salber D, Matusch A, Becker JS, Palm C (2011) IMAGENA: image generation and analysis—an interactive software tool handling LA-ICP-MS data. Int J Mass Spectrom 307:232–239
42. LA-iMageS Software. http://www.la-images.net/index.html. Accessed Nov 2016
43. Lunsford KA, Peter GF, Yost RA (2011) Direct matrix-assisted laser desorption/ionization mass spectrometric imaging of cellulose and hemicellulose in Populus tissue. Anal Chem 83:6722–6730
44. Bishop DP, Clases D, Fryer F, Williams E, Wilkins S, Hare DJ, Cole N, Karst U, Doble PA (2016) Elemental bio-imaging using laser ablation-triple quadrupole-ICP-MS. J Anal Atom Spectrom 31:197–202
45. Barst BD, Bridges K, Korbas M, Roberts AP, Van Kirk K, McNeel K, Drevnick PE (2015) The role of melano-macrophage aggregates in the storage of mercury and other metals: an example from yelloweye rockfish (Sebastes ruberrimus). Environ Toxicol Chem 34:1918–1925
46. Chan Q, Afton SE, Caruso JA (2010) Selenium speciation profiles in selenite-enriched soybean (Glycine Max) by HPLC-ICPMS and ESI-ITMS. Metallomics 2:147–153
47. Bierla K, Bianga J, Ouerdane L, Szpunar J, Yiannikouris A, Lobinski R (2013) A

comparative study of the Se/S substitution in methionine and cysteine in Se-enriched yeast using an inductively coupled plasma mass spectrometry (ICP MS)-assisted proteomics approach. J Proteomics 87:26–39

48. Kyriakopoulos A, Rothlein D, Pfeifer H, Bertelsmann H, Kappler S, Behne D (2000) Detection of small selenium-containing proteins in tissues of the rat. J Trace Elem Med Biol 14:179–183

Part IV

Selenoprotein Overexpression, Purification, and Activity

Chapter 17

Overexpression of Recombinant Selenoproteins in *E. coli*

Qing Cheng and Elias S.J. Arnér

Abstract

Expression of selenoproteins necessitates a process of decoding of a UGA codon from termination of translation to insertion of selenocysteine. The mechanisms of this process pose major challenges with regards to recombinant selenoprotein production in *E. coli*, which however can be overcome especially if the Sec residue is located close to the C-terminal end, as is the case for several naturally found selenoproteins. This chapter summarizes a method to achieve such a production.

Key words Selenocysteine, Selenoprotein, Recombinant, SelB, RF2 (prfB), SECIS element, Thioredoxin reductase, TGR

1 Introduction

Selenoproteins are found in all kingdoms of life, albeit not in all organisms [1–7], containing at least one co-translationally inserted selenocysteine (Sec) residue. The Sec residue is incorporated at the position of a UGA codon that needs to be redefined from its normal role of terminating translation (in *E. coli* by the action of Release Factor 2, RF2, encoded by *prfB*) to a sense codon for Sec. This process requires a Sec-dedicated elongation factor, a unique tRNA species originally acylated with Ser that is converted to Sec as it is bound to the tRNA, accessory factors facilitating this process, and a secondary structure in the selenoprotein-encoding mRNA named a Sec Insertion Sequence (SECIS) element. These overall features are shared between all species expressing selenoproteins [2, 5, 8–13]. However, the exact nature of the factors involved differs between different domains of life and the *E. coli* selenoprotein synthesis machinery is not compatible with the mammalian types of SECIS elements. This makes it impossible to express a mammalian selenoprotein-encoding gene directly in *E. coli*. Several methods have been developed to overcome this hurdle, including the use of native chemical ligation [14] or directed evolution of the Sec-specific tRNA and/or elongation factors [8, 15–17]. The design of

Laurent Chavatte (ed.), *Selenoproteins: Methods and Protocols*, Methods in Molecular Biology, vol. 1661,
DOI 10.1007/978-1-4939-7258-6_17, © Springer Science+Business Media LLC 2018

an engineered SECIS element compatible with the *E. coli* selenoprotein synthesis machinery has also been successfully used to express mammalian or parasitic thioredoxin reductase variants as well as other proteins containing a Sec residue close to the C-terminal end [18–25]. The methodology and prerequisites for such expression have been described elsewhere in further detail [21, 23, 25–27] and are in the present chapter outlined in the form of a simple step-by-step method, mainly adapted from the approaches described in [21]. Note that this method for recombinant selenoprotein production in *E. coli* may prove to become further improved as several groups are currently working on such improvements.

2 Materials

Buffers, media, *E. coli* cloning and expression strains, plasmids, chemicals, and other materials needed for conventional recombinant protein synthesis can typically be used also for recombinant selenoprotein expression.

2.1 Reagents

1. LB medium and LB agar plates with adequate antibiotics: Carbenicillin (or Ampicillin) and Kanamycin were prepared in distilled water at 50 mg/ml, and the solution was sterilized by passing through a sterile syringe filter with 0.22 μm pore size hydrophilic PVDF membrane. Chloramphenicol was prepared in 95% ethanol at 34 mg/ml. Aliquot and store the antibiotics solutions at −20 °C. For liquid LB medium, dissolve 10 g peptone, 5 g yeast extract, 10 g NaCl in 1 l distilled water. Add additional 15 g agar to 1 l LB medium to prepare the medium for casting LB agar plates. Utilize the standard autoclave procedure to sterile the medium. To prepare the LB agar plates, wait for the medium's temperature drop down to about 50 °C and then add desired antibiotics with 1000-fold dilution into the medium with gentle stir, try to avoid generating foam. Pour about 20 ml medium to a sterilized 9 cm petri dish to cast the LB agar plate. After the medium solidified, store the plates at 4 °C and use these plates within 2 months. Carbenicillin is recommended to be used over Ampicillin for making the LB agar plates for its much longer shelf life at 4 °C. For liquid LB medium, add corresponding antibiotics with 1000-fold dilution right before starting the bacterial culture. Ampicillin can be used here instead of Carbenicillin for economic purposes.
2. Competent *E. coli* cells: Chemically competent *E. coli* BL21 (DE3) cells can be either commercially purchased or prepared in house using standard method. We routinely use Mix & Go *E. coli* Transformation Kit & Buffer Set (Zymo Research, T3001) to prepare highly competent *E. coli* cells.

3. L-cysteine: prepare L-cysteine solution in water at 100 mg/ml and sterilize it using a syringe filter as previously described. Aliquot and store the solution at −20 °C. Do not store the solution at 4 °C for it is easily oxidized followed by precipitation. Dilute the solution 1000-fold upon usage.
4. Selenite: prepare selenite solution in water at 5 mM and sterilize it using a syringe filter as previously described. Aliquot and store the solution at −20 °C. Dilute the solution 1000-fold upon usage.
5. Radioactive selenite. For autoradiography of ^{75}Se-labeled selenoproteins, [^{75}Se]-selenite can be obtained from the Research Reactor Center, University of Missouri, Columbia, USA. Selenium-75 is a radioactive gamma emitter with a 119.8-day half-life. The working concentration is 0.1–0.5 μCi/ml. Follow local regulation on radioactive substance usage to handle and utilize this solution.
6. IPTG: Prepare IPTG solution in water at 1 M and sterilize it using a syringe filter as previously described. Aliquot and store the solution at −20 °C. Dilute the solution 1000-fold upon usage.
7. Tris stock solution: Dissolve 121 g Tris in 800 ml distilled water, adjust pH using 37% HCl to 7.5, compensate the solution to 1 l with distilled water to make 1 M stock solution.
8. EDTA stock solution: Dissolve 84 g EDTA disodium salt in 300 ml distilled water, adjust pH using 10 M NaOH to 8.0, compensate the solution to 500 ml with distilled water to make 0.5 M stock solution.
9. TE buffer: Combine 50 ml 1 M Tris stock solution, 4 ml 0.5 M EDTA stock solution, and 946 ml distilled water to prepare the TE buffer (50 mM Tris–HCl, 2 mM EDTA, pH 7.5).
10. Lysozyme solution: Dissolve 500 mg lysozyme in 10 ml TE buffer to prepare the 50 mg/ml stock solution. Aliquot and store the solution at −20 °C. Dilute 100-fold upon usage.
11. TSS solution: Combine 12.5 ml (40% w/v) PEG 6000, 2.5 ml DMSO, 2.5 ml 1 M $MgCl_2$, 32.5 ml LB medium to prepare 50 ml TSS solution Aliquot and store the solution at −20 °C. [28]
12. pSUABC. For improved selenoprotein yields, overexpression of the *E. coli selA, selB,* and *selC* genes using the pSUABC plasmid is recommended [21, 25].

2.2 Instrumentation

1. Shaking incubator set at 37 °C or room temperature.
2. Spectrophotometer that can be set at 463 and 600 nm for single wavelength measurement, and at 340 and 412 nm for continuous measurement (if mammalian-type Thioredoxin reductases are to be expressed. Other wavelength may be applied depending on the protein of interest).

3. Centrifuge: Conventional refrigerated centrifuge for harvesting bacterial cells (7000 rpm, typically 1–2 l scale, ≈6000 × *g*), and for clarifying bacterial lysate (18,000 rpm, typically 30–50 ml scale, ≈38000 × *g*). Conventional refrigerated benchtop centrifuge for routine centrifuge in Eppendorf tubes.
4. SDS-PAGE system: Any available SDS-PAGE system can be utilized. We routinely use Bolt™ Bis-Tris Plus Mini Gels system from Thermo Fisher Scientific.
5. Autoradiography imager: Typhoon FLA 7000 from GE Healthcare or any equivalent imager can be utilized. Alternatively, X-ray film based blotting system can also be used.

3 Methods

3.1 Design and Validation of a SECIS Element Compatible with E. coli Selenoprotein Production

1. A gene of interest to be expressed as a selenoprotein in *E. coli* following the method described herein first needs to be equipped with a SECIS element compatible with the selenoprotein synthesis machinery of the host cell. Although this has been accomplished for production of selenoproteins with internally located Sec residues, such as a Sec-substituted GST [29], the requirements of the SECIS element typically yield higher yields and a native sequence of the expressed selenoprotein if the Sec residue is located close to the C-terminus. If this is to be obtained, the following criteria typically apply, although additional SECIS variants may also prove to be functional [10]:
   (a) Incorporation of a UGA at the position encoding a penultimate Sec residue, followed by the codon for the last amino acid (typically Gly in mammalian thioredoxin reductases [27] or Sel-tagged proteins [21, 30]), subsequently followed by the genuine termination codon (typically UAA).
   (b) Maintenance of the SelB binding loop region of the SECIS element, consisting of 17 conserved nucleotides, positioned 11 nt away on the 3′-side of the Sec-encoding UGA [22, 27].

   For an illustration of a SECIS element following these criteria, *see* Fig. 1.

2. When the SECIS element of choice has been designed and incorporated into the gene encoding the required selenoprotein to be expressed in *E. coli* (following classical cloning methods not described here), the functionality can best be validated using ^{75}Se-labeling using a small-scale expression from a classical expression plasmid, such as the pET series of vectors (Promega, *see* the manufacturer's manual for use). For this optional validation of selenoprotein expression, no co-transformation

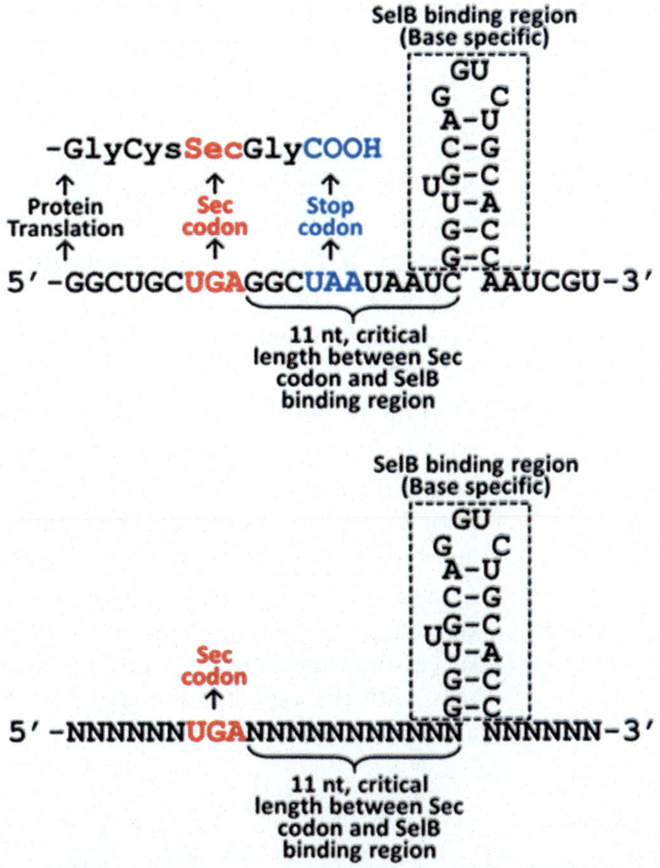

Fig. 1 The top scheme shows the mRNA structure of a SECIS element compatible with Sec insertion in *E. coli* and expression of proteins with the C-terminal motif –GlyCysSecGlyCOOH, such as found in thioredoxin reductases or Sel-tagged proteins. Note that the SelB binding region (*dashed box*) needs to be maintained in terms of both sequence and 11 nucleotide distance from the UGA, while the 11 nucleotides following the UGA codon are flexible in terms of sequence. The lower scheme indicates the requirements of an engineered SECIS element to be functional in *E. coli* to support Sec insertion at the UGA codon, with "N" indicating any nucleotide. Please note, however, that also in the SelB binding region there can be some additional flexibility in terms of exact sequence. See the main text for further information and references to original articles investigating these parameters

with pSUABC is needed to increase yields and no optimization of expression is typically required (*see* **Note 1**). Simply proceed with the following steps:

(a) Transform the plasmid for selenoprotein expression into a host cell compatible with the chosen vector, e.g., BL21 (DE3) for the pET vector series, and screen transformants on LB agar plates containing suitable antibiotics (e.g., 50 μg/ml kanamycin for pET21b). Duplicate colonies picked for expression analyses on a second plate before proceeding with the next step.

(b) Make a small-scale (1–2 ml) expression of the picked colonies using the transformed bacteria grown in classical LB medium containing 0.5–1 μCi [^{75}Se]-selenite.

(c) Separate proteins of whole bacterial lysates with SDS-PAGE (lyse the bacterial pellet directly into reducing SDS-PAGE sample buffer).

(d) Stain the SDS-PAGE gel with Coomassie blue or equivalent and take a picture of the gel before the next step.

(e) Seal the stained gel in a plastic bag and stabilize the bag on a piece of filter paper. Make sure the gel itself is physically secured.

(f) Outline the position of the gel by marking the filter paper with 1–2 μl radioactive sample taken from **step c**.

(g) Make an autoradiography of the marked gel using a Phosphor Screen (GE healthcare) or equivalent equipment for autoradiography, or using an X-ray film.

(h) Compare the autoradiography with the Coomassie blue-stained gel picture to confirm that the ^{75}Se has been successfully incorporated into the expressed selenoprotein, with the expected molecular weight. *See* Fig. 2. for an example autoradiography.

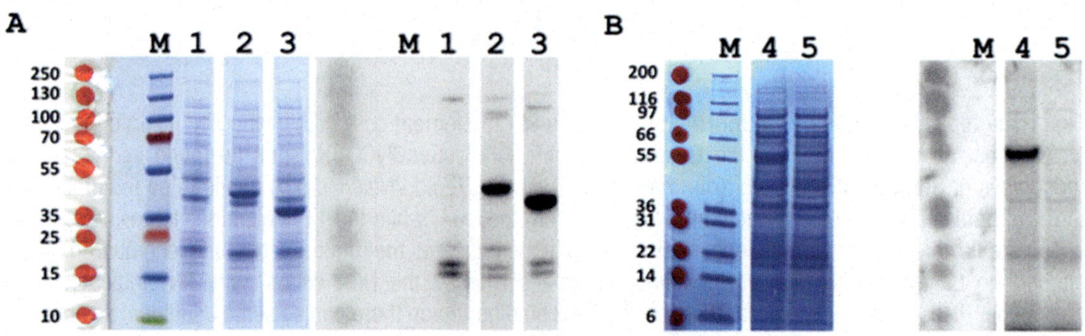

1. **Control with no IPTG induction**
2. **mTrx-GFP-ST (~ 39 kDa)**
3. **AnxA5-ST (~ 37 kDa)**
4. **Rat TrxR (~ 55 kDa)**
5. **Control with no IPTG induction**

Fig. 2 . These Coomassie-stained SDS-PAGE examples (left) with their corresponding autoradiography results (right) illustrate typical results from recombinant selenoprotein expression in *E. coli* and validation using ^{75}Se-labeling. In (**a**) are examples using Sel-tagged proteins functionally characterized elsewhere [19], with *lane 1*: negative control omitting IPTG induction, *lane 2*: Sel-tagged mutant thioredoxin fused to GFP (mTrx-GFP-ST), and *lane 3*: Sel-tagged annexin A5 (AnxA5-ST). In (**b**) examples are shown for expression of recombinant rat thioredoxin reductase (Rat TrxR) [33], using *lane 4*: IPTG induction, and *lane 5*: a negative control without IPTG induction. Shown in both (**a**) and (**b**) are also molecular weight markers (M) with their positions indicated using small aliquots of diluted ^{75}Se-solution and a *red marker pen* (*left*, size markers furthermore indicated in kDa) and as seen in the corresponding autoradiography (*right*)

3.2 Expression of a Recombinant Selenoprotein in *E. coli*

Once a SECIS element has been designed and validated following the steps described above under Subheading 3.1, the following protocol can be followed to achieve optimal yield of selenoprotein expression in *E. coli*, based upon that previously described for expression of variants of thioredoxin reductase or Sel-tagged proteins [21].

1. Grow the expression clone(s) obtained from Subheading 3.1. to turbidity ($OD_{600} = 0.3–0.4$) at 37 °C. Take 1 ml culture and centrifuge to pellet the bacteria, discard the supernatant.
2. For simple co-transformation, add 1 μl pSUABC plasmid (100 ng/μl) on the top of the bacterial pellet and then add 100 μl ice-cold TSS solution for TSS transformation [31] (*see* **Note 2**).
3. Let the tube stand at 4 °C for at least 30 min (not on ice).
4. Add 1 ml fresh LB medium and incubate for 1 h at 37 °C with vigorous shaking (250 rpm).
5. Screen transformants on an LB agar plate containing suitable antibiotics, e.g., both 50 μg/ml kanamycin (for pET 21a) and 34 μg/ml chloramphenicol (for pSUABC).
6. Bacterial clones expressing the target selenoprotein can again be identified and verified by small-scale expression followed by analysis using SDS-PAGE, with or without ^{75}Se-labeling (*see* above). Verified clones of transformed bacteria should be preserved at −80 °C as glycerol stocks.
7. Inoculate a 25 ml overnight culture (LB medium containing required antibiotics for selection, e.g., 50 μg/ml kanamycin or ampicillin for a pET vector with kanamycin or ampicillin resistance and 34 μg/ml chloramphenicol for pSUABC) with the glycerol stock bacteria obtained from **step** 7 at 37 °C in a shaking incubator (200 rpm).
8. Inoculate a 500 ml culture in LB medium containing required antibiotics with the 5 ml overnight culture obtained from **step** 7 at 37 °C in a shaking incubator (200 rpm).
9. Follow the growth rate of the bacterial culture by determining its OD_{600} once in an hour and plot on a linear OD_{600} over time diagram. When the culture growth rate slows down as it reaches late exponential phase (typically after about 7 h after inoculation and OD_{600} would reach 2.4 under culturing circumstance described above; assess this using the growth curve), cool the culture down to room temperature (approx. 20–24 °C) [26] (*see* **Note 3**).
10. Add L-cysteine, selenite and IPTG to the culture at final concentrations of 1 μM, 5 μM, and 0.5 mM, respectively, and continue growth overnight at room temperature in a shaking

incubator (200 rpm). To avoid rapid chemical reaction between concentrated L-cysteine and selenite (selenite being reduced to elementary selenium by L-cysteine), add them separately and mix the culture well in each addition of the chemicals (*see* **Note 4**).

11. Harvest the cells by centrifugation at 6000 × *g* (7000 rpm in a Sorvall GSA rotor) for 30 min at 4 °C.
12. Discard the medium and suspend the bacteria in 25 ml ice-cold TE buffer. If the protein is to be purified over a nickel column, use any suitable binding buffer for IMAC instead (typically: 100 mM Tris–HCl, 250 mM NaCl, 20 mM imidazole, pH 7.5).
13. Add lysozyme to a final concentration of 0.5 mg/ml and incubate the cell suspension at 4 °C for 45 min with end-over-end mixing. A couple of freeze and thaw procedure may help lysis bacteria intensively. Sonication or addition of DNAse may be applied if the sample is too viscous.
14. Centrifuge the sample at 4 °C, 38,000 × *g* (18,000 rpm in a Sorvall SS34 rotor) for 30 min to remove insoluble cell debris. The supernatant is then ready to be subjected to purification of the selenoprotein (*see* **Note 5**).
15. (Optional) Check the protein solubility by analyzing both whole cell lysate, supernatant and pellet with SDS-PAGE.

4 Notes

1. The pSUABC plasmid has a chloramphenicol resistance and p15a origin of replication, whereby it is preferable to choose an expression vector with another origin of replication (e.g., the pET vectors) and another resistance cassette, in order to allow for facile subsequent co-transformation with pSUABC.
2. The pSUABC plasmid is not essential, but typically increases the recombinant selenoprotein yield at least five- to sevenfold [23, 25–27].
3. To increase the yield of the recombinant selenoprotein, we have found it to be optimal to induce protein expression at late exponential phase during slow growth at lower temperature [26]. Induction at stationary phase, however, results in very low yield.
4. Addition of L-cysteine together with selenite to the growth medium prevents unspecific selenium incorporation into normally sulfur containing amino acids (L-Cys and L-Met) through the sulfur metabolism [32], thereby ensuring specific selenium incorporation into the Sec residue of the Sel-tag and only there.

5. Purification methods are not described here, as these will depend upon the choice of selenoprotein to be expressed.

Acknowledgments

The authors acknowledge funding to ESJA from The Swedish Cancer Society, The Swedish Research Council, Swedish Foundation for Strategic Research, Knut and Alice Wallenberg Foundation, and Karolinska Institutet.

References

1. Castellano S, Gladyshev VN, Guigo R, Berry MJ (2008) SelenoDB 1.0: a database of selenoprotein genes, proteins and SECIS elements. Nucleic Acids Res 36:D332–D338
2. Kryukov GV, Castellano S, Novoselov SV, Lobanov AV, Zehtab O, Guigo R, Gladyshev VN (2003) Characterization of mammalian selenoproteomes. Science 300:1439–1443
3. Kryukov GV, Gladyshev VN (2004) The prokaryotic selenoproteome. EMBO Rep 5:538–543
4. Lobanov AV, Fomenko DE, Zhang Y, Sengupta A, Hatfield DL, Gladyshev VN (2007) Evolutionary dynamics of eukaryotic selenoproteomes: large selenoproteomes may associate with aquatic life and small with terrestrial life. Genome Biol 8:R198
5. Lobanov AV, Hatfield DL, Gladyshev VN (2009) Eukaryotic selenoproteins and selenoproteomes. Biochim Biophys Acta 1790:1424–1428
6. Taskov K, Chapple C, Kryukov GV, Castellano S, Lobanov AV, Korotkov KV, Guigo R, Gladyshev VN (2005) Nematode selenoproteome: the use of the selenocysteine insertion system to decode one codon in an animal genome? Nucleic Acids Res 33:2227–2238
7. Zhang Y, Fomenko DE, Gladyshev VN (2005) The microbial selenoproteome of the Sargasso Sea. Genome Biol 6:R37
8. Brocker MJ, Ho JM, Church GM, Soll D, O'Donoghue P (2014) Recoding the genetic code with selenocysteine. Angew Chem Int Ed Engl 53:319–323
9. Yoshizawa S, Böck A (2009) The many levels of control on bacterial selenoprotein synthesis. Biochim Biophys Acta 1790:1404–1414
10. Gursinsky T, Grobe D, Schierhorn A, Jager J, Andreesen JR, Sohling B (2008) Factors and selenocysteine insertion sequence requirements for the synthesis of selenoproteins from a gram-positive anaerobe in Escherichia coli. Appl Environ Microbiol 74:1385–1393
11. Hatfield DL, Carlson BA, Xu XM, Mix H, Gladyshev VN (2006) Selenocysteine incorporation machinery and the role of selenoproteins in development and health. Prog Nucleic Acid Res Mol Biol 81:97–142
12. Gladyshev VN, Kryukov GV (2001) Evolution of selenocysteine-containing proteins: significance of identification and functional characterization of selenoproteins. Biofactors 14:87–92
13. Böck A, Forchhammer K, Heider J, Leinfelder W, Sawers G, Veprek B, Zinoni F (1991) Selenocysteine: the 21st amino acid. Mol Microbiol 5:515–520
14. Hondal RJ (2009) Using chemical approaches to study selenoproteins-focus on thioredoxin reductases. Biochim Biophys Acta 1790:1501–1512
15. Miller C, Brocker MJ, Prat L, Ip K, Chirathivat N, Feiock A, Veszpremi M, Soll D (2015) A synthetic tRNA for EF-Tu mediated selenocysteine incorporation in vivo and in vitro. FEBS Lett 589:2194–2199
16. Haruna K, Alkazemi MH, Liu Y, Soll D, Englert M (2014) Engineering the elongation factor Tu for efficient selenoprotein synthesis. Nucleic Acids Res 42:9976–9983
17. Aldag C, Brocker MJ, Hohn MJ, Prat L, Hammond G, Plummer A, Soll D (2013) Rewiring translation for elongation factor Tu-dependent selenocysteine incorporation. Angew Chem Int Ed Engl 52:1441–1445
18. Xu J, Eriksson SE, Cebula M, Sandalova T, Hedstrom E, Pader I, Cheng Q, Myers CR, Antholine WE, Nagy P, Hellman U, Selivanova G, Lindqvist Y, Arner ES (2015) The conserved Trp114 residue of thioredoxin reductase 1 has a redox sensor-like function triggering oligomerization and crosslinking upon

oxidative stress related to cell death. Cell Death Dis 6:e1616

19. Cheng Q, Lu L, Grafstrom J, Olofsson MH, Thorell JO, Samen E, Johansson K, Ahlzen HS, Stone-Elander S, Linder S, Arner ES (2012) Combining 11C.-AnxA5 PET imaging with serum biomarkers for improved detection in live mice of modest cell death in human solid tumor xenografts. PLoS One 7:e42151
20. Cheng Q, Lu L, Grafstrom J, Olofsson MH, Thorell JO, Samen E, Johansson K, Ahlzen HS, Linder S, Arner ES, Stone-Elander S (2012) Site-specifically 11C-labeled Sel-tagged annexin A5 and a size-matched control for dynamic in vivo PET imaging of protein distribution in tissues prior to and after induced cell death. Biochim Biophys Acta 1830:2562–2573
21. Cheng Q, Stone-Elander S, Arnér ESJ (2006) Tagging recombinant proteins with a Sel-tag for purification, labeling with electrophilic compounds or radiolabeling with carbon-11. Nat Protoc 1:604–613
22. Cheng Q, Johansson L, Thorell JO, Fredriksson A, Samen E, Stone-Elander S, Arner ES (2006) Selenolthiol and dithiol C-terminal tetrapeptide motifs for one-step purification and labeling of recombinant proteins produced in E. coli. Chembiochem 7:1976–1981
23. Johansson L, Chen C, Thorell JO, Fredriksson A, Stone-Elander S, Gafvelin G, Arner ES (2004) Exploiting the 21st amino acid-purifying and labeling proteins by selenolate targeting. Nat Methods (1):61–66
24. Gromer S, Johansson L, Bauer H, Arscott LD, Rauch S, Ballou DP, Williams CH Jr, Schirmer RH, Arnér ESJ (2003) Active sites of thioredoxin reductases — why selenoproteins? Proc Natl Acad Sci U S A 100:12618–12623
25. Arnér ESJ, Sarioglu H, Lottspeich F, Holmgren A, Böck A (1999) High-level expression in *Escherichia coli* of selenocysteine-containing rat thioredoxin reductase utilizing gene fusions with engineered bacterial-type SECIS elements and co-expression with the selA, selB and selC genes. J. Mol. Biol 292:1003–1016
26. Rengby O, Johansson L, Carlson LA, Serini E, Vlamis-Gardikas A, Kårsnäs P, Arnér ESJ (2004) Assessment of production conditions for efficient use of Escherichia coli in high-yield heterologous recombinant Selenoprotein synthesis. Appl Environ Microbiol 70:5159–5167
27. Arnér ESJ (2002) Recombinant expression of mammalian selenocysteine-containing thioredoxin reductase and other selenoproteins in Escherichia coli. Methods Enzymol 347:226–235
28. Chen NY, Zhang JJ, Paulus H (1989) Chromosomal location of the Bacillus Subtilis aspartokinase II gene and nucleotide sequence of the adjacent genes homologous to uvrC and trx of Escherichia coli. J Gen Microbiol
29. Jiang Z, Arnér ESJ, Mu Y, Johansson L, Shi J, Zhao S, Liu S, Wang R, Zhang T, Yan G, Liu J, Shen J, Luo G (2004) Expression of selenocysteine-containing glutathione S-transferase in Escherichia coli. Biochem Biophys Res Commun 321:94–101
30. Wallberg H, Grafstrom J, Cheng Q, Lu L, Martinsson Ahlzen HS, Samen E, Thorell JO, Johansson K, Dunas F, Olofsson MH, Stone-Elander S, Arner ES, Stahl S (2012) HER2-positive tumors imaged within 1 hour using a site-specifically 11C-labeled Sel-tagged affibody molecule. J Nucl Med 53:1446–1453
31. Chung CT, Niemela SL, Miller RH (1989) One-step preparation of competent Escherichia coli: transformation and storage of bacterial cells in the same solution. Proc Natl Acad Sci U S A 86:2172–2175
32. Müller S, Heider J, Böck A (1997) The path of unspecific incorporation of selenium in Escherichia coli. Arch Microbiol 168:421–427
33. Cheng Q, Sandalova T, Lindqvist Y, Arnér ESJ (2009) Crystal structure and catalysis of the selenoprotein thioredoxin reductase 1. J Biol Chem 284:3998–4008

Chapter 18

Preparation of Selenocysteine-Containing Forms of Human SELENOK and SELENOS

Zhengqi Zhang, Jun Liu, and Sharon Rozovsky

Abstract

Selenoprotein K (SELENOK) and Selenoprotein S (SELENOS) are the members of the endoplasmic-reticulum-associated degradation (ERAD) complex, which is responsible for translocating misfolded proteins from the endoplasmic reticulum (ER) to the cytosol for degradation. Besides its involvement in the ERAD, SELENOK was shown to bind and stabilize the palmitoyl transferase DHHC6, and thus contributes to palmitoylation. SELENOK and SELENOS reside in the ER membrane by the way of a single transmembrane helix. Both contain an intrinsically disordered region with a selenocysteine (Sec) located one or two residues away from the C-terminus. Here, we describe the preparation of the Sec-containing forms of SELENOS and SELENOK. SELENOK, which contains no native cysteines, was prepared in an *E. coli* cysteine auxotroph strain by exploiting the codon and the insertion machinery of Cys for the incorporation of Sec. In contrast, the preparation of SELENOS, which contains functionally important cysteine residues, relied on *E. coli*'s native Sec incorporation mechanism.

Key words Selenoprotein, SELENOK, SELENOS, Membrane selenoproteins, Membrane enzymes, Selenocysteine, SelK, SelS

1 Introduction

Eight of the 25 human selenoproteins are integral membrane proteins [1]. Two of these, SELENOS (previously known as SelS [2]) and SELENOK (previously known as SelK [2]), participate in the degradation of misfolded proteins, a critical component of proteostasis [3]. The two proteins share a similar organization with a short luminal N-segment, a single pass transmembrane α-helix anchored to the ER, and a cytosolic intrinsically disordered region with the Sec one or two residues away from the C-terminus. Both the proteins bind to derlins, which are transmembrane proteins and essential constituents of the ERAD complex (Fig. 1) [4–6]. SELE NOS recruits the valosin-containing p97, an ATPase that provides the energy to pull protein substrates out of the ERAD complex [7, 8]. It also interacts with additional p97 adaptors, substrate

Laurent Chavatte (ed.), *Selenoproteins: Methods and Protocols*, Methods in Molecular Biology, vol. 1661,
DOI 10.1007/978-1-4939-7258-6_18, © Springer Science+Business Media LLC 2018

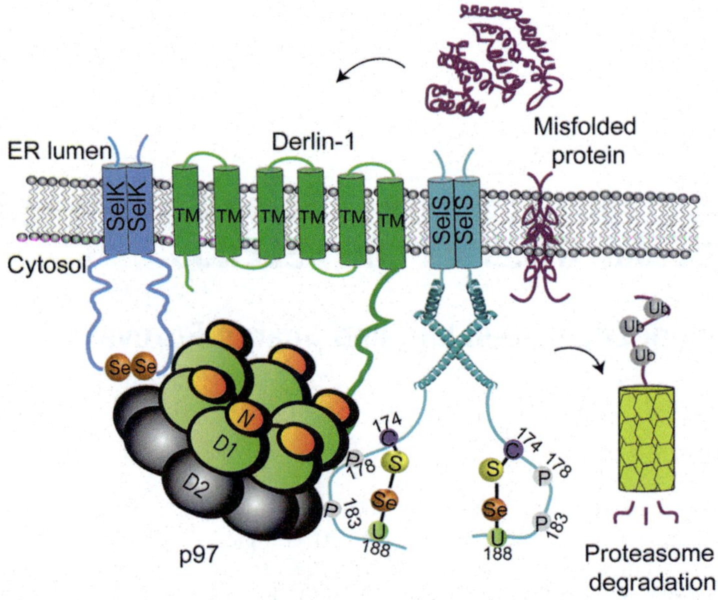

Fig. 1 SELENOK and SELENOS interact with multiple proteins in the ERAD complex to coordinate degradation of misfolded proteins

recruiters, and partners [9]. In contrast to SELENOS, SELENOK's specific role in the ERAD pathway is not yet known [1, 10]. In addition to its involvement in the ERAD, SELENOK, however, was shown to bind and stabilize the palmitoyl transferase DHHC6 [11]. SELENOK thus contributes to palmitoylation, i.e., the reversible attachment of fatty acids onto cysteine residues [12].

Since both SELENOS and SELENOK possess a transmembrane α-helix, their preparation requires their stabilization in a detergent environment that mimics the native cellular membrane [13]. Their expression is also nonstandard, as heterologous expression of membrane proteins is well known to be restricted by the host's limited membrane area, and by improper folding or lack of stability in the non-native membrane [14]. Furthermore, the two proteins are intrinsically disordered and as such they are prone to proteolytic cleavage in host organisms [15]. All these challenges were addressed by expressing both SELENOK and SELENOS in the robust host *E. coli* using a fused protein partner to increase solubility and prevent integration into *E. coli* membranes [10, 16–18]. For both the proteins the incorporation of Sec is essential to maintain their functional integrity [10, 11, 17, 18]. Because SELENOK contains no cysteine, but SELENOS has two cysteines, one of which is critical for its function, different strategies were employed to incorporate the single Sec in the respective protein. In cases of Cys-less proteins, such as SELENOK, there is no need to use the low efficiency, cellular Sec-insertion machinery to incorporate Sec. Instead, a higher yield can be obtained by exploiting the native Cys incorporation machinery for Sec insertion. To that end the Sec codon is mutated to Cys and a cysteine auxotroph *E. coli* strain is employed to prevent Cys biosynthesis [19]. The cysteine auxotroph is then grown in defined growth medium that is depleted of Cys and enriched in L-selenocystine [20]. Misloading the cysteinyl-tRNA with Sec instead of Cys leads then to successful Sec incorporated into the protein. To maximize the yield while minimizing the undesired incorporation of Cys the selenocystine is added after inducing expression. This also allows keeping Sec's concentration in the growth medium low in order to minimize the impact of selenocystine's cytotoxicity. However, this approach cannot be used for SELENOS because here the Sec is paired with a Cys and both must be incorporated in a site-specific manner for SELENOS to retain functionality. Therefore, a genetic incorporation method that harnesses *E. coli*'s native Sec incorporation machinery was used [16]. This requires that an RNA element, termed Sec insertion sequence (SECIS), is inserted 11 nucleotides after the UGA codon into the mRNA [21]. If this stem-loop structure occurs in the coding region of the mRNA, this element will dictate the amino acid sequence downstream from the Sec. However, in SELENOS the Sec is located at the C-terminus, and thus the SECIS element can be placed in the 3′ untranslated region. In addition, co-expression with the selA, selB, and selC genes (encoding selenocysteine synthase, a Sec-specific elongation factor and tRNASec, respectively) further increases the efficiency of Sec insertion [22]. It is also beneficial to induce protein expression at a late exponential phase, in which the level of release factor 2 is decreased [23]. The lower concentration of release factor 2 reduces the frequency of wrongful termination of protein translation at the UGA codon because of its misreading as a stop codon. This approach allowed the expression and purification of the cytosolic segment of SELENOS, which was abbreviated as cSELENOS [17, 18]. In contrast, the expression of full-length SELENOS at a late exponential phase led to a low yield. Strategies for the preparation of the full-length SELENOS involve native chemical ligation similar to the one described in the chapter about the preparation of SELENOM. In the case of cSELENOS, the genetic incorporation yielded a large fraction of truncated protein (over 95% of all proteins) in which the Sec was not integrated because protein translation was prematurely terminated. This truncated form, cSELENOS 188Δ, differs from cSELENOS by only two amino acids. The truncated form contains a single cysteine and thus forms intermolecular disulfide bonds between proteins. Fortunately, this leads to the aggregation of the truncated form, which greatly aids the purification of the homodimeric, Sec-containing cSELENOS via size exclusion chromatography. The expression and purification strategies for both cSELENOS and SELENOK are summarized in Fig. 2.

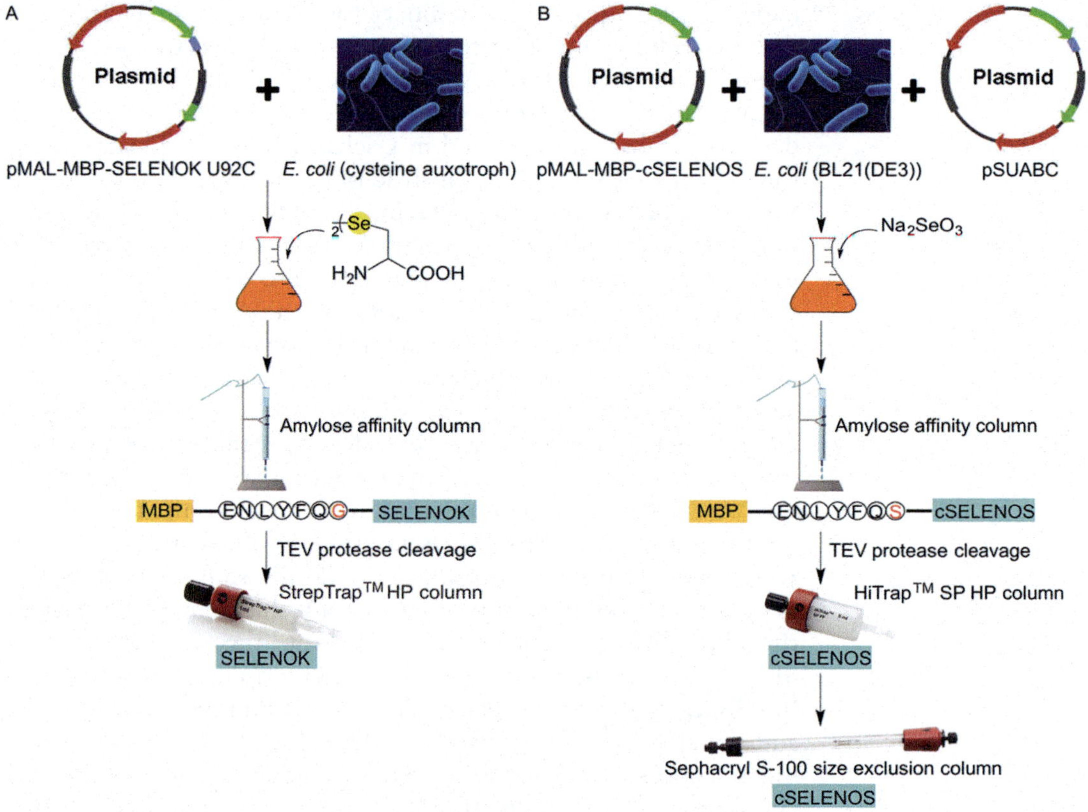

Fig. 2 Schematics illustrating the strategies for expression and purification of SELENOK and SELENOS. (**a**) SELENOK is expressed by supplementing the growth medium with selenocystine and purified by amylose and StrepTrapTM HP affinity columns. (**b**) cSELENOS is expressed using Sec genetic incorporation machinery and purified by amylose, HiTrapTM SP cation exchange, and HiPrepTM 16/60 Sephacryl S-100 HR size exclusion columns

2 Materials

2.1 General Materials

1. LB Medium: 10 g/L tryptone, 5 g/L yeast extract, 5 g/L NaCl and 2 g/L glucose.
2. Isopropyl-β-D-galactoside (IPTG).
3. Antibiotics: ampicillin, kanamycin, and chloramphenicol.
4. Protease inhibitors: phenylmethylsufonyl fluoride (PMSF) and benzamidine.
5. Detergents: Triton X-100 and n-dodecyl-β-D-maltoside (DDM).
6. Ethylenediaminetetraacetic acid (EDTA).
7. Maltose.
8. D-desthiobiotin.

9. Chromatography media: Amylose resin (NEB), StrepTrap™ HP column, HiTrap™ SP HP column, HiPrep™ 16/60 Sephacryl S-100 HR size exclusion column, Superdex™ 200 10/300 GL size exclusion column, and HiTrap™ desalting column (all from GE Healthcare Life Sciences).

2.2 Stock Solution for the Defined Medium (Adapted from Studier's MDAG Medium [24])

1. 1 M $MgSO_4$: Dissolve 12 g $MgSO_4$ in 100 mL water and then autoclave.
2. 1000× trace metal: stock solutions of different trace metals should be made separately and mixed together before use. The specific components of the trace metal solutions are listed in Table 1. $FeCl_3$ is dissolved in 0.12 M HCl. All other metals are dissolved in water.
3. 40% Glucose Solution: dissolve 40 g glucose in 72 mL water and adjust the final volume to 100 mL, then perform filter sterilization using steriflip (with a 0.22 μm pore size).
4. 25% Aspartate Solution: dissolve 25 g aspartic acid and 8 g NaOH in 84 mL water and adjust the final volume to 100 mL and the pH to 7.0, then autoclave.
5. 50× M (Salt Solution): dissolve 17.75 g Na_2HPO_4, 17 g KH_2PO_4, 13.4 g NH_4Cl, and 3.55 g Na_2SO_4 in 80 mL water and adjust the final volume to 100 mL and the pH to 7.0, then autoclave.
6. 17 Amino Acids Mixture (except cysteine, tyrosine, methionine):10 mg of each amino acid/mL. To make 100 mL stock of the amino acid solution, dissolve 1 g of each amino acid (Na-Glu, Asp, Lys-HCl, Arg-HCl, His-HCl, Ala, Pro, Gly, Thr, Ser, Gln, Asn, Val, Leu, Ile, Phe, Trp) in water. The amino acid solution is sterilized using steriflip (with a 0.22 μm pore size).
7. 2.5% Methionine Solution: dissolve 2.5 g methionine in 100 mL water, then autoclave.
8. 1000× Vitamin Stock: prepare the mixture from individual vitamin stock solutions and then sterilized it using steriflip (with 0.22 μm pore size). The specific components of individual vitamin solutions are listed in Table 2.

2.3 Defined Growth Medium

Mix together different amount of stock solutions (as tabulated in Table 3) with autoclaved water to make defined growth medium immediately prior to use. *See* Table 3 for specific formulations.

2.4 Buffers for SELENOK Purification

1. Lysis buffer: 50 mM Na_3PO_4, 200 mM NaCl, 1 mM EDTA, 0.2% (v/v) Triton X-100, 1 mM PMSF, and 1 mM benzamidine, pH 7.5.

Table 1
Trace metal solutions necessary to prepare 100 mL of 1000× trace metal mixture [24]

Components	Stock concentration (M)	Volume to add (mL)
$FeCl_3$	0.1	50
$CaCl_2$	1	2
$MnCl_2$	1	1
$ZnSO_4$	1	1
$CoCl_2$	0.2	1
$CuCl_2$	0.1	2
$NiCl_2$	0.2	1
Na_2MoO_4	0.1	2
Na_2SeO_3	0.1	2
H_3BO_3	0.1	2
H_2O	–	36

Table 2
Vitamin solutions for preparing 100 mL of 1000× vitamin mixture [24]

Components	Stock concentration (mM)	Volume to add (mL)
Nicotinic acid	10	2
Pyridoxine	10	2
Thiamine	10	2
p-Aminobenzoic acid	10	2
Patothenate	10	2
Folic acid	0.1	5
Riboflavin	0.1	5
Vitamin B12	5	4
H_2O	–	76

2. Amylose wash buffer: 50 mM Na_3PO_4, pH 7.5, 200 mM NaCl, 1 mM EDTA, and 0.067% DDM.
3. Amylose elution Buffer: 50 mM Na_3PO_4, 200 mM NaCl, 20 mM maltose, 1 mM EDTA, and 0.067% DDM, pH 7.5.
4. StrepTrap wash buffer: 50 mM Na_3PO_4, 200 mM NaCl, and 0.067% DDM, pH 7.5.

Table 3
Composition of 1 L of defined growth medium

Medium component	Volume to add (mL)
1 M $MgSO_4$	2
1000× trace metal	0.2
40% Glucose solution	10
2.5% Methionine solution	8
25% Aspartate solution	12
17 amino acids	20
50× M	20
1000× vitamin	1
H_2O	926.8

5. StrepTrap elution buffer: 50 mM Na_3PO_4, 200 mM NaCl, 2.5 mM D-desthiobiotin, and 0.067% DDM, pH 7.5.
6. Trypsin digestion buffer: 50 mM Tris–HCl, 20 mM $CaCl_2$, and 0.067% DDM, pH 8.0.

2.5 Buffers for cSELENOS Purification

1. Lysis buffer: 50 mM Na_3PO_4, 200 mM NaCl, 1 mM EDTA, 1 mM PMSF, and 1 mM benzamidine, pH 7.5.
2. Amylose wash buffer and SP column buffer A: 50 mM Na_3PO_4, 200 mM NaCl, and 1 mM EDTA, pH 7.5.
3. Amylose elution buffer: 50 mM Na_3PO_4, 200 mM NaCl, 20 mM maltose, and 1 mM EDTA, pH 7.5.
4. SP column buffer B: 50 mM Na_3PO_4, 1 M NaCl, and 1 mM EDTA, pH 7.5.
5. Trypsin digestion buffer: 50 mM Tris–HCl, and 20 mM $CaCl_2$, pH 8.0.

3 Methods

3.1 Expression and Purification of SELENOK

3.1.1 Clone SELENOK U92C into pMAL-c5X Expression Vector

1. The gene of *Homo sapiens* SELENOK (UniProtKB-Q9Y6D0) is codon optimized for expression in *E. coli*.
2. The expression vector pMAL-c5X is designed to produce fusion proteins with the cytoplasmic maltose binding protein (MBP) and was obtained from New England Biolabs (NEB).
3. In order to facilitate purification, a hexahistidine tag was introduced into MBP between I3 and E4 by standard site directed mutagenesis [16].

4. A linker sequence (NSSS) and a tobacco etch virus (TEV) protease cleavage site (ENLYFQG) were inserted immediately after MBP using AvaI and BamHI restriction sites. The TEV cleavage site was followed by a Met to introduce an NcoI restriction site (CCATGG).
5. SELENOK U92C was cloned into this modified pMAL-c5X vector after the TEV protease cleavage site using NcoI and BamHI restriction sites.
6. To assist in purification, an eight amino acids StrepII tag (WSHPQFEK) was inserted between the TEV protease cleavage site and SELENOK U92C. This vector was named pMAL-MBP-SELENOK U92C.
7. The final pMAL-MBP-SELENOK U92C vector is used to transform the *E. coli* cysteine auxotroph strain BL21(DE3) with the genotype selB::kan cys51E, referred to as BL21 (DE3)cys (*see* **Note 1**) [19].

3.1.2 Expression of SELENOK

1. For optimal yield, freshly transform pMAL-MBP-SELENOK U92C expression vector into BL21(DE3)cys cells. Plate the transformed cells evenly onto LB agar plates supplemented with 100 μg/mL ampicillin, 25 μg/mL kanamycin, and 50 μg/mL L-cysteine, and incubate overnight at 37 °C (*see* **Note 1**).
2. Pick up 2–3 colonies from the plate to grow overnight in 200 mL LB medium supplemented with 100 μg/mL ampicillin, 25 μg/mL kanamycin, and 50 μg/mL L-cysteine at 37 °C.
3. The next morning, transfer 10 mL of overnight culture to 1 L fresh LB medium supplemented with 100 μg/mL ampicillin, 25 μg/mL kanamycin, and 50 μg/mL L-cysteine.

   The optical density (OD) at 600 nm wavelength of the growth medium after the addition of overnight culture should not exceed 0.1.
4. Grow cells at 37 °C while shaking at 250 rpm.
5. Harvest cells when the OD at 600 nm reaches 1.5 (about 6 h) by centrifugation at 4000 × *g* for 20 min at 4 °C.
6. After discarding the supernatant, resuspend the cell pellet in the centrifuge bottles with 100 mL of ice cold, autoclaved water per bottle. This step is necessary for changing the growth medium into a defined mixture and removing residual Cys in the medium. The resuspension process should be carried out on ice.
7. Centrifuge for 10 min at 4000 × *g* at 4 °C.
8. Resuspend the cell pellet with 100 mL cold water on ice as detailed in **step 6** (*see* **Note 2**).
9. Centrifuge for 15 min at 4000 × *g* at 4 °C and discard the supernatant.

10. Collect the cell pellet after centrifugation. Dissolve the pellet in 100 mL of defined medium on ice.
11. Transfer 100 mL of the resuspended cell pellet into 900 mL of defined medium, which has been supplemented with 100 μg/mL ampicillin and 25 μg/mL kanamycin.
12. Grow cells at 37 °C for 30 min and then decrease the temperature to 18 °C. Allow the cells to recuperate at that temperature for an additional 30 to 60 min. The OD at 600 nm should be close to 1.5 at the end of the incubation period.
13. Add 0.5 mM IPTG and 200 μM L-selenocystine to the growth medium to induce expression (*see* **Note 3**).
14. Grow cells overnight at 18 °C while shaking at 250 rpm (between 20 and 24 h).
15. Harvest cells by the centrifugation at 4000 × *g* for 20 min at 4 °C and decant the supernatant.
16. Immediately flash freeze the cells in liquid N_2, unless you intend to purify the protein straightaway.

3.1.3 Purification of SELENOK

SELENOK purification is carried out using two chromatographic steps (Fig. 2a). First, the fusion protein MBP-SELENOK is purified by amylose affinity chromatography. Then, the fusion partner MBP is cleaved off by TEV protease. Finally, MBP and TEV protease are removed from SELENOK by strep tag affinity purification.

1. Resuspend the cell pellet with 30–50 mL of lysis buffer (add PMSF just before you lyse the cells). Lyse the cells using a high pressure homogenizer (e.g., EmulsiFlex-C5, Avestin) operating continuously at 9000 psi performing at least 5 cycles. Keep the homogenizer on ice to minimize SELENOK denaturation (*see* **Notes 4** and **5**).
2. Immediately afterward centrifuge the cell lysate for 60 min at 12,000 × *g* or higher. Collect the supernatant.
3. To prepare the amylose affinity column preequilibrate the column with 5 column volumes (CV) of amylose loading buffer. Subsequently, load the column with the supernatant and collect the flow-through. The binding capacity for amylose resin is around 3 mg MBP-SELENOK fusion protein per milliliter of resin. For 1 L of the medium, 10 mL of amylose resin will suffice (*see* **Note 6**).
4. Wash the amylose affinity column with 10 CV of amylose washing buffer.
5. Elute MBP-SELENOK fusion protein with 3 CV amylose elution buffer.
6. Estimate the concentration of eluted fusion protein MBP-SELENOK by measuring its absorbance at 280 nm and using an extinction coefficient of 88,810 M^{-1} cm^{-1}.

7. Now perform TEV protease cleavage of MBP-SELENOK. The molar ratio of TEV protease to MBP-SELENOK fusion protein should be around 1:20. Add the desired amount of TEV protease to the eluted protein (in elution buffer). Let the TEV cleavage reaction occur overnight (12–14 h) at 4 °C while gently stirring (*see* **Notes 7–9**).
8. Evaluate the purity and percentage of successful TEV cleavage by running a 16% Tris-Glycine gel. A typical result is shown in Fig. 3.
9. If the TEV cleavage is successful, proceed to run a 5 mL StrepTrap™ HP column to remove MBP and TEV. The binding capacity of StrepTrap™ HP column is around 7 mg of protein per milliliter of resin.
10. Before loading the StrepTrap™ HP column with the TEV cleavage mixture, remove aggregates by centrifuging the mixture at 10,000 × *g* for 10 min or filtering using a 0.22 μm syringe filter.
11. To prepare the StrepTrap™ HP column preequilibrate it with 5 CV of StrepTrap washing buffer.

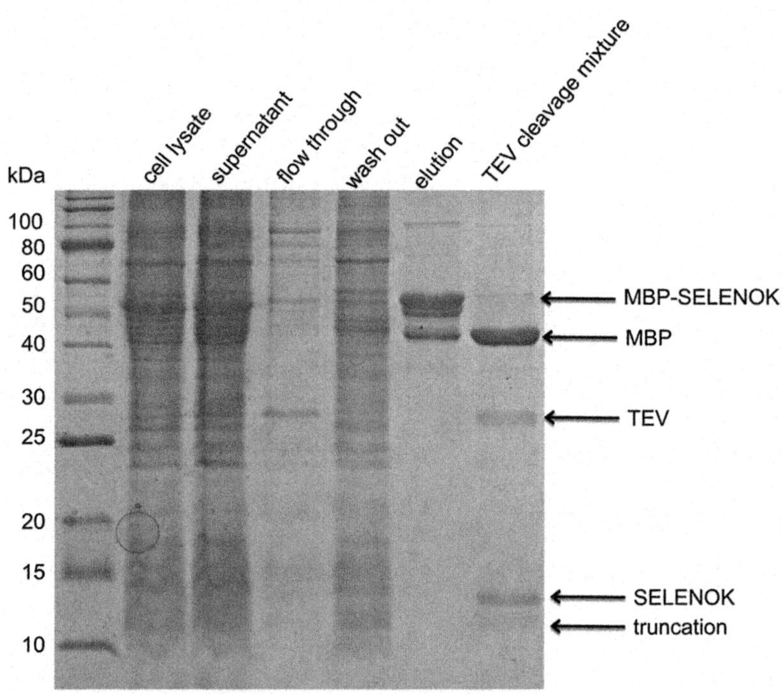

Fig. 3 Purification of MBP-SELENOK by amylose affinity chromatography and its subsequent cleavage by TEV protease as monitored by 16% reducing Tris-Glycine SDS-PAGE

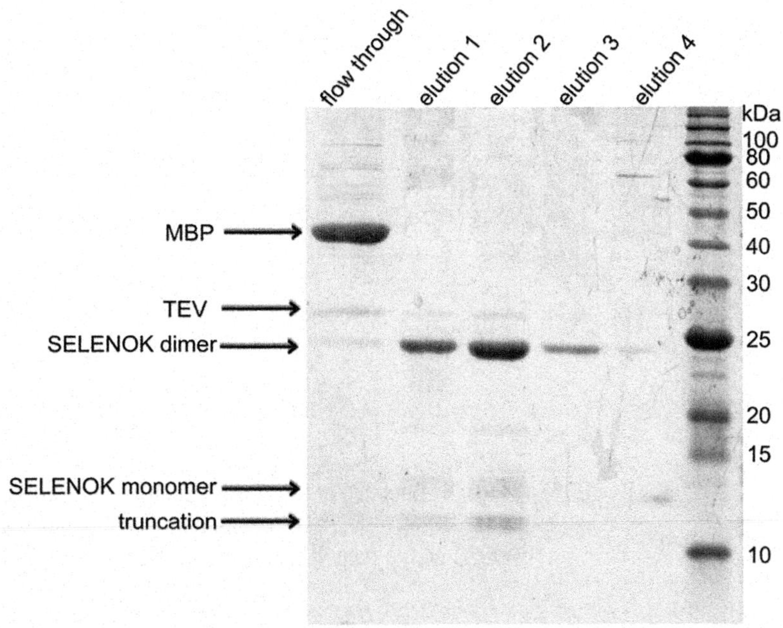

Fig. 4 Purification of SELENOK by StrepTrap™ HP column monitored by 16% nonreducing Tris-Glycine SDS-PAGE

12. Load the supernatant onto the StrepTrap™ HP column and collect the flow-through.
13. Wash the StrepTrap™ HP column with 10 CV of StrepTrap washing buffer.
14. Elute SELENOK using 3 CV StrepTrap elution buffer. The yield of SELENOK is around 2 mg proteins from 1 L defined growth medium.
15. Estimate the purity of the protein by SDS-PAGE. Typically, it is higher than 95% at this point (*see* **Note 10**). A typical result is shown in Fig. 4.
16. If necessary, perform size exclusion chromatography. To that end, concentrate SELENOK to 5 mg/mL and then filter the sample using a 0.22 μm syringe filter. Load the sample onto a HiPrep™ 16/60 Sephacryl S-100 HR size exclusion column or a Superdex™ 200 10/300 GL size exclusion column and elute with amylose wash buffer. The molecular weight of the SelK DDM complex is around 90 kDa. Collect the fractions containing SELENOK and check the purity by running a 16% Tris-Glycine SDS-PAGE.
17. Approximate the Sec incorporation ratio by intact protein mass spectrometry (*see* **Notes 11** and **12**). To this end, first reduce SELENOK (10 μM) by adding 5 mM dithiothreitol (DTT) and incubate at room temperature for 20 min. Inject the protein sample into the mass spectrometer (e.g., a Xevo G2-S

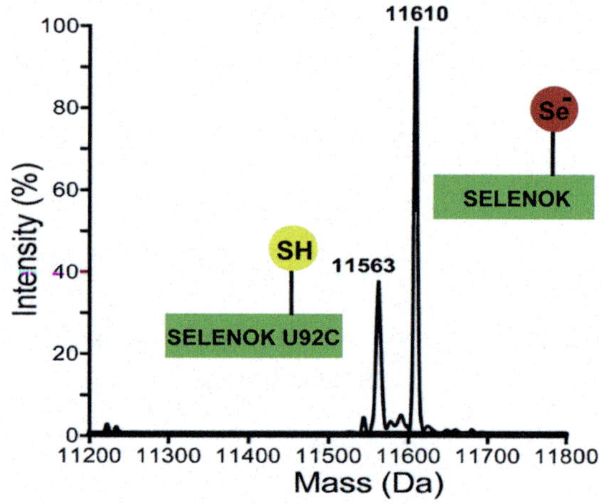

Fig. 5 Deconvoluted mass spectrum of reduced SELENOK. In addition to SELENOK (11,609 Da calculated weight), also a small amount of SELENOK U92C with a calculated molecular mass of 11,563 Da is present

QTOF, Waters), which is coupled with a reverse phase HPLC column such as an ACQUITY UPLC Protein BEH C4 reverse-phase column (300 Å, 1.7 μm, 2.1 mm × 150 mm). Use an acetonitrile gradient from 5 to 95% with 0.1% formic acid, over a running time of 5 min at a constant flow rate of 0.5 mL/min. Acquire the data from m/z 350 to 2000, at a rate of 1 s/scan. Typically, about 70–80% of the protein contains Sec at position 92 while the remaining contains Cys. A typical, deconvoluted mass spectrum is shown in Fig. 5 (*see* **Note 13**).

18. Concentrate SELENOK (using for example an Amicon ultra centrifugal filter unit with a membrane molecular weight limit of a 100 kDa) to 5 mg/mL and then add 10% glycerol and flash freeze in liquid N_2. For long-term storage, keep the protein at −80 °C (*see* **Note 14**).

3.1.4 Tandem Mass Sequencing of SELENOK

To confirm the presence of Sec in SELENOK, it is digested by trypsin protease and subjected to peptide sequencing.

1. Exchange the buffer of the SELENOK solution into the trypsin digestion buffer using a 5 mL HiTrapTM desalting column.
2. Reduce SELENOK by adding 5 mM DTT and incubate at room temperature for 20 min.
3. Boil at 95 °C for 5 min. Cool down the sample to room temperature, add 20 mM iodoacetamide (IAM), and incubate for 20 min to alkylate the Sec.
4. Add 0.5 μg trypsin into 100 μL of alkylated SELNOK (1 mg/mL) from **step 3** and digest at 37 °C for 16 h.

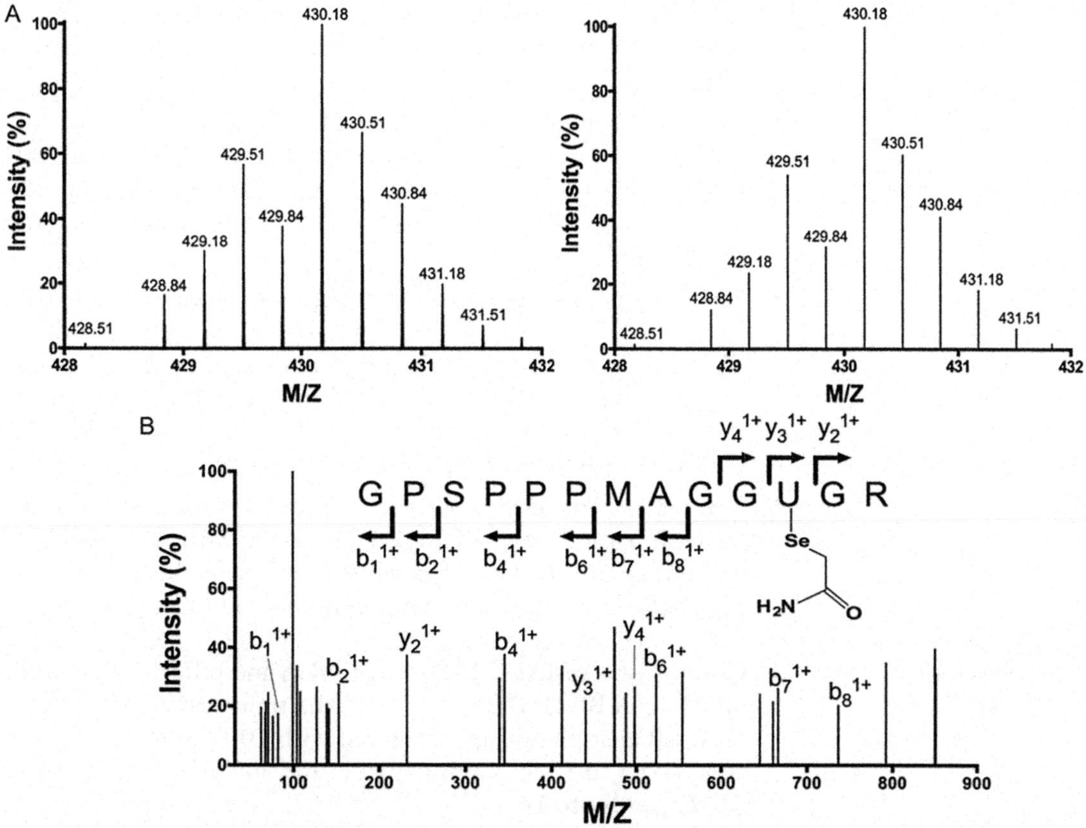

Fig. 6 Identification of Sec in SELENOK. (**a**) Mass spectrum of Sec-containing peptide from SELENOK with the sequence GPSPPPMAGGUU(IAM)GR, where IAM stands for iodoacetamide and U for Sec. The pattern is indicative of selenium isotopic abundance. The theoretical *M*/*Z* of the triply charged peptide GPSPPPMAG-GUU(IAM)GR is shown on *right* and the experimental data on *left*. (**b**) MS/MS spectrum of the peptide from panel **a** confirms Sec is present in this peptide

5. Sequence the Sec-containing peptide using, for example, a Q Exactive Orbitrap interfaced with Ultimate 3000 LC system. 5 μL of trypsin-digested protein from **step 4** are loaded on an Ace ultracore super C18 reverse-phase column (75 mm × 2.1 mm) via an autosampler. The samples are analyzed using a 45 min gradient going from 5% B to 95% B at a flow rate of 0.2 mL/min. Solvent A is water containing 0.1% formic acid and solvent B is acetonitrile containing 0.1% formic acid. A typical result is shown in Fig. 6.

3.2 Expression and Purification cSELENOS

3.2.1 Design of Constructs for Expression

1. The gene of *Homo sapiens* SELENOS (UniProtKB-Q9BQE4) is codon optimized for expression in *E. coli*. The Sec codon must remain UGA.
2. The expression vector pMAL-c5X, designed to produce fusion proteins with the MBP, was obtained from NEB.

3. A hexahistidine tag was introduced into MBP between I3 and E4 to facilitate purification by standard site directed mutagenesis [16].
4. A linker sequence NSSS and TEV protease cleavage site (ENLYFQS) was inserted immediately after MBP using AvaI and BamHI restriction sites. The TEV cleavage site was followed by a Met to introduce an NcoI restriction site (CCATGG).
5. The cytoplasmic portion of SELENOS (residues 52–189, abbreviated as cSELENOS) was cloned into this modified pMAL-c5X vector using NcoI and BamHI sites. To allow Sec encoding, an *E. coli* formate dehydrogenase H SECIS element was also included immediately after the stop codon [25]. This vector was named pMAL-MBP-cSELENOS.
6. The plasmid of pMAL-MBP-cSELENOS is used for cotransformation of *E. coli* BL21(DE3), along with the pSUABC plasmid (*see* **Note 15**) expressing *E. coli* SelA, SelB, and SelC under the control of their endogenous promoters [22].

3.2.2 Expression of cSELENOS

1. Cotransform pMAL-MBP-cSELENOS and pSUABC plasmids into E. coli BL21(DE3) and plate the transformed cells evenly onto LB agar plates supplemented with 100 μg/mL ampicillin and 34 μg/mL chloramphenicol, and incubate overnight at 37 °C (*see* **Note 16**).
2. Pick up 2–3 colonies from the agar plate to grow overnight in the 200 mL LB medium supplemented with 100 μg/mL ampicillin with 34 μg/mL chloramphenicol at 37 °C.
3. The following morning, transfer 10 mL overnight culture to the 1 L defined medium supplemented with 100 μg/mL ampicillin and 34 μg/mL chloramphenicol. The OD at 600 nm of the medium after the addition of overnight culture should not exceed 0.1.
4. Grow cells at 37 °C while shaking at 250 rpm.
5. Lower the growth temperature to 18 °C when the OD at 600 nm reaches 2.0. Then let the cells recover for 30–60 min at 18 °C.
6. To induce expression, add 0.5 mM IPTG, 10 μM Na_2SeO_3, and 100 μg/mL L-cysteine to the growth medium. A typical OD at the point of induction is 2.2–2.4.
7. Grow cells at 18 °C while shaking at 250 rpm for 20–24 h.
8. Harvest cells by a 15 min centrifugation at 5000 × *g* at a temperature of 4 °C and subsequently discard the supernatant.
9. Immediately freeze cells in liquid N_2, unless you intend to purify the protein straightaway.

3.2.3 Purification of cSELENOS

cSELENOS purification is carried out using three chromatographic steps (Fig. 2b). First, the fusion protein MBP-cSELENOS is isolated by amylose affinity chromatography. Next, the fusion partner MBP is cleaved off by TEV protease. MBP, TEV, and aggregated truncated SELENOS (cSELENOS 188Δ) are then removed by cation exchange chromatography. Finally, the purity with respect to cSELENOS is increased by separating it from the truncated form, cSELENOS 188Δ, using size exclusion chromatography.

1. Resuspend the cell pellet with 30–50 mL of lysis buffer (add PMSF just before you lyse the cells). Lyse the cells using a high pressure homogenizer (EmulsiFlex-C5, Avestin) operating continuously at 9000 psi for at least 5 cycles. Keep the homogenizer on ice to minimize cSELENOS denaturation.
2. Immediately centrifuge the cell lysate for 60 min at 12000 × *g* and collect the supernatant.
3. To prepare the amylose affinity column preequilibrate it with 5 CV of amylose loading buffer. Subsequently, load the supernatant onto the column and collect flow-through. The binding capacity for amylose resin is typically around 7 mg MBP-cSELENOS fusion protein per milliliter of resin. For 1 L of the medium, 30 mL of amylose resin should thus suffice.
4. Wash the amylose affinity column with 10 CV of amylose washing buffer.
5. Elute MBP-cSELENOS with 3 CV of amylose elution buffer.
6. Estimate the concentration of the eluted MBP-cSELENOS by measuring the UV absorbance at 280 nm using an extinction coefficient of 81,945 M^{-1} cm^{-1}.
7. Cleave MBP-cSELENOS using TEV protease. The molar ratio of TEV protease to MBP-cSELENOS fusion protein should be around 1:20. Add the desired amount of TEV protease to the eluted protein (in elution buffer). Perform TEV cleavage reaction overnight (12–14 h) at 4 °C while gently stirring.
8. Evaluate the purity and success of TEV cleavage by running a reducing 16% Tris-Glycine gel. A typical result is depicted in Fig. 7.
9. When the cleavage by the TEV protease is complete, proceed to remove MBP and TEV protease using a 5 mL HiTrapTM SP HP column. The binding capacity for HiTrapTM SP HP column is about 40 mg cSELENOS per milliliter of resin.
10. Centrifuge the TEV cleavage mixture at 10,000 × *g* for 10 min or filter it using a 0.22 μm syringe filter prior to loading it onto the HiTrapTM SP HP column.
11. To prepare the HiTrapTM SP HP column preequilibrate it with 5 CV of SP column buffer A.

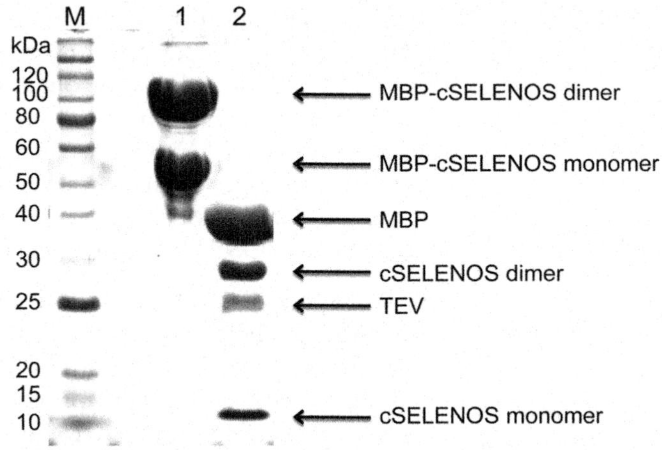

Fig. 7 Purification of MBP-cSELENOS by amylose affinity chromatography and subsequent cleavage by TEV protease as monitored by 16% nonreducing Tris-Glycine SDS-PAGE. *Lane 1*: Amylose column elution; *Lane 2*: TEV protease cleavage mixture

12. Load the supernatant onto the HiTrapTM SP HP column and collect the flow-through.
13. Wash the HiTrapTM SP HP column with 10 CV of SP column buffer A.
14. Elute cSELENOS with a gradient of SP column buffer B from 0 to 100% over 15 CV.
15. Evaluate the purity of different fractions by nonreducing 16% Tris-Glycine SDS-PAGE. A typical result is shown in Fig. 8. Note that cSELENOS elutes at a lower salt concentration than the truncated form cSELENOS 188Δ (*see* **Notes 17** and **18**).
16. Approximate the ratio of cSELENOS to cSELENOS 188Δ by intact protein mass spectrometry. First, reduce cSELENOS (10 μM) by adding 5 mM DTT and incubate at room temperature for 20 min. Inject the protein sample into the mass spectrometer (e.g., a Xevo G2-S QTOF, Waters), which is coupled with a reverse phase HPLC column such as an ACQUITY UPLC Protein BEH C4 reverse-phase column (300 Å, 1.7 μm, 2.1 mm × 150 mm). Use an acetonitrile gradient from 5 to 95% with 0.1% formic acid, over a running time of 5 min at a constant flow rate of 0.5 mL/min. Acquire the data from *m*/*z* 350–2000, at a rate of 1 s/scan.
17. Combine fractions that contain cSELENOS, concentrate them to 5 mg/mL, and filter the sample with a 0.45 μm syringe filter. Load the sample onto a HiprepTM 16/60 Sephacryl S-100 HR size exclusion column and elute with SP column buffer A. Collect fractions containing the cSELENOS and evaluate the purity by 16% Tris-Glycine SDS-PAGE. cSELENOS is usually eluted at a later time than the truncated form cSELENOS

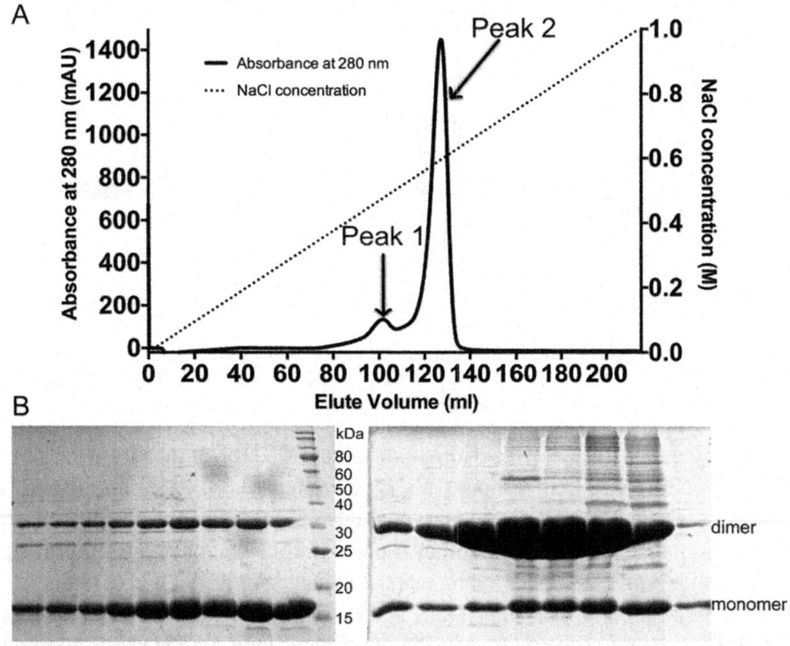

Fig. 8 Purification of cSELENOS by cation exchange chromatography. (**a**) Elution profiles of cSELENOS from a HitrapTM SP HP ion exchange column under nonreducing conditions. The mixture of MBP-cSELENOS following cleavage by TEV protease was loaded to remove the TEV protease and MBP. Only cSELENOS and cSELENOS188Δ bind the column. Peak 1, which elutes with 300 mM NaCl, contains the majority of the Sec-containing form. Peak 2, which elutes with 600 mM NaCl, contains cSELENOS 188Δ. cSELENOS 188Δ bind tighter because it forms oligomers by mixed disulfide bonds. (**b**) 16% nonreducing Tris-Glycine SDS PAGE analysis of cSELENOS/cSELENOS188Δ mixtures following separation by Hitrap SP FF column. The gel on the *left* shows different fractions under the first peak; the gel on the *right* shows different fractions under the second peak. Peak 1 contains less of the higher oligomeric forms

188Δ. A typical result is depicted in Fig. 9. About 0.6 mg of cSELENOS can be obtained from 1 L of growth medium (*see* **Note 19**).

18. Approximate the ratio of cSELENOS to cSELENOS 188Δ in different fractions by mass spectrometry as detailed in **step 16**. A typical result is shown in Fig. 10 (*see* also **Notes 20** and **21**).
19. Collect all fractions containing cSELENOS and concentrate to 5 mg/mL and then add 10% glycerol and flash freeze with liquid N_2. For long-term storage, store the protein at −80 °C.

3.2.4 Tandem Mass Sequencing of cSELENOS

To confirm the presence of Sec in cSELENOS, it is digested by trypsin protease and subjected to peptide sequencing.

1. Exchange cSELENOS into the trypsin digestion buffer using a 5 mL HiTrapTM desalting column.

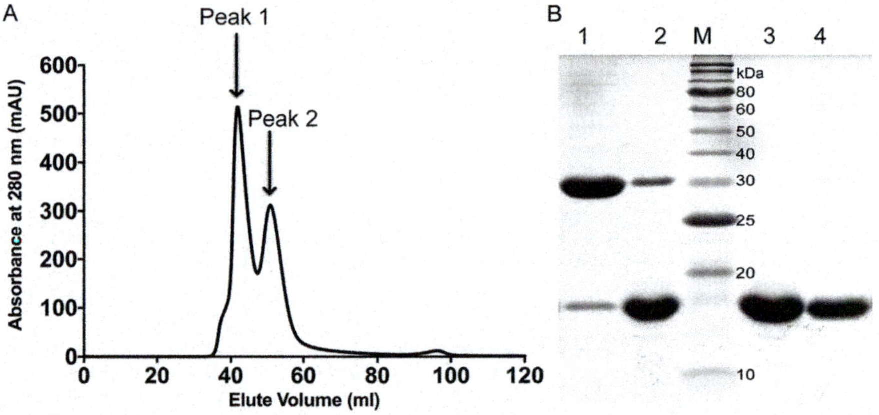

Fig. 9 Separation of the cSELENOS from cSELENOS 188Δ by size exclusion chromatography. (**a**) Elution profiles of cSELENOS from a HiprepTM 16/60 Sephacryl S-100 HR size exclusion column under nonreducing conditions. Peak 2, which eluted last, contains cSELENOS. (**b**) SDS-PAGE analysis of cSELENOS/cSELENOS 188Δ mixtures following separation by HiprepTM 16/60 Sephacryl S-100 HR size exclusion chromatography. *Lanes 1* and *2* were run under nonreducing conditions: *lane 1*, cSELENOS first peak; *lane 2*, cSELNEOS second peak; *lane M*, protein molecular mass standards (the molecular mass in kilodaltons is noted on the *right*). *Lanes 3* and *4* were run under reducing conditions: *lane 4*, cSELENOS first peak; *lane 5*, cSELENOS second peak. The image was modified from ref. 17

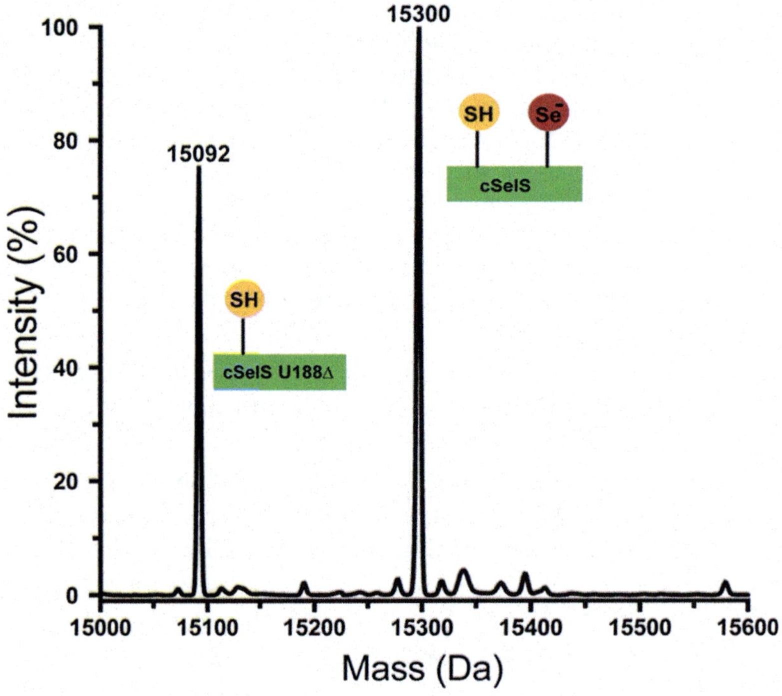

Fig. 10 Deconvoluted mass spectrum of reduced cSELENOS after the purification by HiprepTM 16/60 Sephacryl S-100 HR size exclusion column. cSELENOS was enriched from cSELENOS 188Δ to over 50%. The calculated molecular mass of cSELENOS 188Δ is 15,093 Da and that of cSELENOS is 15,300 Da

2. Reduce cSELENOS by adding 5 mM DTT and incubate the mixture at room temperature for 20 min.
3. Boil at 95 °C for 5 min. Cool down the sample to room temperature, add 20 mM IAM, and incubate for 20 min to alkylate the Sec and Cys.
4. Add 0.5 μg trypsin into 100 μL alkylated cSELENOS (1 mg/mL) solution from **step 3** and digest at 37 °C for 9–10 h.
5. Sequence the Sec-containing peptide using a Q Exactive Orbitrap interfaced with Ultimate 3000 LC system. Load 5 μL of trypsin-digested protein on an Ace ultracore super C18 reverse-phase column (75 mm × 2.1 mm) via an autosampler. The samples are analyzed using a 45 min gradient going from 5% B to 95% B at a flow rate of 0.2 mL/min. Solvent A is water containing 0.1% formic acid and solvent B is acetonitrile containing 0.1% formic acid. A typical result from this procedure is shown in Fig. 11.

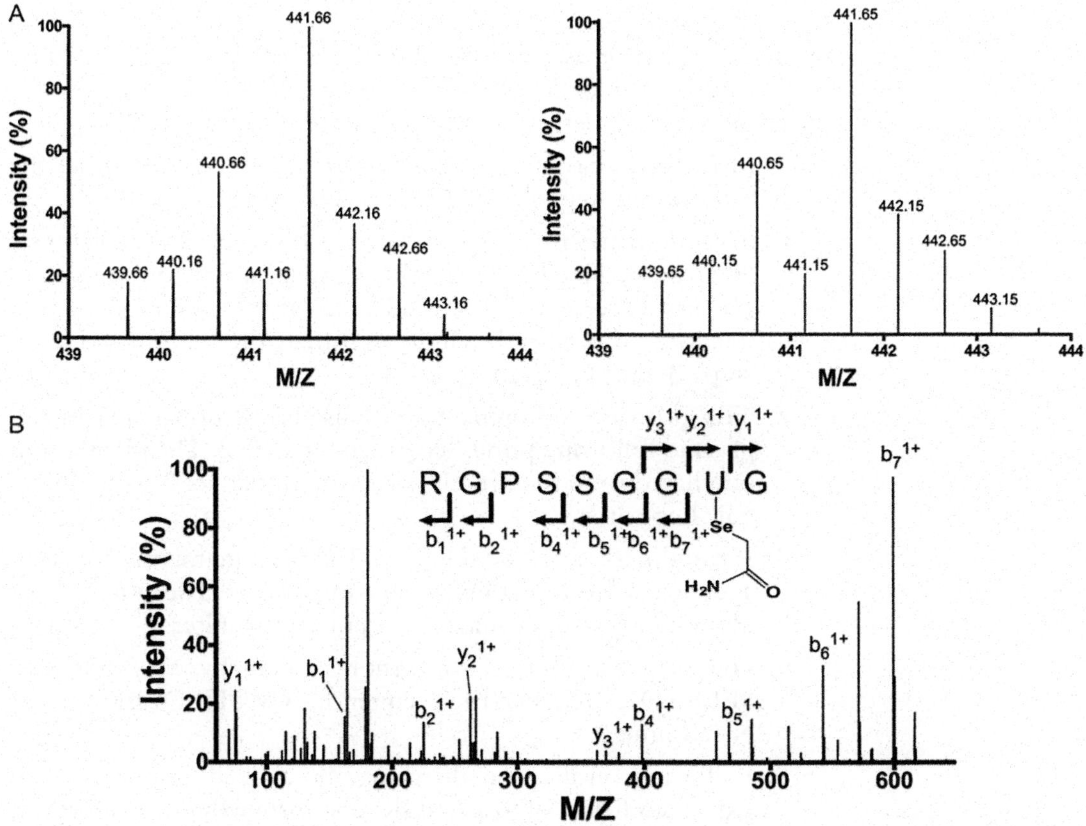

Fig. 11 Identification of Sec in SELENOS. (**a**) Mass spectrum of Sec-containing peptide from cSELENOS with the sequence RGPSSGGU(IAM)G, where IAM stands for iodoacetamide and U for Sec. The pattern is indicative of selenium isotopic abundance. The theoretical *M/Z* of the triply charged peptide RGPSSGGU(IAM)G is shown on *right* and the experimental data on *left*. (**b**) MS/MS spectrum of the peptide from panel **a** confirms Sec is present in this peptide

4 Notes

Notes Regarding SELENOK Expression and Purification

1. BL21(DE3)cys cells require 25 μg/mL kanamycin and 50 μg/mL L-cysteine for growth. The strain can be requested from Dr. Marie-Paule Strub at the NIH laboratory of molecular biophysics. Other cysteine auxotroph *E. coli* strains are available from the *E. coli* Genetic Stock Center. A regular BL21(DE3) *E. coli* strain can also be employed instead of BL21(DE3)cys, but the resulting sample contains equal amounts of cSELENOK and cSELENOK U92C.
2. Washing the cells with water when transferring from LB to defined growth medium is essential for removing residual Cys and reaching high Sec incorporation levels.
3. L-selenocystine is not soluble at neutral pH. It could be dissolved in 1 M HCl or 1 M NaOH to prepare a 100 mM stock. Alternatively, the powder can be directly added to the growth medium.
4. Add the protease inhibitor PMSF just before cell lysis. PMSF is easily degraded in aqueous solutions, so the stock solution is prepared in methanol or isopropanol. Beware of toxicity.
5. SELENOK has autoproteolytic activity and therefore all the purification steps should be carried out on ice or at 4 °C.
6. Save all fractions (flow-through, wash out, and elution) until you have verified by gel that the binding capacity of the column was not exceeded. If you find that there is fusion protein in fractions other than the elution, estimate the amount and rerun **steps 3–8** of Subheading 3.1.3.
7. TEV protease was prepared in house using a pMHTDelta238 plasmid following published procedures [26]. The plasmid was purchased from the Protein Structure Initiative: Biology Materials Repository.
8. If cloudiness is observed in the TEV cleavage mixture, remove aggregates by centrifuging the mixture at 10,000 × *g* for 10 min or filtering using a 0.22 μm syringe filter.
9. Following cleavage of the fusion protein by TEV protease, SELENOK retains the sequence GWSHPQFEK in its N-terminus.
10. Following purification there may be a small amount of two truncated forms of SELENOK: one form exhibiting a truncation between H79 and L80 with a molecular weight of 10,128 Da and another form that terminates even earlier between P66 and P67 with a molecular weight of 8647 Da. Both truncated proteins appear to be caused in vivo by protease activity in *E. coli*.

11. Inductively Coupled Plasma Mass Spectrometry (ICP-MS) is the preferred method to quantitatively determine the concentration of selenium in the protein sample.
12. The ratio of SELENOK to SELENOK U92C is estimated from their relative intensities in the mass spectrum. For reliable estimation the ion counts must be kept well below the saturation level of the mass detector.
13. If Sec incorporation is low, the most likely reason is that residual Cys is not completely removed when the cells are transferred from the LB medium to defined medium or that the cells were not allowed sufficient time to consume the internal Cys pool. To improve the ratio of Sec over Cys incorporation the concentration of L-selenocystine in the defined medium can be increased up to 300 mM; however this will reduce yield.
14. When concentrating membrane proteins avoid sharp gradients in detergents and proteins. Frequently mix the concentrate in concentrator devices. Do not exceed a concentration of 8 mg/mL.

Notes Regarding cSELENOS Expression and Purification

15. The pSUABC plasmid is available upon request from Prof. E. S. J. Arner at the Karolinska Institutet [22, 27].
16. The pMAL-MBP-cSELENOS and pSUABC plasmids can be co-transformed into electrocompetent BL21(DE3) cells following standard protocols for electro transformation. Alternatively, it can be done sequentially by transforming pSUABC plasmid into chemically competent BL21(DE3) cells first, making the cells containing pSUABC plasmid to be chemically competent and then transforming them with the pMAL-MBP-cSELENOS plasmid [27].
17. cSELENOS 188Δ tends to aggregate in the absence of reducing agents because of the formation of intermolecular disulfide bonds. It binds the HiTrapTM SP HP cation exchange column tighter and thus requires for elution at higher concentration of salt than cSELENOS.
18. Only 5% of the total expressed protein is the cSELENOS. After the purification by ion exchange chromatography, cSELENOS can be enriched up to 20% of the total protein.
19. We found that compared to the TB growth medium, the use of the LB medium results in a higher yield.
20. Since cSELENOS 188Δ aggregates it either elutes though the size exclusion column's void volume or runs ahead of the cSELENOS. Consequently, cSELENOS can be separated from the truncated form cSELENOS 188Δ based on size and can be enriched up to 70–90% after size exclusion chromatography.

21. Only cSELENOS, but not cSELENOS 188Δ, is enzymatically active. Thus, it is possible to test its presence in different fractions by using an insulin reductase assay instead of employing mass spectrometry [17].

Acknowledgments

We thank Prof. E. S. J. Arner from the Karolinska Institutet for the generous donation of the pSUABC plasmid and Dr. Marie-Paule Strub at the NIH laboratory of molecular biophysics for the donation of BL21(DE3)cys cells. The Delaware COBRE program supported this research with grants from the National Institute of General Medical Sciences under awards P30 GM110758-02 and P20 GM104316. We acknowledge the use of instrumentation made accessible through the National Science Foundation under Grant No. CHE-1337449. This work was primarily supported by the National Science Foundation under Grant No. MCB-1054447 "CAREER: Reactivity of Selenoproteins" and Grant No. MCB-1616178.

References

1. Liu J, Rozovsky S (2015) Membrane-bound selenoproteins. Antioxid Redox Signal 23(10):795–813. doi:10.1089/ars.2015.6388
2. Gladyshev VN, Arner ES, Berry MJ, Brigelius-Flohe R, Bruford EA, Burk RF, Carlson BA, Castellano S, Chavatte L, Conrad M, Copeland PR, Diamond AM, Driscoll DM, Ferreiro A, Flohe L, Green FR, Guigo R, Handy DE, Hatfield DL, Hesketh J, Hoffmann PR, Holmgren A, Hondal RJ, Howard MT, Huang K, Kim HY, Kim IY, Kohrle J, Krol A, Kryukov GV, Lee BJ, Lee BC, Lei XG, Liu Q, Lescure A, Lobanov AV, Loscalzo J, Maiorino M, Mariotti M, Sandeep Prabhu K, Rayman MP, Rozovsky S, Salinas G, Schmidt EE, Schomburg L, Schweizer U, Simonovic M, Sunde RA, Tsuji PA, Tweedie S, Ursini F, Whanger PD, Zhang Y (2016) Selenoprotein gene nomenclature. J Biol Chem 291(46):24036–24040. doi:10.1074/jbc.M116.756155
3. Lilley BN, Ploegh HL (2005) Multiprotein complexes that link dislocation, ubiquitination, and extraction of misfolded proteins from the endoplasmic reticulum membrane. Proc Natl Acad Sci U S A 102(40):14296–14301. doi:10.1073/pnas.0505014102
4. Shchedrina VA, Novoselov SV, Malinouski MY, Gladyshev VN (2007) Identification and characterization of a selenoprotein family containing a diselenide bond in a redox motif. Proc Natl Acad Sci U S A 104(35):13919–13924. doi:10.1073/pnas.0703448104
5. Lee JH, Kwon JH, Jeon YH, Ko KY, Lee SR, Kim IY (2014) Pro178 and Pro183 of selenoprotein S are essential residues for interaction with p97(VCP) during endoplasmic reticulum-associated degradation. J Biol Chem 289(20):13758–13768. doi:10.1074/jbc.M113.534529
6. Lee JH, Park KJ, Jang JK, Jeon YH, Ko KY, Kwon JH, Lee SR, Kim IY (2015) Selenoprotein S-dependent selenoprotein K binding to p97(VCP) protein is essential for endoplasmic reticulum-associated degradation. J Biol Chem 290(50):29941–29952. doi:10.1074/jbc.M115.680215
7. Ye YH, Shibata Y, Yun C, Ron D, Rapoport TA (2004) A membrane protein complex mediates retro-translocation from the ER lumen into the cytosol. Nature 429(6994):841–847. doi:10.1038/nature02656
8. Lilley BN, Ploegh HL (2004) A membrane protein required for dislocation of misfolded proteins from the ER. Nature 429(6994):834–840. doi:10.1038/nature02592
9. Turanov AA, Shchedrina VA, Everley RA, Lobanov AV, Yim SH, Marino SM, Gygi SP, Hatfield DL, Gladyshev VN (2014) Selenoprotein S is involved in maintenance and transport

of multiprotein complexes. Biochem J 462 (3):555–565. doi:10.1042/bj20140076

10. Liu J, Zhang Z, Rozovsky S (2014) Selenoprotein K form an intermolecular diselenide bond with unusually high redox potential. FEBS Lett 588(18):3311–3321. doi:10.1016/j.febslet.2014.07.037
11. Fredericks GJ, Hoffmann FW, Rose AH, Osterheld HJ, Hess FM, Mercier F, Hoffmann PR (2014) Stable expression and function of the inositol 1,4,5-triphosphate receptor requires palmitoylation by a DHHC6/selenoprotein K complex. Proc Natl Acad Sci U S A 111(46):16478–16483. doi:10.1073/pnas.1417176111
12. Fredericks GJ, Hoffmann PR (2015) Selenoprotein K and protein palmitoylation. Antioxid Redox Signal 23(10):854–862. doi:10.1089/ars.2015.6375
13. Hardy D, Bill RM, Jawhari A, Rothnie AJ (2016) Overcoming bottlenecks in the membrane protein structural biology pipeline. Biochem Soc Trans 44(3):838–844. doi:10.1042/bst20160049
14. Lyons JA, Shahsavar A, Paulsen PA, Pedersen BP, Nissen P (2016) Expression strategies for structural studies of eukaryotic membrane proteins. Curr Opin Struct Biol 38:137–144. doi:10.1016/j.sbi.2016.06.011
15. Christianson JC, Olzmann JA, Shaler TA, Sowa ME, Bennett EJ, Richter CM, Tyler RE, Greenblatt EJ, Harper JW, Kopito RR (2011) Defining human ERAD networks through an integrative mapping strategy. Nat Cell Biol 14 (1):93–105. doi:10.1038/ncb2383
16. Liu J, Srinivasan P, Pham DN, Rozovsky S (2012) Expression and purification of the membrane enzyme selenoprotein K. Protein Expr Purif 86(1):27–34. doi:10.1016/j.pep.2012.08.014
17. Liu J, Li F, Rozovsky S (2013) The intrinsically disordered membrane protein selenoprotein S is a reductase in vitro. Biochemistry 52 (18):3051–3061. doi:10.1021/bi4001358
18. Liu J, Rozovsky S (2013) Contribution of selenocysteine to the peroxidase activity of selenoprotein S. Biochemistry 52(33):5514–5516. doi:10.1021/bi400741c
19. Muller S, Senn H, Gsell B, Vetter W, Baron C, Bock A (1994) The formation of diselenide bridges in proteins by incorporation of selenocysteine residues - biosynthesis and characterization of (Se)(2)-thioredoxin. Biochemistry 33(11):3404–3412. doi:10.1021/bi00177a034
20. Strub MP, Hoh F, Sanchez JF, Strub JM, Bock A, Aumelas A, Dumas C (2003) Selenomethionine and selenocysteine double labeling strategy for crystallographic phasing. Structure 11 (11):1359–1367. doi:10.1016/j.str.2003.09.014
21. Yoshizawa S, Böck A (2009) The many levels of control on bacterial selenoprotein synthesis. Biochim Biophys Acta 1790(11):1404–1414. doi:10.1016/j.bbagen.2009.03.010
22. Arnér ESJ, Sarioglu H, Lottspeich F, Holmgren A, Bock A (1999) High-level expression in Escherichia coli of selenocysteine-containing rat thioredoxin reductase utilizing gene fusions with engineered bacterial-type SECIS elements and co-expression with the selA, selB and selC genes. J Mol Biol 292(5):1003–1016. doi:10.1006/jmbi.1999.3085
23. Rengby O, Johansson L, Carlson LA, Serini E, Vlamis-Gardikas A, Karsnas P, Arner ESJ (2004) Assessment of production conditions for efficient use of Escherichia coli in high-yield heterologous recombinant selenoprotein synthesis. Appl Environ Microbiol 70 (9):5159–5167. doi:10.1128/AEM.70.9.5159-5167.2004
24. Studier FW (2005) Protein production by auto-induction in high-density shaking cultures. Protein Expr Purif 41(1):207–234. doi:10.1016/j.pep.2005.01.016
25. Cheng Q, Johansson L, Thorell JO, Fredriksson A, Samen E, Stone-Elander S, Arner ESJ (2006) Selenolthiol and dithiol C-terminal tetrapeptide motifs for one-step purification and labeling of recombinant proteins produced in E. coli. ChemBioChem 7(12):1976–1981. doi:10.1002/cbic.200600326
26. Blommel PG, Fox BG (2007) A combined approach to improving large-scale production of tobacco etch virus protease. Protein Expr Purif 55(1):53–68. doi:10.1016/j.pep.2007.04.013
27. Cheng Q, Stone-Elander S, Arner ESJ (2006) Tagging recombinant proteins with a Sel-tag for purification, labeling with electrophilic compounds or radiolabeling with C-11. Nat Protoc 1(2):604–613. doi:10.1038/nprot.2006.87

Chapter 19

Selenocysteine-Mediated Expressed Protein Ligation of SELENOM

Jun Liu, Qingqing Chen, and Sharon Rozovsky

Abstract

A sizeable fraction of the selenoproteome encodes oxidoreductases possessing a thioredoxin fold, a structural motif that is shared among a diverse group of enzymes. In these oxidoreductases, the active site is comprised of a cysteine and a selenocysteine separated by one to two amino acids. In a subset of these selenoproteins, such as human SELENOH, SELENOM, SELENOT, SELENOV, SELENOW, and SELENOF, this redox motif is positioned immediately after the first β-sheet in a short loop, and is essential for interactions with its substrate or partners. Here, we describe the preparation of a representative member of this group, SELENOM, by selenocysteine-driven expressed protein ligation. The preparation employs a peptide bond formation between two protein fragments expressed recombinantly in *E. coli*. This method can be employed to prepare other selenoproteins.

Key words Selenoproteins, SELENOM, Selenoprotein M, Expressed protein ligation, Selenocysteine-mediated expressed protein ligation

1 Introduction

A large percentage of the selenoproteome is composed of selenoproteins whose catalytic Sec resides in a thioredoxin (Trx) fold. The fundamental Trx fold is composed of four stranded β-sheets, packed on one side flanked by two α-helices, although variations to this basic unit are common [1]. This fold encompasses several superfamilies of enzymes with functions from detoxification, to the formation and reduction of disulfide bonds in proteins [2]. In the majority of enzymes possessing the Trx fold, the redox-active site is typically positioned right before an α helix, capping it (Fig. 1). In contrast, in a subset of selenoproteins, the redox active motif Cys-X-X-Sec (where X stands for any amino acid and Sec stands for selenocysteine) is positioned immediately after the first β-sheet [3]. A conformational switch following the reduction of the intramolecular selenylsulfide bond was proposed to trigger a signaling event or binding of protein partners [4]. In humans, SELENOH,

Laurent Chavatte (ed.), *Selenoproteins: Methods and Protocols*, Methods in Molecular Biology, vol. 1661,
DOI 10.1007/978-1-4939-7258-6_19, © Springer Science+Business Media LLC 2018

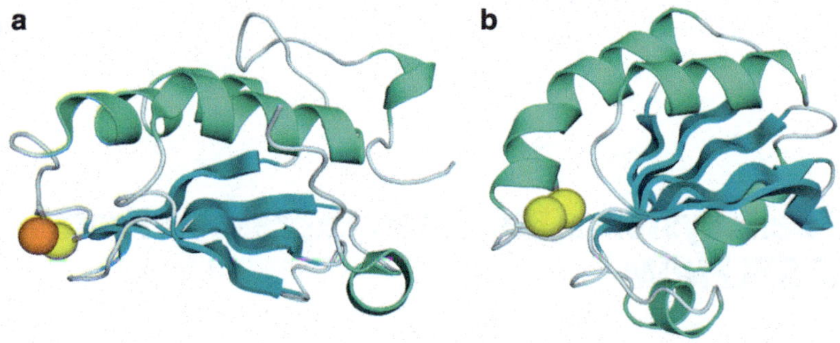

Fig. 1 Selenoproteins with a minimal thioredoxin fold position their redox motif next to the first β-sheet. (**a**) The structure of *Mus musculus* SELENOM (PDB entry 2A2P) is representative of this class. (**b**) Thioredoxin and related proteins position their redox motif immediately before an α-helix. The structure of human thioredoxin (PDB entry 1ERU) exemplifies this class. The catalytic selenylsulfide bond of SELENOM and disulfide bond of Trx are depicted as *spheres*. Selenium is in *orange* and sulfur in *yellow*

SELENOM, SELENOT, SELENOV, SELENOW, and SELENOF exhibit this common Trx fold and placement of the selenoredox motif [5, 6]. This chapter details the preparation of a representative member of this group, SELENOM, in a manner that can be generalized to other selenoproteins with a minimal Trx fold [7].

Like many selenoproteins, SELENOM (UniProt Q8WWX9) is tied to the management of cellular oxidative stress [8–14]. SELENOM resides in the endoplasmic reticulum [15] and its overexpression governs the expression of antioxidant enzymes like glutathione peroxidase and superoxide dismutase [10]. SELENOM knock-out mice showed increased weight gain and changes in metabolism [16]. It was proposed to regulate energy metabolism through participation in leptin signaling [17]. The NMR-derived structure of *Mus musculus* SELENOM U48C mutant was reported [4] (Fig. 1a), but the redox properties and precise function of the wild-type SELENOM with its CGGU motif remain undetermined.

This chapter presents a versatile chemical approach based on native chemical ligation (NCL) to prepare the wild-type selenium-containing SELENOM [18]. In Sec-driven NCL, a Sec residue mediates the formation of the peptide bond from two respective protein fragments to form the target protein [19–22]. As shown in Fig. 2, an N-terminal Sec from one fragment initiates the nucleophilic attack of the C-terminal thioester from the complementary fragment and subsequently undergoes a Se→N acyl shift to generate the native amide bond (Fig. 2). The ligation rate of Sec-mediated NCL is faster and more pH tolerant than Cys-mediated ligation, and the resulting ligation efficiency is often higher than that of Cys-mediated reactions [19, 23]. Sec-mediated NCL is an attractive method for preparing selenoproteins because it is possible

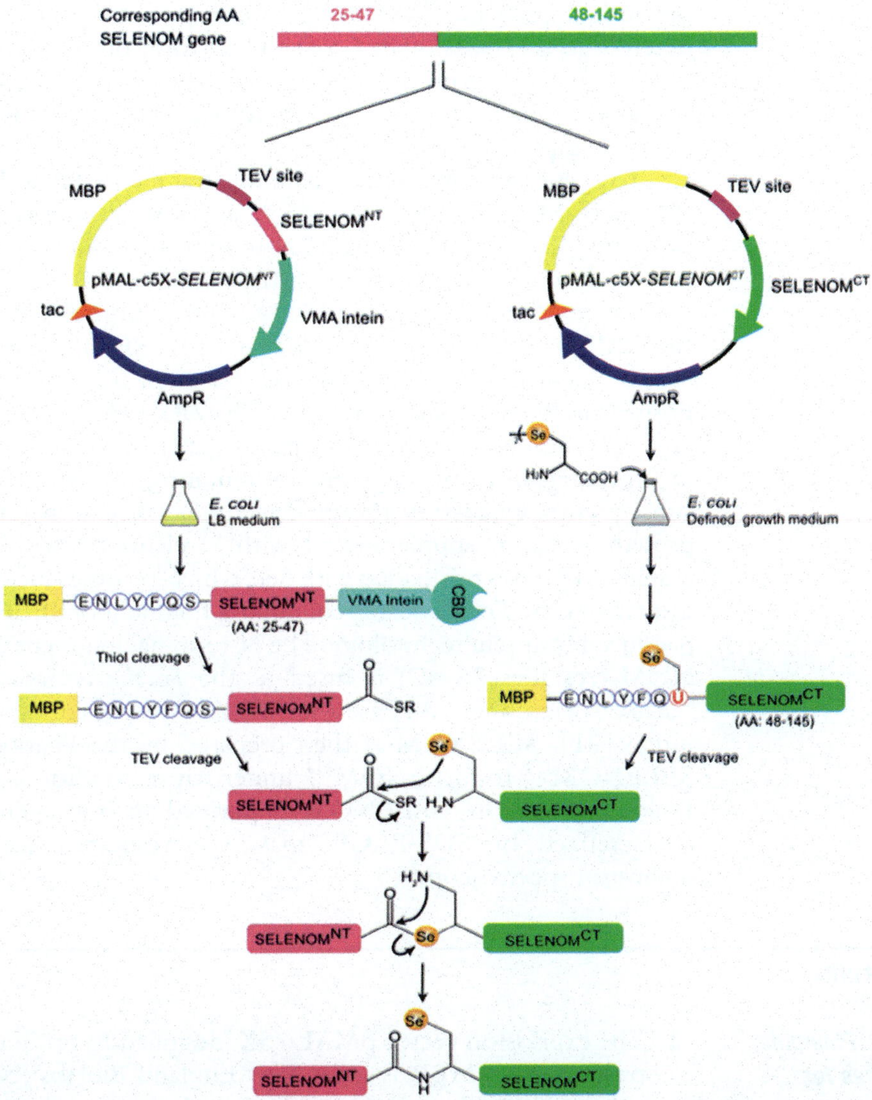

Fig. 2 Scheme of Sec-mediated expressed protein ligation of SELENOM. *E. coli* bearing the expression vector pMAL-c5X-MBP-SELENOM^{CT} (residues 48–145 of SELENOM) is grown in defined growth medium supplemented with L-selenocystine for the incorporation of Sec into the protein. Following TEV cleavage, the exposed selenolate of SELENOM^{CT} initiates the nucleophilic attack on SELENOM^{NT}-thioester. SELENOM^{NT}-thioester is prepared by intein technology (*see* Fig. 3). The native peptide bond is formed through the attack of selenolate onto the thioester, followed by a Se→N shift

to achieve high yield, and in addition it is compatible with introducing posttranslational modifications or chemical tags into the protein. However, since the preparation of selenium-containing protein fragments has traditionally relied on chemical synthesis, this method was restricted by limitations on the size of the Sec-containing peptide fragment and its solubility [21, 24–27]. The production of SELENOM discussed here is distinguished from previous approaches [28] in that all fragments are prepared by heterologous expression in *E. coli* and rely only on standard protein expression and purification [7]. This reduces cost and simplifies preparation by alleviating the need to synthesize the seleno-containing part of the protein.

To generate the Sec-containing fragment of SELENOM (SELENOM^{CT} (residues 48–145)), a cytosolic maltose binding protein (MBP) is coupled to its N-terminus for enhanced expression and solubility, generating MBP-SELENOM^{CT}. A Tobacco Etch Virus (TEV) protease cleavage site is introduced between the MBP and SELENOM^{CT} so that SELENOM^{CT} can be efficiently released since TEV protease has good tolerance at its P1′ position in the recognition site ENLYFQ/P1′ (where P1′ stands for any canonical amino acid except P) [29]. The Sec residue is incorporated into the fragment by mutating the Sec to Cys and subsequently expressing the protein in *E. coli* grown on defined growth medium supplemented with L-selenocystine [30]. The cysteinyl-tRNA is misloaded with Sec, which becomes incorporated instead of Cys. The thioester-containing fragment of SELENOM is produced by intein technology. The N-terminal fragment of SELENOM (residues 25–47) is fused to the *Saccharomyces cerevisiae* Vacuolar Membrane ATPase (Sce VMA1; 454 residues; 56 kDa) intein [31]. SELENOM is then prepared by the ligation of the SELENOM^{CT} to SELENOM^{NT} under native conditions (Fig. 2). Following ligation, SELENOM is purified to homogeneity and characterized by SDS-PAGE, mass spectrometry, and circular dichroism spectroscopy.

2 Materials

2.1 Gene, Plasmids, and Enzymes for Molecular Cloning

1. The expression vector pMAL-c5X, designed to produce fusion proteins with MBP, is from New England Biolabs (NEB). To aid purification, a general cloning vector was prepared by introducing a hexahistidine tag to the N-terminus of MBP [32]. In addition, a TEV protease site was inserted at the C-terminus of MBP to facilitate efficient release of the target protein or protein fragment. The VMA intein was introduced following MBP and the TEV cleavage site allowing cloning of the protein of interest fused to MBP, the VMA intein or both. This pMAL-c5X-VMA cloning vector is available through Addgene (plasmid #86590) [7].
2. The gene of *Homo sapiens* SELENOM (accession number: NP_536355.1) optimized for *E. coli* expression can be obtained from Addgene (plasmid #86579). The gene was inserted into a pMAL vector to allow expression of SELENOM U48C fused to hexahistidine-tagged MBP creating the plasmid

pMAL-c5X-SELENOM^{U48C}. A TEV protease cleavage site exists between MBP and SELENOM U48C so that SELENOM U48C can be efficiently released from MBP [7].

3. Restriction enzymes SapI and NdeI are from NEB.
4. Q5™ site-directed mutagenesis kit is from NEB.

2.2 Buffers and Chemicals for Protein Expression and Purification

1. LB medium: 10 g/L tryptone, 5 g/L yeast extract, 5 g/L NaCl, 2 g/L glucose. Autoclave the medium and allow it to cool to room temperature prior to adding antibiotics.
2. LB agar plate: 10 g/L tryptone, 5 g/L yeast extract, 10 g/L NaCl, 15 g/L agar, autoclaved. Supplement with 100 μg/mL of ampicillin (Amp) once the solution temperature is below 50 °C but before solidification, and pour into individual petri dishes. Seal the LB agar plate with Parafilm™ after complete solidification, and store at 4 °C for up to 1 week.
3. Lysis buffer: 50 mM sodium phosphate, 200 mM sodium chloride (NaCl), 2 mM ethylenediaminetetraacetic acid (EDTA), 1 mM phenylmethylsulfonyl fluoride (PMSF), and 1 mM benzamidine, pH 7.5.
4. Amylose binding buffer: 50 mM sodium phosphate, 200 mM NaCl, and 2 mM EDTA, pH 7.5.
5. Amylose elution buffer: 50 mM sodium phosphate, 200 mM NaCl, 20 mM maltose, and 2 mM EDTA, pH 7.5.
6. IMAC binding buffer: 50 mM sodium phosphate, 200 mM NaCl, and 10 mM imidazole, pH 7.5.
7. IMAC elution buffer: 50 mM sodium phosphate, 200 mM NaCl, and 1 M imidazole, pH 7.5.
8. Chitin binding buffer: 25 mM 2-(*N*-morpholino)ethanesulfonic acid (MES), 200 mM NaCl, and 2 mM EDTA, pH 6.5.
9. Chitin cleavage buffer: 25 mM MES, 200 mM NaCl, 75 mM sodium 2-mercaptoethanesulfonate (MESNA), and 2 mM EDTA, pH 6.5.
10. Gel filtration buffer: 50 mM sodium phosphate, 200 mM NaCl, 2 mM EDTA, and 5 mM 1,4-dithio-DL-threitol (DTT), pH 7.5.

2.3 Protein Expression and Purification

1. Isopropyl 1-thio-β-D-galactopyranoside (IPTG).
2. L-selenocystine (*see* **Note 1** for preferred vendors of L-selenocystine).
3. Amylose resin and chitin resin from NEB.
4. HisTrap™ HP, HiPrep™ 26/10, and HiPrep™ 16/60 Sephacryl™ S-100 HR columns from GE Healthcare Life Sciences.

2.4 Expressed Selenoprotein Ligation

1. Sodium 2-mercaptoethanesulfonate (MESNA).
2. 4-Mercaptophenylacetic acid (MPAA).
3. pH meter with a micro probe.

2.5 Mass Spectrometry Characterization

1. Mass spectrometry autosampler vials (Waters).
2. Trypsin-ultra, mass spectrometry grade (NEB).
3. Iodoacetamide (IAM).
4. Spin desalting columns (Pierce).
5. Buffer for trypsin digestion: 50 mM Tris–hydrochloride, and 20 mM calcium chloride, pH 8.0.

2.6 Circular Dichroism Spectroscopy

1. Circular dichroism spectropolarimeter (such as Jason J-810 spectropolarimeter).
2. 10 mM potassium phosphate buffer, 50 mM sodium sulfate (Na_2SO_4), pH 7.5.
3. Superdex™ 75 10/300 GL (GE Healthcare Life Sciences).

3 Methods

3.1 Molecular Cloning of MBP-SELENOM^{NT}-VMA and MBP-SELENOM^{CT}

1. Clone MBP-SELENOM^{NT} gene, which includes MBP, TEV cleavage site, and SELENOM residues 25–47, into vector pMAL-c5X-VMA using restriction sites NdeI and SapI to create the vector pMAL-c5X-*SELENOM^{NT}* for the expression of SELENOM^{NT} fragment (Fig. 2). A TEV protease cleavage site assists the efficient release of SELENOM^{NT} fragment from the fusion construct.
2. The expression vector for SELENOM^{CT} is prepared by deleting residues 25–47 from pMAL-c5X-*SELENOM^{U48C}* to generate pMAL-c5X-*SELENOM^{CT}* (SELENOM residues 48–145, beginning with U48 mutated to C48). Mutagenesis can be accomplished with the Q5™ Site-Directed Mutagenesis Kit from NEB.

3.2 Protein Expression of MBP-SELENOM^{NT} Thioester

1. Transform pMAL-c5X-SELENOM^{NT} into *E. coli* strain BL21 (DE3) or BL21. Incubate on ice 30 ng of plasmid DNA with 50 μL of chemically competent BL21(DE3) in a 1.5 mL microcentrifuge tube for 30 min. Heat shock in a water bath at 42 °C for 50 s. Allow the cells to recover on ice for an additional 2–5 min, then add 400 μL of autoclaved LB medium and incubate with shaking at 250 rpm for 1 h at 37 °C. Spin down the cells by centrifugation at 2500 × *g* and discard 400 μL of supernatant. Gently resuspend the cell pellet in the remainder of the LB medium and plate cells evenly onto LB agar plates supplemented with 100 μg/mL Amp, and incubate overnight at 37 °C.

2. Begin an overnight culture by inoculating 100 mL of autoclaved LB medium, supplemented with 100 μg/mL Amp, with a few colonies from a freshly transformed plate of MBP-SELENOM^{NT}-VMA. Incubate with shaking at 250 rpm overnight at 37 °C.
3. The next morning, transfer 10 mL of overnight culture to 1 L of fresh autoclaved LB medium supplemented with 100 μg/mL Amp and grow at 37 °C with good aeration. When the optical density (OD) at 600 nm reaches 0.5, the temperature is lowered to 18 °C, and the cells are allowed to shake at the lower temperature for an additional 0.5 h. Protein expression is induced with 0.5 mM of IPTG.
4. The cells are harvested after 18–20 h by centrifugation at 5000 × *g* at 4 °C for 15 min. Decant as much supernatant as possible and resuspend the cell paste in the lysis buffer.

3.3 Protein Purification of MBP-SELENOM^{NT} Thioester

The MBP-SELENOM^{NT} thioester is obtained via two chromatographic steps. First, the MBP-SELENOM^{NT}-VMA fusion protein is obtained by amylose affinity chromatography through MBP binding. Then while immobilized through the chitin binding domain on the chitin column, the VMA fusion partner is cleaved off using MESNA, releasing the MBP-SELENOM^{NT} fragment in the chitin column flow-through while generating the C terminal thioester.

1. Lyse the cells by a high-pressure homogenizer (EmulsiFlex™-C5, Avestin) on ice, operating continuously at 9000 psi.
2. Immediately remove cell debris by a 1 h centrifugation of the cell lysate at 5000 × *g* at 4 °C. Collect the supernatant.
3. Carry out amylose purification at 4 °C. Equilibrate the amylose column with 5 column volumes (CV) of amylose binding buffer. Load the supernatant onto the column by gravity.
4. Wash with 5 CV of amylose binding buffer.
5. Elute with 5 CV amylose elution buffer and collect the eluent fusion protein MBP-SELENOM^{NT}-VMA.
6. Equilibrate the chitin column with 4 CV of chitin binding buffer. Load the previous eluent from **step 4** onto the chitin column.
7. Wash with 4 CV of chitin binding buffer.
8. Wash with 2 CV of chitin column cleavage buffer, allowing the resin to become fully soaked. Drain the column by gravity until there is about 1 in. of cleavage buffer left above the resin bed, cap the bottom of the column and allow intein cleavage to proceed at 25 °C overnight.
9. Elute with 2 CV of chitin column binding buffer, and collect the fractions that contain MBP-SELENOM^{NT} thioester.

10. Repeat **steps** 7 and **8** to fully cleave any MBP-SELENOM^{NT}-VMA still bound to the chitin column in order to obtain the highest yield of the thioester fragment (*see* **Note 2** regarding the stability of the thioester).
11. Run SDS-PAGE and mass spectrometry to confirm intein cleavage and thioester formation. A typical mass spectrum of the MBP-SELENOM^{NT} with thioester is shown in Fig. 3.
12. Measure protein concentration with an extinction coefficient of 74,830 M^{-1} cm^{-1}; note that excessive thiol reagents may interfere with UV measurements. Concentrate the MBP-SELENOM^{NT} with thioester to 10 mg/mL for later use in expressed selenoprotein ligation of SELENOM.

3.4 Protein Expression of MBP-SELENOM^{CT}

The MBP-SELENOM^{CT} fusion protein is obtained via single step purification by amylose affinity chromatography.

1. Transform pMAL-c5X-SELENOM^{CT} into *E. coli* strain BL21 (DE3) using the procedure described in Subheading 3.3.
2. Begin a small overnight culture by inoculating 10 mL of autoclaved LB medium, supplemented with 100 μg/mL Amp, with a few colonies from a freshly transformed plate of MBP-SELENOM^{CT}. Incubate with shaking at 250 rpm overnight at 37 °C.
3. Collect 1 mL of the above overnight culture and pellet the cells at 2500 × *g* for 5 min. Resuspend the pellet in 1 L of modified MDAG growth medium (based on Studier original MDAG [33]) as listed in Table 1, step 1. Grow cells at 37 °C with shaking at 250 rpm until OD at 600 nm exceeds 1.5. This typically takes up to 14 h.
4. Supplement the growth medium as detailed in Table 1, step 2 (*see* **Notes 3** and **4** on improving Sec incorporation), continuously shake for an additional 10 min at 37 °C.
5. Lower the temperature to 18 °C, allow cells to equilibrate for an additional 10 min, and induce protein expression by adding 0.5 mM IPTG.
6. Continue protein expression for 18–20 h post-induction at 18 °C, then harvest by centrifugation at 5000 × *g* at 4 °C for 15 min. Decant as much of the supernatant as possible and resuspend the cell pellet in the lysis buffer.

3.5 Protein Purification of MBP-SELENOM^{CT}

1. Lyse the cells by a high-pressure homogenizer on ice, operating continuously at 9000 psi.
2. Immediately remove cell debris by a 1 h centrifugation of the cell lysate at 5000 × *g* at 4 °C. Collect the supernatant.
3. Carry out amylose purification at 4 °C. Equilibrate the amylose column with 4 CV of amylose binding buffer. Load the supernatant onto the amylose column, and let it drain by gravity.

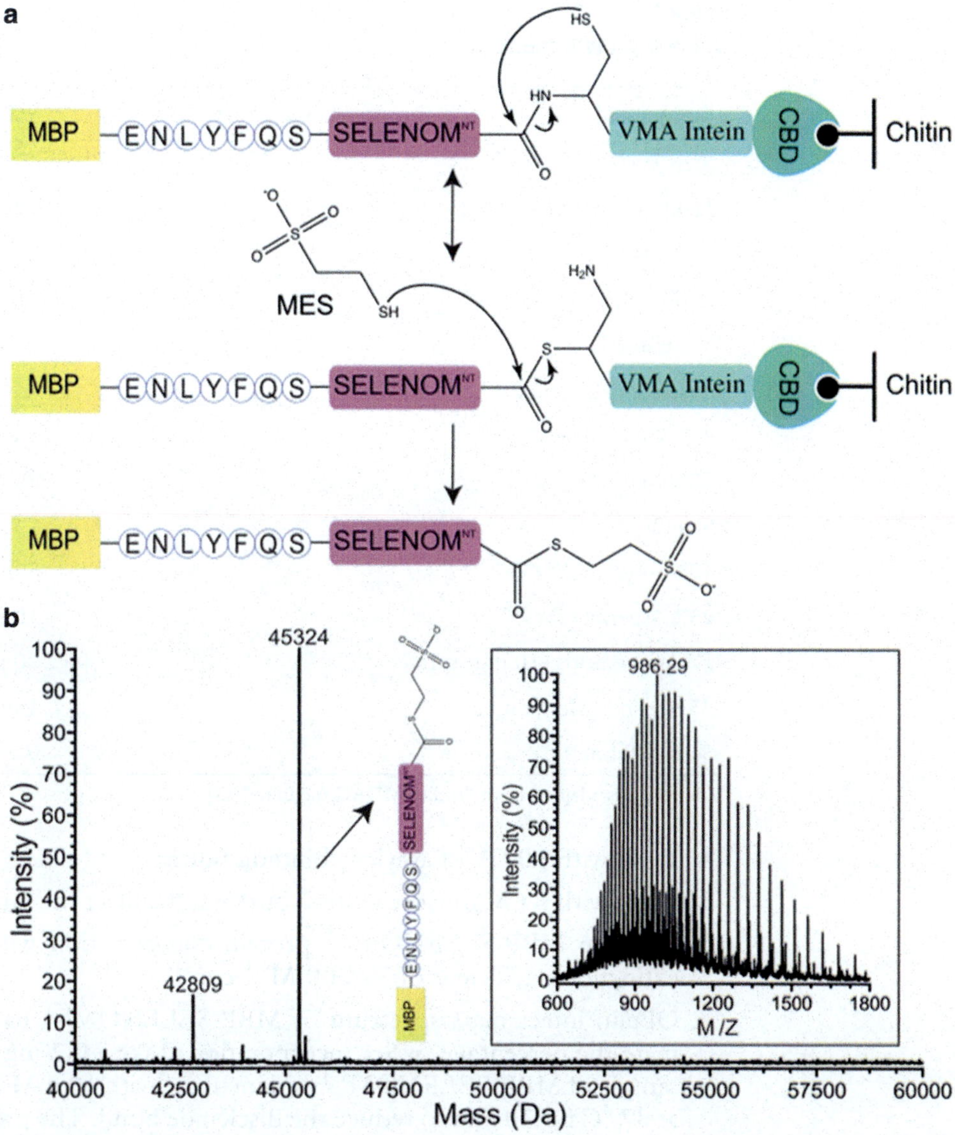

Fig. 3 MBP-SELENOM^{NT} thioester characterization by mass spectrometry. (**a**) Scheme demonstrating MBP-SELENOM^{NT} thioester production. MBP-SELENOM^{NT}-VMA fusion is purified by amylose affinity chromatography and then bound to a chitin column through the chitin binding domain fused at the C-terminal of intein. During cleavage of intein, the thioester form of MBP-SELENOM^{NT} (MBP-SELENOM^{NT}-MES) is formed and eluted from the column with the addition of 75 mM MESNA. (**b**) MBP-SELENOM^{NT} thioester detected by mass spectrometry. The theoretical mass of the MBP-SELENOM^{NT} thioester is 45,322 Da, and measured as 45,324 Da. The 42,809 Da peak arises from a truncation. *Inset* shows the *M*/*Z* spectrum. Reprinted with permission from Liu J, Chen Q, Rozovsky S (2017) Utilizing Selenocysteine for Expressed Protein Ligation and Bioconjugations. J Am Chem Soc. doi:10.1021/jacs.6b10991. Copyright 2017 American Chemical Society

Table 1
Defined growth medium

Component	Volume
Step 1	
H_2O (autoclaved)	926 mL
1 M $MgSO_4$	2 mL
1000× metals[a]	0.2 mL
Vitamin mix[a]	1 mL
Ampicillin 100 mg/mL	1 mL
40% Glucose (W/V)	5 mL
50× M[a]	20 mL
Step 2	
40% Glucose (W/V)	5 mL
25% Aspartate (W/V)	10 mL
17 Amino acids (10 mg/mL)[a]	20 mL
25 mg/mL Methionine	4 mL
Selenocystine	0.1 g

[a]The recipe is adapted from Studier MDAG medium [33]

4. Wash with 10 CV of amylose binding buffer.
5. Elute with 5 CV amylose elution buffer and collect the sample.
6. Measure MBP-SELENOM^{CT} protein concentration with an extinction coefficient of 79,300 M^{-1} cm^{-1}.
7. Obtain intact mass spectrum of MBP-SELENOM^{CT} to estimate the percentage of Sec incorporation. Take a 0.2 mg/mL sample of MBP-SELENOM^{CT} and incubate with 10 mM DTT at 37 °C for 30 min to reduce the diselenide bond. The protein can then be alkylated by adding excess IAM or analyzed reduced. Inject the sample onto a reversed-phase liquid chromatography (RPLC) column such as a Waters Xbridge C4 that is appropriate for protein analysis. Perform an acetonitrile elution gradient from 5 to 95% supplemented with 0.1% formic acid. MBP-SELENOM^{CT} elutes close to halfway of the gradient. Sec incorporation is estimated by quantitating the peak intensity of the Sec-containing form over that of the Cys- and Sec-containing protein forms together (*see* **Note 4**).

   With the expression strategy described in Subheading 3.4, higher than 90% incorporation of Sec is typically achieved for Sec-mediated NCL (*see* **Notes 5** and **6** for Sec enrichment). A typical mass spectrum of reduced MBP-SELENOM^{CT} is shown in Fig. 4a.

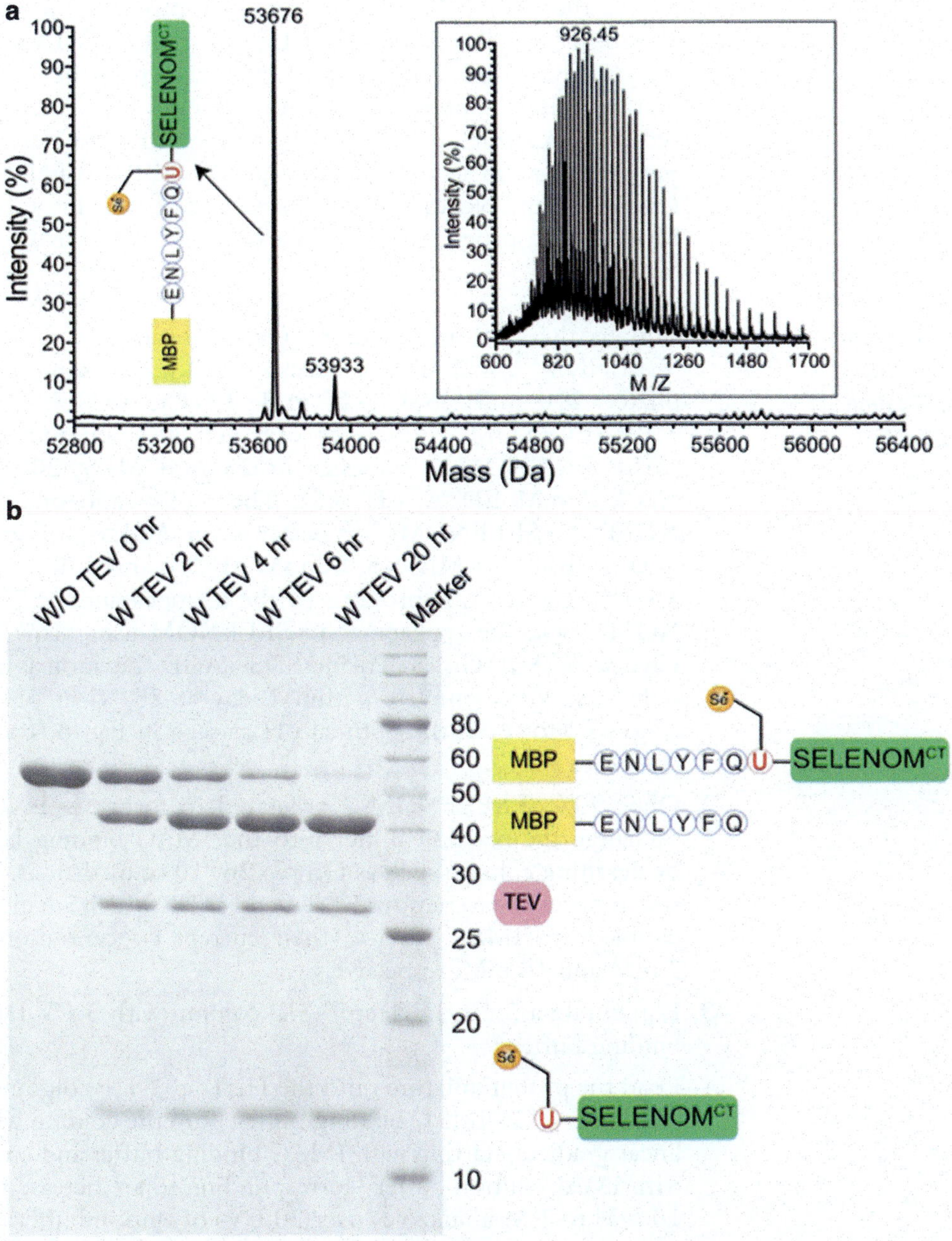

Fig. 4 MBP-SELENOM^{CT} purification and the release of SELENOM^{CT} from MBP by TEV protease cleavage. (**a**) Deconvoluted electrospray ionization mass spectrum of reduced MBP-SELENOM^{CT}. The calculated molecular mass of MBP-SELENOM^{CT} is 53,674 Da. The 53,933 Da form is due to a 6-phosphogluconolactone modification, most likely on the His$_6$-tag at the N-terminal of MBP [34]. *Inset* shows the *M*/*Z* spectrum. (**b**) MBP-SELENOM^{CT} can be efficiently cleaved by TEV protease in the presence of a reductant (here DTT). Proteins are identified by 16% Tris-Glycine SDS-PAGE under reducing conditions. MBP-SELENOM^{CT} is incubated with TEV protease at a molar ratio of 5:1 in 50 mM sodium phosphate, 200 mM NaCl, 2 mM EDTA, and 10 mM DTT, pH 7.5. Reprinted with permission from Liu J, Chen Q, Rozovsky S (2017) Utilizing Selenocysteine for Expressed Protein Ligation and Bioconjugations. J Am Chem Soc. doi:10.1021/jacs.6b10991. Copyright 2017 American Chemical Society

8. Cleave the MBP-SELENOM^{CT} by TEV protease, purified as previously described [32, 35], at 1:5 molar ratio of TEV protease to MBP-SELENOM^{CT} at 25 °C for 16 h in the amylose elution buffer supplemented with 10 mM DTT. A strong reducing reagent such as DTT, and room temperature is required for efficient TEV cleavage of MBP-SELENOM^{CT} (*see* Fig. 4b and **Notes 7** and **8** on selection of reducing agents). Monitor the progress of TEV cleavage by 16% Tris-Glycine SDS-PAGE.

3.6 Expressed Protein Ligation and Purification of SELENOM

1. Add to the cleavage mixture from Subheading 3.5 (MBP-SELENOM^{CT} with TEV protease in 50 mM sodium phosphate, 200 mM NaCl, 2 mM EDTA, pH 7.5), the MBP-SELENOM^{NT} thioester obtained in Subheading 3.3 (MBP-SELENOM^{NT} thioester in 25 mM MES, 200 mM NaCl, 2 mM EDTA, pH 6.5). The molar ratio of SELENOM^{CT} to SELENOM^{NT} thioester is 1:5. Add to the ligation mixture 100 mM MESNA, 50 mM MPAA, and 5 mM DTT. Monitor the reaction progress by the disappearance of SELENOM^{CT} and the appearance of SELENOM using 16% Tris-Glycine SDS-PAGE or Tricine-SDS-PAGE. Ligation is typically over 90% complete within 1 day at 25 °C at pH 7.0 (Fig. 5). However, the optimal pH as seen in Fig. 6 is at 6.5 (*see* **Note 9**).
2. Once 90% of SELENOM^{CT} is ligated to form SELENOM, exchange the reaction buffer into the IMAC binding buffer by desalting column such as HiPrep 26/10 or by dialysis. This step is necessary to remove EDTA and DTT, which are incompatible with IMAC resin at their current concentrations of 2 mM and 10 mM, respectively.
3. Equilibrate a 5 mL HisTrap™ HP column with 5 CV IMAC binding buffer.
4. Load the protein mixture onto the HisTrap™ HP column and wash with 5 CV IMAC binding buffer. Run the column with a linear gradient starting with IMAC binding buffer and ending with IMAC elution buffer (corresponding to an increase from 10 mM to 1 M imidazole) over 20 CVs of elution buffer.
5. Collect the fractions containing SELENOM, and concentrate using a centrifugal filter unit (Amicon Ultra-15 Centrifugal Filter Units, 3 kDa cutoff) to 5 mg/mL. Then load the sample onto an appropriate size exclusion column suitable for small proteins (<20 kDa) such as Superdex™ 75 10/300 GL or Sephacryl™ S-100, preequilibrated with gel filtration buffer. Collect fractions containing SELENOM and verify their purity by visualizing on 16% Tris-Glycine SDS-PAGE.

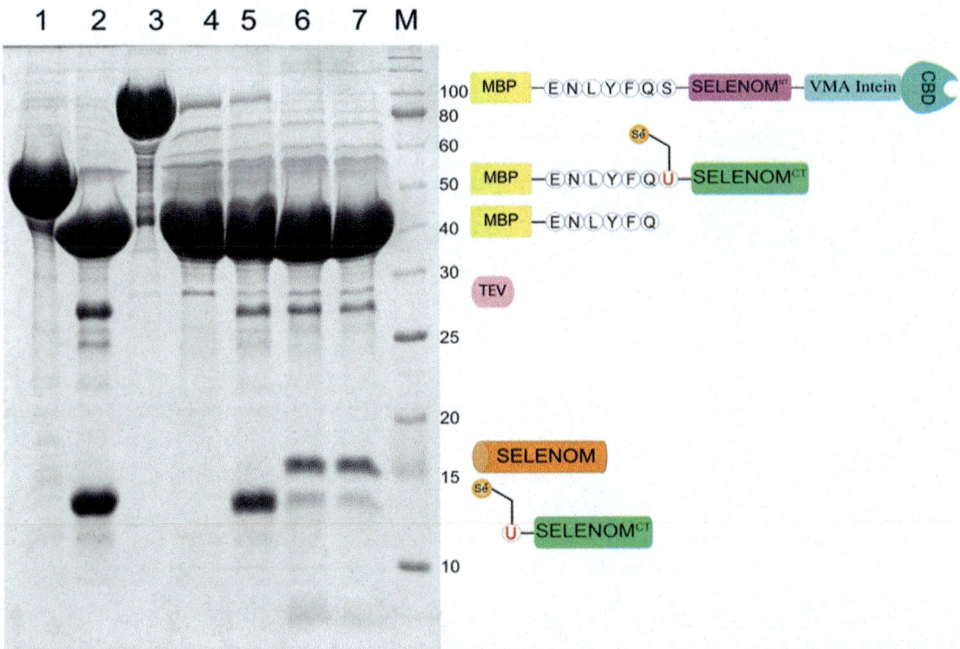

Fig. 5 SELENOM ligation as monitored by SDS-PAGE: *Lane 1*: MBP-SELENOM^{CT}; *lane 2*: same as *lane 1* but after TEV protease cleavage; *lane 3*: MBP-SELENOM^{NT}-VMA; *lane 4*: MBP-SELENOM^{NT} thioester; *lanes 5–7*: the ligation of SELENOM monitored on days 0, 1, and 2. *Lane M*: molecular mass standards. After TEV protease cleavage, SELENOM^{NT} thioester cannot be detected because of its 2 kDa mass. Reprinted with permission from Liu J, Chen Q, Rozovsky S (2017) Utilizing Selenocysteine for Expressed Protein Ligation and Bioconjugations. J Am Chem Soc. doi:10.1021/jacs.6b10991. Copyright 2017 American Chemical Society

6. Determine SELENOM protein concentration by UV-Vis spectroscopy with an extinction coefficient of 18,450 M^{-1} cm^{-1}.
7. Obtain intact mass spectrum of purified SELENOM according to Subheading 3.5, **step** 7 in order to further characterize sample purity. A representative spectrum of SELENOM at the final purifications stage is shown in Fig. 7.
8. Co-purified cellular thiols may present an obstacle to downstream applications. *See* **Note 10** on removing small thiol species from purified SELENOM.

3.7 Tandem Mass Sequencing of SELENOM

1. Reduce 6 μg of SELENOM in 60 μL of trypsin digest buffer by the addition of 5 mM DTT. Incubate for 5 min at 95 °C to achieve full reduction and denaturation. Allow the sample to cool to room temperature, then immediately add 12 mM IAM and incubate for another 20 min to fully alkylate exposed Sec and Cys residues, as well as the remaining DTT.
2. Add 0.3 μg of the trypsin enzyme into the above mixture and digest at 37 °C for 16 h. Employ a 15 μm, 200 Å pore size ZipTip™ to desalt the trypsin digested protein sample using the vendor procedure.

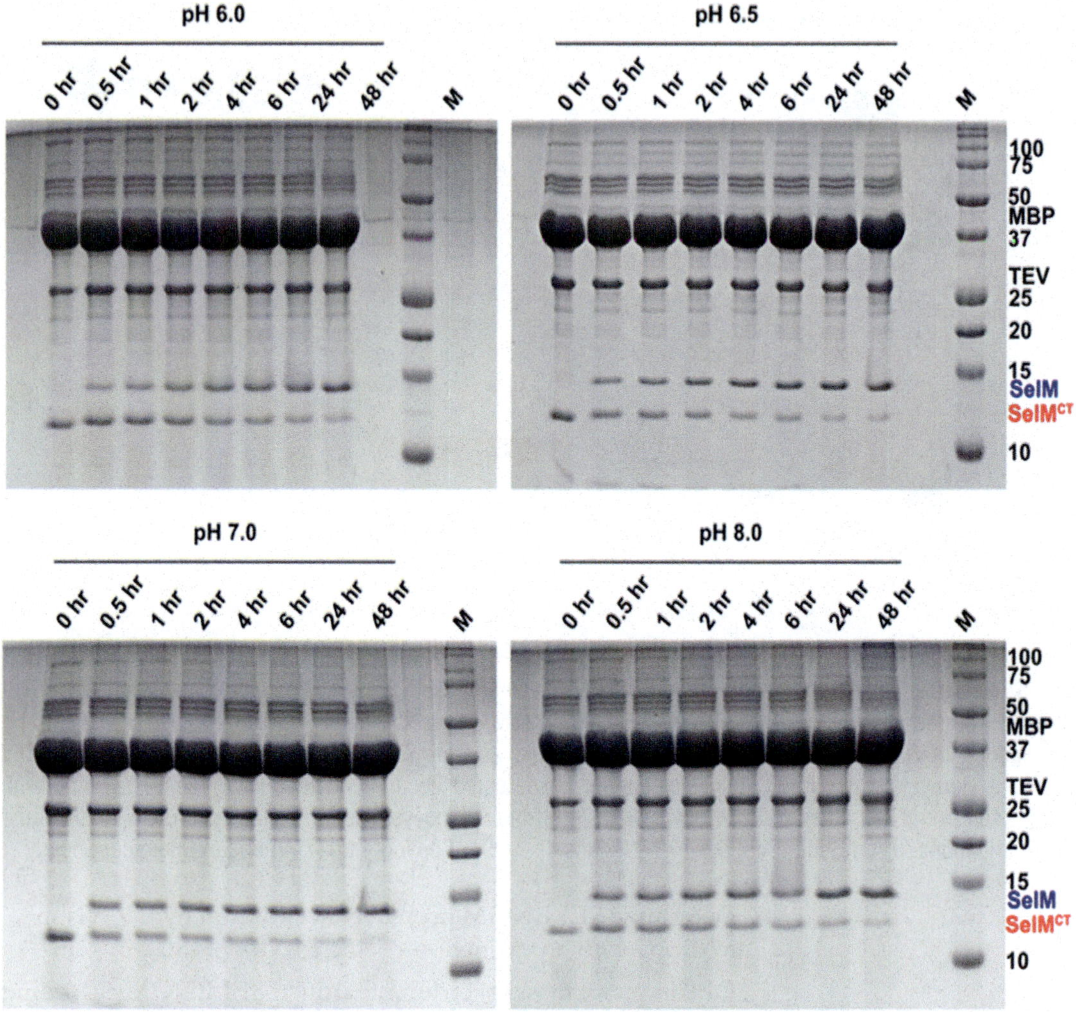

Fig. 6 The rate of SELENOM ligation at 25 °C at pH 6–8. MBP-SELENOM^{NT} was incubated with TEV cleaved SELENOM^{CT} at a molar ratio of 5:1 in 50 mM sodium phosphate, 200 mM NaCl, and 2 mM EDTA, supplemented with 100 mM MESNA, 50 mM MPAA, and 5 mM DTT. The reaction was monitored for 48 h. Proteins are identified on a 16% Tricine-SDS-PAGE with reducing agents. The reaction yield is higher than 90% at pH 6 and above as judged by the disappearance of SELENOM^{CT}. Reprinted with permission from Liu J, Chen Q, Rozovsky S (2017) Utilizing Selenocysteine for Expressed Protein Ligation and Bioconjugations. J Am Chem Soc. doi:10.1021/jacs.6b10991. Copyright 2017 American Chemical Society

3. The position of the Sec residue is confirmed by sequencing the Sec-containing peptide using a Q Exactive Orbitrap mass spectrometer interfaced with an Ultimate 3000 LC system (or equivalent setup). Load 1 μg of trypsin digested SELENOM onto a reverse-phase LC column such as C18 suitable for small peptide analysis. Perform an acetonitrile gradient supplemented with 0.1% formic acid over a run time of at least 45 min to

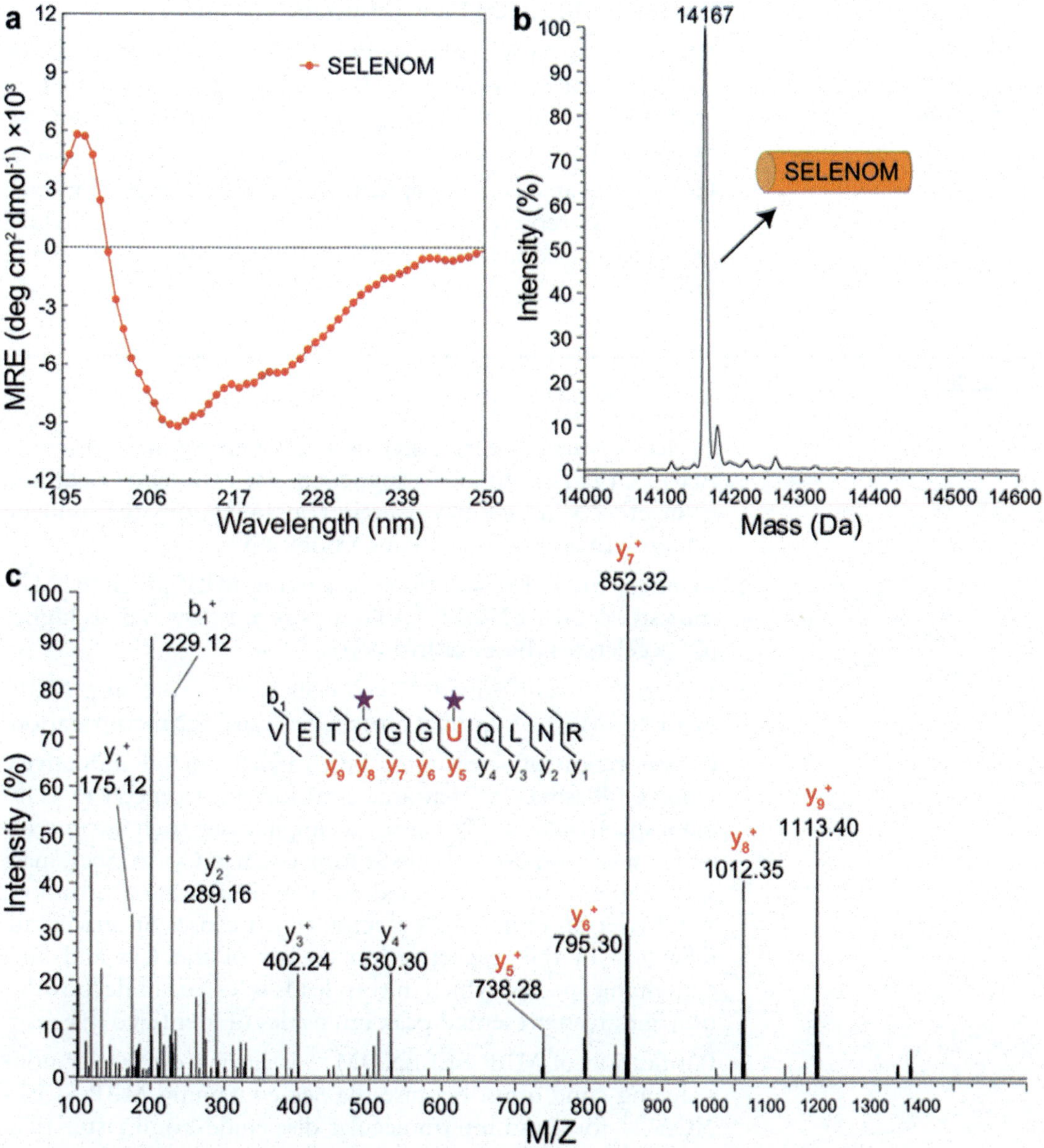

Fig. 7 Characterization of SELENOM. (**a**) CD spectrum of SELENOM prepared by Sec-EPL. (**b**) Deconvoluted ESI-MS of intact SELENOM. (**c**) Tandem MS sequencing of the peptide from panel **b** confirms Sec's presence in SELENOM. Sec and Cys residues were alkylated with iodoacetamide (*purple star*). Fragment ions that contain Sec are colored *red*. Adapted with permission from Liu J, Chen Q, Rozovsky S (2017) Utilizing Selenocysteine for Expressed Protein Ligation and Bioconjugations. J Am Chem Soc. doi:10.1021/jacs.6b10991. Copyright 2017 American Chemical Society

ensure complete elution of all digested fragments. Acquire MS data using a stepped normalized collision energy of 28, 30, 35 eV. A representative sequencing of SELENOM Sec-containing peptide is shown in Fig. 7.

3.8 Circular Dichroism Spectroscopy of SELENOM

1. Exchange SELENOM buffer into the CD buffer.
2. Record the far-UV spectra using a 1 mm path-length cell for the 190–250 nm region at 20 °C. Take three accumulation scans for baseline, take average of eight accumulation scans for SELENOM.
3. Plot the mean residual elipticity against wavelength for each sample. A representative CD spectrum for SELENOM prepared by Sec-mediated native chemical ligation is shown in Fig. 7.

4 Notes

1. L-selenocystine is best purchased through Sigma-Aldrich or Acros Organics. In our experience L-selenocystine acquired from other vendors may contain contaminations that impede cell growth and reduce Sec incorporation.
2. Maintain the pH of solutions containing MBP-SELENOM^{NT} thioester or SELENOM^{NT} thioester at or below 6.5, as higher pH accelerates thioester hydrolysis.
3. Use freshly transformed *E. coli* cells when expressing MBP-SELENOM^{CT} for optimal protein yield and Sec incorporation.
4. Incorporation of selenium is best carried out by Inductively Coupled Plasma (ICP)-atomic emission spectrometry or ICP-mass spectrometry. However, in simple cases such as proteins with a single amino acid substitution (such as Cys vs. Sec), mass spectrometry provides a good estimation. During acquisition, avoid oversaturating the detector which causes an inaccurate reflection in the respective ion counts of the Cys and Sec-containing species, which in turn leads to errors in deconvolution and misrepresented peak intensities of identified masses.
5. Purification of MBP-SELENOM^{CT} is carried out under non-reducing conditions, because the Sec-containing MBP-SELENOM^{CT} forms an intermolecular diselenide bond faster than the formation of disulfide bond by the Cys-containing variant. Extended wash helps to enrich the fraction of Sec-containing MBP-SELENOM^{CT} due to tighter binding of the dimeric diselenide-containing form to affinity columns over that of the monomeric Cys-containing form.
6. Excess L-selenocystine in the growth medium can lead to decreased yield due to cellular toxicity. In our experience, 0.1 g L-selenocystine per L growth medium resulted in incorporation of over 90% Sec into target proteins. In the cases where the percent of Sec incorporation is low the concentration of L-selenocystine can be increased to 0.15 g per L growth medium at the cost of about 20–30% decrease in protein yield.

7. MBP-SELENOM^{CT} forms a diselenide bond that prevents efficient TEV protease cleavage. Therefore, reducing reagents such as DTT or tris(2-carboxyethyl)phosphine (TCEP) is required to break the diselenide bond during TEV protease cleavage. Similarly, the diselenide bond between SELENOM^{CT} must be broken during ligation using reducing agents.
8. When possible, avoid the use of TCEP and tris(hydroxypropyl) phosphine (THP) which can promote selenium elimination from Sec.
9. Hydrolysis of the thioester increases with increased pH. Sec-mediated ligation can be carried out at pH lower than 7.0 to stabilize the thioester (Fig. 6).
10. An estimated 0.1–1 mM of cellular-free thiols is usually present in the purified SELENOM protein sample after size exclusion chromatography. It is recommended that this free thiol population be removed to avoid downstream interference with protein activity. Reduce the protein with 10 mM DTT for 2 h at 25 °C or overnight at 4 °C. Then extensively dialyze the protein against 50 mM sodium phosphate, 200 mM NaCl, 2 mM EDTA, pH 7.5, until the free thiol count of the protein is undetectable using 5,5′-dithiobis-(2-nitrobenzoic acid) or DTNB assay [36].

Acknowledgments

This project was supported by the Delaware COBRE program, with grants from the National Institute of General Medical Sciences—NIGMS (5 P30 GM110758-02 and 5 P20 GM104316) from the National Institutes of Health. Qingqing Chen acknowledges support by chemical biology training grant award T32GM008550 from the National Institute of General Medical Sciences. This work was primarily supported by the National Science Foundation under Grant No. MCB-1054447 "CAREER: Reactivity of Selenoproteins" and Grant No. MCB-1616178.

References

1. Qi Y, Grishin NV (2005) Structural classification of thioredoxin-like fold proteins. Proteins 58(2):376–388. doi:10.1002/prot.20329
2. Atkinson HJ, Babbitt PC (2009) An atlas of the thioredoxin fold class reveals the complexity of function-enabling adaptations. PLoS Comput Biol 5(10):e1000541. doi:10.1371/journal.pcbi.1000541
3. Aachmann FL, Fomenko DE, Soragni A, Gladyshev VN, Dikiy A (2007) Solution structure of selenoprotein W and NMR analysis of its interaction with 14-3-3 proteins. J Biol Chem 282(51):37036–37044. doi:10.1074/jbc.M705410200
4. Ferguson AD, Labunskyy VM, Fomenko DE, Araç D, Chelliah Y, Amezcua CA, Rizo J, Gladyshev VN, Deisenhofer J (2006) NMR

structures of the selenoproteins Sep15 and SelM reveal redox activity of a new thioredoxin-like family. J Biol Chem 281 (6):3536–3543. doi:10.1074/jbc. M511386200
5. Hatfield DL, Tsuji PA, Carlson BA, Gladyshev VN (2014) Selenium and selenocysteine: roles in cancer, health, and development. Trends Biochem Sci 39(3):112–120. doi:10.1016/j. tibs.2013.12.007
6. Labunskyy VM, Hatfield DL, Gladyshev VN (2014) Selenoproteins: molecular pathways and physiological roles. Physiol Rev 94 (3):739–777. doi:10.1152/physrev.00039. 2013
7. Liu J, Chen Q, Rozovsky S (2017) Utilizing selenocysteine for expressed protein ligation and bioconjugations. J Am Chem Soc. doi:10. 1021/jacs.6b10991
8. Reeves MA, Bellinger FP, Berry MJ (2010) The neuroprotective functions of selenoprotein m and its role in cytosolic calcium regulation. Antioxid Redox Signal 12(7):809–818. doi:10.1089/ars.2009.2883
9. Hwang DY, Cho JS, Oh JH, Shim SB, Jee SW, Lee SH, Seo SJ, Lee SK, Lee SH, Kim YK (2005) Differentially expressed genes in transgenic mice carrying human mutant presenilin-2 (N141I): correlation of selenoprotein M with Alzheimer's disease. Neurochem Res 30 (8):1009–1019. doi:10.1007/s11064-005-6787-6
10. Hwang DY, Sin JS, Kim MS, Yim SY, Kim YK, Kim CK, Kim BG, Shim SB, Jee SW, Lee SH, Bae CJ, Lee BC, Jang MK, Cho JS, Chae KR (2008) Overexpression of human selenoprotein M differentially regulates the concentrations of antioxidants and H_2O_2, the activity of antioxidant enzymes, and the composition of white blood cells in a transgenic rat. Int J Mol Med 21(2):169–179. doi:10.3892/ijmm.21. 2.169
11. Goo JS, Kim YN, Choi KM, Hwang IS, Kim JE, Lee YJ, Kwak MH, Shim SB, Jee SW, Lim CJ, Seong JK, Hwang DY (2013) Proteomic analysis of kidneys from selenoprotein M transgenic rats in response to increased bioability of selenium. Clin Proteomics 10(1):10. doi:10. 1186/1559-0275-10-10
12. Chen P, Wang RR, Ma XJ, Liu Q, Ni JZ (2013) Different forms of selenoprotein M differentially affect A beta aggregation and ROS generation. Int J Mol Sci 14(3):4385–4399. doi:10. 3390/ijms14034385
13. Garcia-Triana A, Gomez-Jimenez S, Peregrino-Uriarte AB, Lopez-Zavala A, Gonzalez-Aguilar G, Sotelo-Mundo RR, Valenzuela-Soto EM, Yepiz-Plascencia G (2010) Expression and silencing of selenoprotein M (SelM) from the white shrimp Litopenaeus vannamei: effect on peroxidase activity and hydrogen peroxide concentration in gills and hepatopancreas. Comp Biochem Physiol 155(2):200–204. doi:10.1016/j.cbpa.2009. 10.037
14. Reeves MA, Berry MJ (2012) Selenoprotein M. In: Hatfield DL, Berry MJ, Gladyshev VN (eds) Selenium: its molecular biology and role in human health, 3rd edn. Springer, New York. doi:10.1007/978-1-4614-1025-6_15
15. Korotkov KV, Novoselov SV, Hatfield DL, Gladyshev VN (2002) Mammalian selenoprotein in which selenocysteine (Sec) incorporation is supported by a new form of Sec insertion sequence element. Mol Cell Biol 22 (5):1402–1411. doi:10.1128/mcb.22.5. 1402-1411.2002
16. Pitts MW, Reeves MA, Hashimoto AC, Ogawa A, Kremer P, Seale LA, Berry MJ (2013) Deletion of selenoprotein M leads to obesity without cognitive deficits. J Biol Chem 288 (36):26121–26134. doi:10.1074/jbc.M113. 471235
17. Ogawa-Wong AN, Berry MJ, Seale LA (2016) Selenium and metabolic disorders: an emphasis on type 2 diabetes risk. Forum Nutr 8(2):80. doi:10.3390/nu8020080
18. Dawson PE, Muir TW, Clark-Lewis I, Kent SB (1994) Synthesis of proteins by native chemical ligation. Science 266(5186):776–779. doi:10. 1126/science.7973629
19. Hondal RJ, Nilsson BL, Raines RT (2001) Selenocysteine in native chemical ligation and expressed protein ligation. J Am Chem Soc 123 (21):5140–5141. doi:10.1021/ja005885t
20. Quaderer R, Sewing A, Hilvert D (2001) Selenocysteine-mediated native chemical ligation. Helv Chim Acta 84(5):1197–1206. doi:10.1002/1522-2675(20010516) 84:5<1197::aid-hlca1197>3.0.co;2-#
21. Gieselman MD, Xie L, van Der Donk WA (2001) Synthesis of a selenocysteine-containing peptide by native chemical ligation. Org Lett 3 (9):1331–1334. doi:10.1021/ol015712o
22. Malins LR, Mitchell NJ, Payne RJ (2014) Peptide ligation chemistry at selenol amino acids. J Pept Sci 20(2):64–77. doi:10.1002/psc.2581
23. McGrath NA, Raines RT (2011) Chemoselectivity in chemical biology: acyl transfer reactions with sulfur and selenium. Acc Chem Res 44(9):752–761. doi:10.1021/ar200081s
24. Hondal RJ (2005) Incorporation of selenocysteine into proteins using peptide ligation. Protein Pept Lett 12(8):757–764. doi:10.2174/ 0929866054864319

25. Metanis N, Keinan E, Dawson PE (2006) Synthetic seleno-glutaredoxin 3 analogues are highly reducing oxidoreductases with enhanced catalytic efficiency. J Am Chem Soc 128(51):16684–16691. doi:10.1021/ja0661414
26. Casi G, Roelfes G, Hilvert D (2008) Selenoglutaredoxin as a glutathione peroxidase mimic. Chembiochem 9(10):1623–1631. doi:10.1002/cbic.200700745
27. Berry SM, Gieselman MD, Nilges MJ, van Der Donk WA, Lu Y (2002) An engineered azurin variant containing a selenocysteine copper ligand. J Am Chem Soc 124(10):2084–2085. doi:10.1021/ja0169163
28. Dery L, Reddy PS, Dery S, Mousa R, Ktorza O, Talhami A, Metanis N (2017) Accessing human selenoproteins through chemical protein synthesis. Chem Sci. doi:10.1039/c6sc04123j
29. Kapust RB, Tozser J, Copeland TD, Waugh DS (2002) The P1′ specificity of tobacco etch virus protease. Biochem Biophys Res Commun 294 (5):949–955. doi:10.1016/s0006-291x(02)00574-0
30. Muller S, Senn H, Gsell B, Vetter W, Baron C, Bock A (1994) The formation of diselenide bridges in proteins by incorporation of selenocysteine residues - biosynthesis and characterization of (Se)(2)-thioredoxin. Biochemistry 33 (11):3404–3412. doi:10.1021/bi00177a034
31. Chong S, Williams KS, Wotkowicz C, Xu MQ (1998) Modulation of protein splicing of the Saccharomyces cerevisiae vacuolar membrane ATPase intein. J Biol Chem 273 (17):10567–10577. doi:10.1074/jbc.273.17.10567
32. Liu J, Li F, Rozovsky S (2013) The intrinsically disordered membrane protein selenoprotein S is a reductase in vitro. Biochemistry 52 (18):3051–3061. doi:10.1021/bi4001358
33. Studier FW (2005) Protein production by auto-induction in high-density shaking cultures. Protein Expr Purif 41(1):207–234. doi:10.1016/j.pep.2005.01.016
34. Geoghegan KF, Dixon HB, Rosner PJ, Hoth LR, Lanzetti AJ, Borzilleri KA, Marr ES, Pezzullo LH, Martin LB, LeMotte PK, McColl AS, Kamath AV, Stroh JG (1999) Spontaneous alpha-N-6-phosphogluconoylation of a "His tag" in Escherichia coli: the cause of extra mass of 258 or 178 Da in fusion proteins. Anal Biochem 267(1):169–184. doi:10.1006/abio.1998.2990
35. Blommel PG, Fox BG (2007) A combined approach to improving large-scale production of tobacco etch virus protease. Protein Expr Purif 55(1):53–68. doi:10.1016/j.pep.2007.04.013
36. Ellman GL (1959) Tissue sulfhydryl groups. Arch Biochem Biophys 82(1):70–77. doi:10.1016/0003-9861(59)90090-6

Chapter 20

Monitoring of Methionine Sulfoxide Content and Methionine Sulfoxide Reductase Activity

Lionel Tarrago, Emmanuel Oheix, Zalán Péterfi, and Vadim N. Gladyshev

Abstract

The sulfur-containing amino acid methionine (Met) plays critical roles in protein synthesis, methylation, and sulfur metabolism. Both in its free form and in the form of an amino acid residue, it can be oxidized to the *R* and *S* diastereomers of methionine sulfoxide (MetO). Organisms evolved methionine sulfoxide reductases (MSRs) to reduce MetO to Met, with the MSRs type A (MSRA) and type B (MSRB) being specific for the *S* and *R* forms of MetO, respectively. In mammals, the selenoprotein MSRB1 plays an important protein repair function, and its expression is tightly regulated by dietary selenium. In this chapter, we describe a protocol for determining the concentration of protein-based Met-*R*-O and its analysis in HEK293 cells using a genetically encoded ratiometric fluorescent biosensor MetROx. We also describe the procedure for quantifying MSR activities in cell extracts using specific substrates and a reverse phase HPLC-based method.

Key words Fluorescent microscopy, Genetically encoded fluorescent sensors, HPLC, Methionine sulfoxide reductase, Oxidative stress, Protein oxidation

1 Introduction

Living cells are constantly exposed to reactive oxygen species (ROS) and other oxidative insults. Under physiological conditions, ROS levels are tightly controlled, and these species are actively involved in cell signaling. However, abnormally high ROS levels can lead to deleterious oxidation of macromolecules [1]. The amino acid methionine (Met), either as a metabolite or as an amino acid residue, is sensitive to oxidation and is converted by ROS to Met sulfoxide (MetO), which exist in the form of two diastereomers, *rectus* (*R*) (Met-*R*-O) and *sinister* (*S*) (Met-*S*-O). These oxidized forms can be reduced back to Met by specialized enzymes, methionine sulfoxide reductases of type A (MSRA) and type B (MSRB), which are specific for Met-*S*-O and Met-*R*-O, respectively. MSRs use thiol/selenothiol chemistry to reduce their substrates using the reducing power generally provided by the thioredoxin system [2].

Laurent Chavatte (ed.), *Selenoproteins: Methods and Protocols*, Methods in Molecular Biology, vol. 1661, DOI 10.1007/978-1-4939-7258-6_20, © Springer Science+Business Media LLC 2018

MSRA reduces the free form of Met-*S*-O and the respective oxidized amino acid residue with similar efficiency, whereas MSRB specializes in the reduction of protein-based Met-*R*-O [3]. Bacteria and unicellular eukaryotes also evolved dedicated enzymes for the reduction of the free form of Met-*R*-O, but such enzymes are absent in higher eukaryotes [4, 5]. As a result, mammals cannot efficiently reduce the free Met-*R*-O [6]. Mammalian genomes have a single MSRA and three MSRB genes. MSRB1 is located in the cytosol and nucleus, MSRB2 and MSRB3 are in mitochondria and the endoplasmic reticulum, respectively, whereas alternative first exon splicing allows producing, from a single gene, various MSRA forms that are localized in different cellular compartments [7]. MSRB1 is the only selenocysteine-containing MSR in mammals and was initially identified as selenoprotein R (SelR) and selenoprotein X (SelX) by searching for putative SECIS element structures in EST databases using bioinformatics tools [8, 9]. Knockout mouse models revealed that MSRB1 is a nonessential selenoprotein but it has a major antioxidant function in kidney and liver [10]. Moreover, the expression of MSRB1 is regulated by Se availability and declines as a function of age [10, 11]. In kidney cells, MSRB1 was shown to reactivate the TRPM6 magnesium-channel which can be inactivated by the oxidation of its Met1755 residue [12]. An important function was found in macrophages, in which MSRB1 plays a key role, together with MICAL monooxygenases, in the regulation of actin polymerization through cyclic oxidation and reduction of two key Met residues [13]. As MSRB1 is intricately linked with Se metabolism, studies on Se and selenoproteins may require quantification of Met-*R*-O levels and MSR activity in cells. In this chapter, we describe the procedures that allow determining Met-*R*-O levels and its changes in living cells using the genetically encoded ratiometric fluorescent biosensor MetROx [14, 15]. We also provide protocols for quantification of MSRA, MSRB, and total MSR activities in protein extracts using specific substrates and an HPLC-based procedure.

2 Materials

In all materials and methods described in this article, the terms methionine, methionine sulfoxide, or any of their abbreviations always refer to the L-enantiomers. All solutions should be prepared in ultrapure water (sensitivity of 18 MΩ cm at 25 °C).

2.1 Monitoring Met-R-O Concentration in Living Mammalian Cells

1. HEK293T/17 cells.
2. Plasmids for the expression of MetROx, its inactive version (C129S MetROx or C69S/C129S MetROx) in the appropriate cell compartment, and pEGFP-C3 plasmid (*see* **Note 1**).
3. Dulbecco's Modified Eagle Medium (D-MEM).

4. Opti-MEM without phenol red and without serum.
5. Fetal Bovine Serum (FBS).
6. Penicillin and streptomycin antibiotics solution. Stock solution contains 10,000 U/mL penicillin and 10 mg/mL streptomycin and is diluted 1:100 when used in cultured cells.
7. Transfection reagent (e.g., FuGENE HD from Promega).
8. 35 mm glass-bottom dishes.
9. Live cell imaging solution. Supplement 500 mL of Live Cell Imaging Solution (Life technologies) with 0.45 g glucose, sterilize by filtration through 0.2 μm filter and store at 4 °C.
10. A motorized inverted microscope (e.g., Olympus IX81) equipped with a cooled-CCD camera (e.g., Hamamatsu Orca ER) in conjunction with a zero-drift focus compensation system with the following filters: excitation (420/40 and 500/16), emission (535/30), and the acquisition software (e.g., MetaMorph from Universal Imaging, Downingtown, PA).
11. Dithiothreitol solution. To prepare 1 M solution, add 1.54 g of DTT (MW = 154.25 g/mol, min 98%) to 10 mL of water. Mix well by vortexing. Store solution at −20 °C.
12. L-methionine-*R*,*S*-sulfoxide (MW = 165.21 g/mol, min 99%). To prepare 100 mM solution, add 1.65 g to 100 mL of water. Mix well by vortexing. Store solution at room temperature or below.
13. Software for image analysis (e.g., ImageJ [16] or Fiji [17]).

2.2 Monitoring Methionine Sulfoxide Reductase Activity in Cell Lysates

1. Dabsyl-L-Methionine (Dabsyl-Met), MW = 436.55 g/mol, min 95%.
2. Dimethyl sulfoxide (DMSO), min 99%.
3. Hydrogen peroxide (H_2O_2), 30% (8.8 M).
4. Disposable SEP-PAK® C18 Cartridges (Waters).
5. HPLC solvent B: Acetonitrile HPLC grade.
6. HPLC solvent A: Acetate buffer. To make 29 mM solution, slowly add 1.665 mL of glacial acetic acid (min 99%) to 700 mL water in 1-L graduated cylinder or a glass beaker. Adjust the pH to 4.16 with 5 N NaOH. Adjust the volume to 1000 mL with water in a volumetric flask. Pass through the 0.2 μm filter and degas. Store at room temperature.
7. Tris–HCl buffers. Dissolve 12.1 g of Trizma® base (MW = 121.14 g/mol, min 99%) to 800 mL in a 1-L cylinder or a glass beaker. *Adjust pH at 8.0 with concentrated HCl and adjust the volume to 1000 mL with water in a volumetric flask to obtain a 0.1 M solution.* Filter (0.2 μm). To obtain 30 mM solution, dilute 333 mL of the above solution in 500 mL water

and adjust the volume to 1000 mL with water in a volumetric flask. Filter (0.2 μm) and store at 4 °C or room temperature.

8. For separation of *R*- and *S*-diastereomers of MetO, specific equipment includes: heating stirring plate, heating mantle or oil bath, condenser, glass-sintered funnel, Büchner flask, Schlenck line equipped with high vacuum pump or rotary evaporator, and polarimeter (e.g., Anton Paar Modular Circular Polarimeter 200, $\lambda = 589$ nm) or Nuclear Magnetic Resonance spectrometer (e.g., Bruker Avance III 400 MHz).
9. L-methionine-*R*,*S*-sulfoxide (MW = 165.21 g/mol, min 98%, Sigma-Aldrich).
10. Picric acid crystalline or moistened with water (MW = 229.10 g/mol, min 98%).
11. Acetone (min 99.5%).
12. Triethylamine (MW = 101.19 g/mol, min 99%). To prepare 5 mL of 1 M solution, add 0.70 mL triethylamine using a pipette in a 5 mL volumetric flask, and complete with water to 5 mL. To prepare 50 mL of 0.175 M triethylamine, add 1.23 mL triethylamine in a 50 mL volumetric flask using a pipette and complete with water to 50 mL.
13. 4-*N*,*N*-dimethylaminoazobenzene-4′sulfonyl chloride (dabsyl chloride) (MW = 323.80 g/mol). Dissolve 200 mg of dabsyl chloride in 50 mL acetone to obtain a 4 mg/mL solution.
14. Sodium carbonate buffer. Dissolve 12.88 g of $Na_2CO_3{\cdot}10H_2O$ (MW = 286.14 g/mol) in 50 mL of water. To obtain 0.9 M $NaHCO_3$, pH 9.0 solution, dissolve 3.78 g of $NaHCO_3$ (MW = 84.01 g/mol) in 40 mL of water and adjust slowly the pH to 9.0 with the $Na_2CO_3{\cdot}10H_2O$ solution.
15. Dithiothreitol solution. To prepare 1 M solution, add 1.54 g of DTT (MW = 154.25 g/mol, min 98%) to 10 mL of water. Mix well by vortexing. Store the solution at −20 °C.
16. Reverse-phase High Performance Liquid Chromatography (HPLC) system. The HPLC device must be equipped with a two-solvent mixing pump (e.g., Waters 1525) and UV-visible detectors (e.g., Waters PDA 996 PhotoDiode Array), and optionally with an autosampler (e.g., Waters Autosampler 717 plus). Dabsyl-Met and dabsyl-MetO are detected at 466 nm. The column used is a SunFire™ C18 3.5 μm 3.0 × 50 mm (Waters). The program runs for 8 min and starts with 70% solvent A (acetate buffer) and 30% solvent B (acetonitrile) with a 1 mL/min flow rate. Solvent B percentage increases to 66.7 from 0 to 2 min, then to 100% from 2 to 2.5 min and for a duration of 1 min (3.5 min), then decreases to 30% in 0.2 min (3.7 min) and up to 8 min. Diastereomers of dabsyl-MetO are eluted at 3.5 min and dabsyl-Met is eluted at 4.6 min (*see* **Note 2**).

3 Methods

3.1 Monitoring Met-R-O Concentration in Living Mammalian Cells

3.1.1 Transfection

1. Seed HEK293 cells onto 35 mm glass-bottom dishes (10^5 cells/dish) in D-MEM supplemented with 10% FBS and penicillin (100 U/mL)/streptomycin (100 μg/mL). Culture the cells for 24 h at 37 °C, 5% CO_2.
2. The next day, add 100 μL serum-free Opti-MEM to a sterile 1.5 mL tube.
3. Mix well 1 μg of the sensor-encoding expression vector in the medium (*see* **Note 3**).
4. Add 3 μL of transfection reagent. Mix rapidly by vortexing, and allow 15 min for the liposome-DNA complex to form.
5. Add the content of the tube to the media covering the cells in the glass-bottom dish. Mix gently.

3.1.2 Imaging

1. 16–24 h after transfection, examine effectiveness of transfection by checking fluorescence using your cell culture microscope. Use the GFP filter set on the microscope (*see* **Note 4**).
2. 30 min prior to the experiment, change the medium to the live cell imaging solution supplemented with 5 mM glucose and allow the dish to stay at room temperature (*see* **Note 5**).
3. Place the dish in the microscope mount, and find a group of transfected cells by using the visual mode of the microscope. Once the cells are in focus, immediately close the lamp shutter to avoid unwanted bleaching of the probe.
4. Switch the microscope to the recording mode and activate the following beam path settings (*see* **Note 6**): for excitation, use 420/40 (channel 1) and 500/16 (channel 2) band-pass excitation filters and 535/30 band-pass emission filter. Live imaging should be carried out in time series or xyt mode.
5. Start time series recording. Typically, emission is acquired every 15–30 s for 10–15 min.
6. Remove 0.1 mL of incubation medium and add chemicals in the same volume of buffer using a regular handhold pipette. To determine the ratio of the fully reduced sensor, prepare dithiothreitol solution in the incubation medium, add it to the cells at 1 mM final concentration and incubate until the fluorescent signal is stable (for 1–5 min). To determine the ratio of the fully oxidized sensor, proceed similarly by adding 10–100 mM L-methionine-*R*,*S*-sulfoxide to the cells and incubate until the fluorescent signal is stable (for 5–10 min). Repeat these assays with the cells expressing the inactive sensor.

3.1.3 Data Analysis

The fluorescence intensity analysis procedure described here is made using the software ImageJ/Fiji.

1. Open the stack in the software. Depending on the software used for the acquisition you might need to install a specific plugin or convert the file to work with the format.
2. Open "*ROI Manager*."
3. Set your first region of interest (ROI) on the first cell you would like to measure. Press "t" (or click "*Add*" on the *ROI Manager*) for multiple ROIs. Include non-fluorescent ROIs for background correction. Repeat until all cells are selected.
4. To check all your ROIs, click "*Show All*" on the *ROI Manager*.
5. Use Time series Analyzer to calculate all average intensities in the image stack.
6. Calculate the sensor fluorescence ratio (R) by dividing the background subtracted fluorescence in channel 2 ($F_{500\ \mathrm{nm}}$) by the background subtracted fluorescence in channel 1 ($F_{420\ \mathrm{nm}}$).
7. Use the following equation to calculate the oxidized fraction of the sensor:

$$\text{Oxidized fraction} = \frac{R_{\mathrm{Reduced}} - R}{\mathrm{R}_{\mathrm{reduced}} - R_{\mathrm{Oxidized}}}$$

where R is the measured $F_{500\ \mathrm{nm}}/F_{420\ \mathrm{nm}}$ ratio; R_{Reduced} is the $F_{500\ \mathrm{nm}}/F_{420\ \mathrm{nm}}$ ratio of fully reduced MetROx, and R_{Oxidized} is the $F_{500\ \mathrm{nm}}/F_{420\ \mathrm{nm}}$ ratio of fully oxidized MetROx.

8. To avoid nonspecific changes in the sensor signal, analyze cells expressing the inactive sensors (C129S MetROx or C69S/C129S MetROx), and divide the signal of the active sensor by that obtained with the inactive sensor (*see* **Note 7**).

3.2 Monitoring Methionine Sulfoxide Reductase Activity in Protein Extracts

The activity assay measurement is made using MetO derivatized with dabsyl, which is a derivatization reagent that allows detection of amino acids in visible light using a reverse-phase HPLC procedure. We propose two protocols for the preparation of dabsylated MetO substrates. The first is an easy procedure to obtain a racemic mixture of dabsyl-Methionine-*R*,*S*-sulfoxide (dabsyl-Met-*R*,*S*-O), which allows measuring total MSR activity of a protein extract. The second protocol describes the preparation of individual forms of MetO, i.e., dabsyl-Methionine-*R*-sulfoxide (dabsyl-Met-*R*-O) and dabsyl-Methionine-*S*-sulfoxide (dabsyl-Met-*S*-O); this allows measuring MSRB and MSRA activities of a protein extract separately. This protocol consists of the separation of the *R*- and *S*-diastereomers of MetO from a racemic mixture using picric acid precipitation, followed by their derivatization with dabysl chloride. The separation of both diastereomers of MetO should be very useful for other applications (e.g., genetic selection and screening [4, 18]). The protocol for measuring MSR activity is similar whichever substrate is used.

3.2.1 Preparation of the Racemic Mixture of Dabsyl-methionine-R,S-sulfoxide

1. In a 50 mL plastic tube, dissolve 25 mg of dabsyl-Met in 11.46 mL DMSO to obtain a 5 mM solution (*see* **Note 8**). Keep 0.5 mL for the HPLC standard curve (*see* Subheading 3.2.3).
2. To oxidized dabsyl-Met to dabsyl-Met-*R*,*S*-O, add 341 μL of 8.8 M (30%) H_2O_2 and 18.7 mL water (to 30 mL final volume). Mix well by vortexing. Protect from light by wrapping the tube with aluminum foil and incubate overnight at room temperature.
3. Take 10 μL, add 30 μL water (1/4 dilution) and check oxidation by HPLC by injecting 10 μL (*see* **Note 9** and Subheading 3.2.3).
4. Centrifuge at 3250 × *g* for 10 min at 25 °C.
5. Eliminate H_2O_2 and DMSO using the disposable SEP-PAK® C18 cartridge. Proceed by using a 5 mL syringe to avoid overpressure. Rinse the cartridge with 5 mL acetonitrile:acetate buffer (80:20 v:v). Equilibrate the cartridge with 5 mL acetonitrile:acetate buffer (5:95 v:v). Load 3 × 5 mL dabsyl-Met-*R*,*S*-O solution. Wash the cartridge with 5 mL acetonitrile:acetate buffer (5:95 v:v). Elute dabsyl-Met-*R*,*S*-O with 2 × 5 mL acetonitrile:acetate buffer (5:95 v:v). Repeat the procedure for the remaining dabsyl-Met-*R*,*S*-O solution. Pool eluted dabsyl-Met-*R*,*S*-O in a 50 mL plastic tube and freeze at −80 °C (*see* **Note 10**).
6. Make holes in a 50 mL plastic tube cap, put it on the 50 mL plastic tube of frozen dabsyl-Met-*R*,*S*-O solution and dry under vacuum. You should obtain a red powder.
7. Resuspend the powder in 0.5 mL 0.1 M Tris–HCl, pH 8.0. Mix very well by repeated vortexing (*see* **Note 11**). Transfer the solution to a 1.5 mL test tube (e.g., Eppendorf). Centrifuge for 10 min at maximum speed (e.g., 14,000 × *g*) on a table-top microcentrifuge. If a small insoluble pellet is formed, transfer the supernatant to a clean tube. Store dabsyl-Met-*R*,*S*-O solution at 4 °C.

3.2.2 Preparation of Dabsyl-methionine-R-sulfoxide and Dabsyl-methionine-S-sulfoxide

1. In a 25 mL round-bottom flask with a ground-glass joint, add Met-*R*,*S*-O (1.333 g, 8.069 mmol), picric acid (1.849 g, 8.071 mmol) as solids and 8.5 mL of water (*see* **Note 12**). Position the flask in an oil bath or a heating mantle standing on a hotplate stirrer apparatus fitted on an elevator. Add a stirrer bar in the flask and adapt a suitable condenser on the top.

2. Bring the mixture to reflux while stirring the suspension. As soon as the mixture refluxes gently, stop the stirrer and watch out for the remaining/undissolved solid at the bottom of the flask. If there is no solid left, proceed directly to **step 3**. Otherwise, add 0.5 mL of water through the condenser, resume stirring/reflux for ~2 min, and watch out for the remaining solid again. Keep adding 0.5 mL aliquots of water until the solid dissolves completely (*see* **Note 13**).
3. Stop the heating and let the orange solution to cool down as slowly as possible (ideally overnight in an oil bath/heating mantle). Yellow crystals start to form rapidly. These are enriched in the picrate salt of Met-*S*-O and contain small traces of Met-*R*-O. In order to maximize the separation, make sure the mixture returns to room temperature (~25 °C) before you proceed further.
4. Filter the suspension on a Büchner mounted with a glass-sintered funnel and wash the solid with 5 mL of cold water twice. Recover the yellow crystals (Met-*S*-O enriched) and the yellow solution (Met-*R*-O enriched). Purify them separately by successive crystallization.
5. Purification of Met-*R*-O. Collect the filtrate solution together with wash solutions. Reduce the volume of solvent to ~3–4 mL using a high-vacuum pump or a rotary evaporator until a small amount of solid starts to precipitate. Add 1.5 mL of water and agitate the flask manually. Filter on a glass-sintered funnel and wash the solid with a minimal amount of water (~2 mL). Dispose of the remaining solid.
6. Repeat **step 5** entirely. Then dry the solution under vacuum to obtain a yellow solid. Weigh the solid (e.g., 1.354 g, 393.3 g/mol, 3.44 mmol).
7. Dispose the solid in a 10 mL Erlenmeyer flask. Add 1.1 equivalent of triethylamine as a concentrated 1 M solution and a stirrer bar (*see* **Note 14**). Stir the solution until complete dissolution, then transfer to a 500 mL Erlenmeyer flask. Add 200 mL acetone portion-wise while stirring.
8. Filter the solid on a sintered funnel and wash with 20 mL acetone twice. Dry the solid under high vacuum to obtain Met-*R*-O > 95% pure as a white solid (e.g., $m = 0.515$ g).
9. Purification of Met-*S*-O. Recrystallize the solid from **step 4** once from water (estimated quantity: 20 mL water for 1.76 g of the picrate salt), and once from methanol (estimated quantity: 11 mL methanol for 1.48 g of the picrate salt) as described in **steps 2–4**. Dry under vacuum and weigh (e.g., 1.345 g, 393.3 g/mol, 3.42 mmol).

10. Dispose the solid in a 200 mL Erlenmeyer flask. Add 1.1 equivalent of triethylamine as a dilute 0.175 M solution and a stirrer bar (*see* **Note 15**). Stir the solution until complete dissolution. Add 200 mL acetone portion-wise while stirring.
11. Filter the solid on a sintered funnel and wash with 20 mL acetone twice. Dry the white solid under high vacuum to obtain pure Met-*S*-O (e.g., $m = 0.533$ g).
12. Check purity of the sample by NMR (^{1}H and/or ^{13}C) or by the measurement of the specific rotation using a polarimeter. Compare with the values from the literature [18–21] (*see* **Note 16**).
13. Derivatize Met-*R*-O and Met-*S*-O with dabsyl chloride. In a 50 mL plastic tube, dissolve 10 mg Met-*R*-O in 12 mL 0.9 M $NaCO_3$, pH 9. Add 24 mL of 4 mg/mL dabsyl chloride in acetone. Mix well by vortexing. Proceed similarly for Met-*S*-O.
14. Incubate at 70 °C for 10 min in water bath. Protect from light by covering the bath with aluminum foil (*see* **Note 17**). Freeze at −80 °C.
15. Make holes in a 50 mL plastic tube cap, put it on the 50 mL plastic tube of frozen dabsyl-Met-*R*-O (or dabsyl-Met-*S*-O) solution and dry under vacuum. You should obtain a red powder.
16. Resuspend the powder in 0.5 mL 0.1 M Tris–HCl, pH 8.0. Mix very well by repeated vortexing (*see* **Note 11**). Transfer the solution to a 1.5 mL test tube. Centrifuge at maximum speed (e.g., 14,000 × *g*) on a table-top microcentrifuge. If a small insoluble pellet is formed, transfer the supernatant to a clean tube. Store dabsyl-Met-*R*-O and dabsyl-Met-*S*-O solutions at 4 °C.

3.2.3 Determination of Dabsyl-Met-R,S-O, Dabsyl-Met-R-O, and Dabsyl-Met-S-O Concentrations

1. Prepare a standard curve with the dabsyl-Met solution kept in **step 1**. Add 10–90 μL 30 mM Tris–HCl, pH 8.0 (1/10 dilution) to prepare a solution concentrated at 0.5 nmol/μL. Prepare four dilutions as described in Table 1. Inject 50 μL of each in HPLC to obtain a 0.25, 0.5, 1, and 2 nmol standard curve.
2. Determine dabsyl-Met-*R*,*S*-O concentration using the standard curve made as described in Subheading 3.2.3. The final concentration should be ~80–100 mM. Dilute 10 μL in 990 μL 30 mM Tris–HCl, pH 8.0 (1/100 dilution, final concentration ~0.8–1.0 nmol/μL). Prepare four dilutions similar to those described in Table 1 except that dabsyl-Met-*R*,*S*-O solution volumes should be divided by 2 compared to dabsyl-Met solution volumes and 30 mM Tris–HCl, pH 8 volumes modified accordingly.

Table 1
Examples of dilution used

Final quantity for 50 μL injected (nmol)	0.25	0.5	1	2
0.5 nmol/μL dabsyl-Met solution (μL)	3	6	12	24
30 mM Tris–HCl pH 8 (μL)	97	94	88	76
Acetonitrile (μL)	200	200	200	200

3. Determine dabsyl-Met-*R*-O and dabsyl-Met-*S*-O using the standard curve made as described in Subheading 3.2.3. The final concentration should be ~20–40 mM. Dilute 10 μL in 990 μL 30 mM Tris–HCl, pH 8.0 (1/100 dilution, final concentration ~0.2–0.4 nmol/μL). Prepare four dilutions similar to those described in Table 1 replacing the dabsyl-Met solution with the dabsyl-Met-*S*-O solution. Proceed similarly for the dabsyl-Met-*R*-O solution.
4. Inject 50 μL of each dilution of dabsyl-Met for the standard curve and 50 μL of each dilution of the solution to dose (dabsyl-Met-*R*,*S*-O; dabsyl-Met-*R*-O or dabsyl-Met-*S*-O) and in HPLC, run the separation program and proceed to the analysis.
5. Using a spreadsheet software (e.g., Microsoft Excel), plot areas of peak of dabsyl-Met for each of the four solutions prepared versus the quantity injected (0.25, 0.5, 1, and 2 nmol). Determine the slope of the linear regression. For each tube of the solution to dose, determine the concentration of dabsyl-MetO and those of dabsyl-Met in mM by applying the following equation:

$$\text{Conc.} = \frac{\text{Area} \times 10}{\text{Slope} \times 50}$$

where Conc. is the concentration (mM), Area is the area of the peak of interest, 10 is the initial dilution factor, Slope is the slope of the standard curve, and 50 is the volume injected in μL. Average the four tubes to obtain the concentration of dabsyl-Met-*R*,*S*-O; dabsyl-Met-*R*-O or dabsyl-Met-*S*-O solution.

3.2.4 Activity Assays and HPLC Analysis

As it may differ according to the experiment setting, the cell lysis protocol to obtain soluble proteins is not described here. For mammalian cells, you can use your favorite methods (e.g., CelLytic™ M cell lysis reagent (Sigma)) and quantify protein using standard Bradford assay or equivalent.

1. Prepare the substrate dilution. To measure the total MSR activity (MSRA and MSRB), prepare dabsyl-Met-*R*,*S*-O to 5 mM final concentration in 30 mM Tris–HCl, pH 8.0 using an appropriate volume of the concentrated solution. To measure MSRA or MSRB activity, proceed similarly with dabsyl-Met-*S*-O and dabsyl-Met-*R*-O, respectively. It is advised to prepare fresh solutions prior to each experiment. Equilibrate the solution at 37 °C in water bath or in a dry bath with agitation (e.g., Eppendorf ThermoMixer).
2. Prepare the reaction assay. In a 1.5 mL test tube containing the appropriate volume of 30 mM Tris–HCl, pH 8.0, to a final volume of 200 μL, mix 200 μg of protein with 2 μL of 1 M dithiothreitol (20 mM final concentration). Prepare a control tube without protein (*see* **Note 18**). Equilibrate the solution for 5 min at 37 °C. Start the reaction by adding 20 μL of 5 mM dabsyl-Met-*R*,*S*-O for total activity assay, or of individual diastereomer for MSRA or MSRB activity assays (0.5 mM final concentration). Mix well by short-time vortexing and incubate for 30 min at 37 °C (*see* **Note 19**).
3. Stop the reaction by adding 300 μL acetonitrile.
4. Centrifuge at 4 °C for 30 min at 12,000 × *g*.
5. Inject 50 μL of supernatant in HPLC, run the separation program, and proceed to the analysis.
6. Determine the quantity of dabsyl-Met produced and report activity in pmol of dabsyl-Met produced by minute by mg of protein using the following equation:

$$\text{Activity} = \frac{\text{Q Met} \times \frac{\text{Vol.reaction assay}}{\text{Vol.injected}} \times \text{DF}}{\text{Inc.time} \times \text{Q Prot.}}$$

where Q Met is the quantity of dabsyl-Met produced, Vol. reaction assay is a total volume of the reaction (200 μL), Vol injected is the injected volume (50 μL), DF is the dilution factor (200/300 μL), Inc. time is the time of incubation in min (30 min), and Q Prot. is the total quantity of protein by reaction (200 μg).

4 Notes

1. MetROx expression plasmids targeted to cytosol, mitochondria, the endoplasmic reticulum, and nucleus were made by inserting the MetROx chimera gene and an appropriate targeting sequence between the *NheI* and *BamHI* restriction sites of pEGFP-C3 vector. Available plasmids can be obtained upon

Table 2
Available plasmids

Name	Description
pCyto-MetROx	Expression of MetROx in the cytosol of mammalian cells
pCyto-C129S MetROx	Expression of inactive C129S MetROx in the cytosol of mammalian cells
pCyto-C69S/C129S MetROx	Expression of inactive C69S/C129S MetROx in the cytosol of mammalian cells
pMito-MetROx	Expression of MetROx in mitochondria of mammalian cells
pMito-C129S MetROx	Expression of inactive C129S MetROx in mitochondria of mammalian cells
pNucleus-MetROx	Expression of MetROx in the nucleus of mammalian cells
pNucleus-C129S-MetROx	Expression of inactive C129S-MetROx in the nucleus of mammalian cells
pER-MetROx	Expression of MetROx in the ER of mammalian cells
pER-C129S-MetROx	Expression of inactive C129S-MetROx in the ER of mammalian cells

request from the authors and are described in Table 2. These kanamycin-resistant plasmids can be amplified in *Escherichia coli* and purified using standard procedures.

2. The use of the SunFire™ C18 column allows the fast separation of dabsyl-MetO from dabsyl-Met using the described program. Other columns and programs can be used to separate the *R*- and *S*-diastereomers of dabsyl-MetO [22].
3. For a typical measurement, you should use at least one well with the active MetROx sensor and one with the inactive sensor (C129S MetROx or C69S/C129S MetROx).
4. Since it is a single cell measurement, it is not essential to get all of the cells transfected, but the more fluorescent cells are present in a field of view, the more information you are able to gather by imaging. If no fluorescence is detected, make sure that transfection is successful. You may control for transfection efficiency by separately transfecting with the pEGFP-C3 control plasmid. Depending on the microscope setup and the length of the experiment, photobleaching might be an issue. We advise using a wide field (instead of confocal laser scanning) fluorescence microscope setup with high sensitivity detector to record time series and always record a photobleaching curve prior to starting the actual experiments.
5. Short experiments (less than 30 min) can be carried out without CO_2 and temperature control. However, be mindful of the potentially slower kinetics of redox reactions at room temperature. Do not forget to change the bicarbonate-based D-MEM

to a solution with an adequate buffering system (e.g., 10 mM HEPES, pH 7.4) without CO_2 under these conditions.

6. Use equivalent beam path settings depending on the manufacturer of the microscope and the software driving it. Perform several single scans to set the final focus, adjust light power and exposure time. The initial fluorescence intensities in both the channels should be stable. If your system is not equipped with autofocus capabilities, it is important that the microscope focal plane should not change over the experimental period.
7. Alternatively, the ratio of the inactive sensor could be used as the ratio of the fully reduced sensor in the equation. This allows direct correction of the fluorescence ratio. However, it is possible only if the measured ratio is higher than the one measured for the active sensor.
8. Proceed by the successive addition and removal of 1 mL DMSO to the glass bottle to avoid losing the product by weighing. Adjust the volume if starting from a 100 mg-containing glass bottle. Dabsyl-Met solubilized in DMSO can be stored at room temperature or below.
9. Dabsyl-Met-*R*,*S*-O and dabsyl-Met should be eluted at 3.5 min and 4.6 min, respectively. If no oxidation occurred, use a fresh bottle of H_2O_2 solution. Make sure to respect concentrations.
10. The maximum volume of dabsyl-methionine-*R*,*S*-sulfoxide solution to be loaded on the C18 cartridge may vary. To avoid loss of solution, be sure that it does not come out of the column while loading (the solution is red). Adjust the volume accordingly and repeat the procedure. We advise using a new disposable cartridge for each purification. If you prefer to reuse the cartridge, increase the equilibration volume to 3 × 5 mL acetonitrile:acetate buffer (5:95 v:v).
11. We experienced sometimes problems with solubilization, and vortexing might not be sufficient to completely resolubilize the powder. If this is the case, add 5 N NaOH in 5 μL increments and vortex until complete solubilization. Alternatively, solubilization can be made in acetonitrile or 70% ethanol.
12. The quantities of Met-*R*,*S*-O, picric acid, and water can be slightly adjusted while keeping their ratio constant. However, it must be noted that low-scale recrystallization tends to be less effective, whereas the large-scale ones require particular care in handling.
13. An optimal separation is obtained when the solution is saturated and there is no solid left to interfere with the crystal formation. This means the salts should be *entirely* dissolved in the *minimal* amount of water. If the quantity of salts specified

cannot be dissolved in 15 mL of water, consider checking purity of the starting materials or perform an additional hot-filtration to remove excess solid.

14. The quantity of triethylamine is relative to the amount weighed in **step 6**. In the example given, use 3.8 mL of triethylamine 1 M solution.
15. Similarly to **Note 14**, the volume of triethylamine 0.175 M solution is calculated considering the molar quantity of picrate salt measured in **step 9** (e.g., 21.5 mL).
16. Alternatively, purity could be estimated by reverse phase HPLC as previously described [22].
17. Alternatively, incubation can be done overnight at room temperature. Protect from light by wrapping the tube in aluminum foil.
18. The prepared substrate might be contaminated with dabsyl-Met (generally not more than 5%). Use the control tube to correct the area of the dabsyl-Met formed in a protein-containing reaction.
19. Quantity of protein and incubation time might need to be adjusted for optimal activity measurement. To ensure saturation of MSR with substrate, make sure that sufficient amount of substrate remains at the end of the assay; otherwise, the activity will be underestimated. If no substrate remains, decrease the quantity of protein and/or the incubation time.

Acknowledgments

This work was supported by the NIH grant AG021518. Pascal Rey (CEA, DRF, BIAM, Laboratoire d'Ecophysiologie Moléculaire des Plantes, France) is acknowledged for technical advice.

References

1. Ray PD, Huang B-W, Tsuji Y (2012) Reactive oxygen species (ROS) homeostasis and redox regulation in cellular signaling. Cell Signal 24:981–990. doi:10.1016/j.cellsig.2012.01.008
2. Tarrago L, Gladyshev VN (2012) Recharging oxidative protein repair: catalysis by methionine sulfoxide reductases towards their amino acid, protein, and model substrates. Biochemistry (Biokhimiia) 77:1097–1107. doi:10.1134/S0006297912100021
3. Tarrago L, Kaya A, Weerapana E et al (2012) Methionine sulfoxide reductases preferentially reduce unfolded oxidized proteins and protect cells from oxidative protein unfolding. J Biol Chem 287:24448–24459. doi:10.1074/jbc.M112.374520
4. Le DT, Lee BC, Marino SM et al (2009) Functional analysis of free methionine-R-sulfoxide reductase from Saccharomyces cerevisiae. J Biol Chem 284:4354–4364. doi:10.1074/jbc.M805891200
5. Lin Z, Johnson LC, Weissbach H et al (2007) Free methionine-(R)-sulfoxide reductase from Escherichia coli reveals a new GAF domain function. Proc Natl Acad Sci U S A 104:9597–9602. doi:10.1073/pnas.0703774104
6. Lee BC, Le DT, Gladyshev VN (2008) Mammals reduce methionine-S-sulfoxide with MsrA and are unable to reduce methionine-R-

sulfoxide, and this function can be restored with a yeast reductase. J Biol Chem 283:28361–28369. doi:10.1074/jbc.M805059200

7. Kim H-Y, Gladyshev VN (2004) Methionine sulfoxide reduction in mammals: characterization of methionine-R-sulfoxide reductases. Mol Biol Cell 15:1055–1064. doi:10.1091/mbc.E03-08-0629
8. Kryukov GV, Kryukov VM, Gladyshev VN (1999) New mammalian selenocysteine-containing proteins identified with an algorithm that searches for selenocysteine insertion sequence elements. J Biol Chem 274:33888–33897
9. Lescure A, Gautheret D, Carbon P, Krol A (1999) Novel selenoproteins identified in silico and in vivo by using a conserved RNA structural motif. J Biol Chem 274:38147–38154
10. Fomenko DE, Novoselov SV, Natarajan SK et al (2009) MsrB1 (methionine-R-sulfoxide reductase 1) knock-out mice: roles of MsrB1 in redox regulation and identification of a novel selenoprotein form. J Biol Chem 284:5986–5993. doi:10.1074/jbc.M805770200
11. Novoselov SV, Kim H-Y, Hua D et al (2010) Regulation of selenoproteins and methionine sulfoxide reductases A and B1 by age, calorie restriction, and dietary selenium in mice. Antioxid Redox Signal 12:829–838. doi:10.1089/ars.2009.2895
12. Cao G, Lee KP, van der Wijst J et al (2010) Methionine sulfoxide reductase B1 (MsrB1) recovers TRPM6 channel activity during oxidative stress. J Biol Chem 285:26081–26087. doi:10.1074/jbc.M110.103655
13. Lee BC, Péterfi Z, Hoffmann FW et al (2013) MsrB1 and MICALs regulate actin assembly and macrophage function via reversible stereoselective methionine oxidation. Mol Cell 51:397–404. doi:10.1016/j.molcel.2013.06.019
14. Tarrago L, Péterfi Z, Lee BC et al (2015) Monitoring methionine sulfoxide with stereospecific mechanism-based fluorescent sensors. Nat Chem Biol 11:332–338. doi:10.1038/nchembio.1787
15. Péterfi Z, Tarrago L, Gladyshev VN (2016) Practical guide for dynamic monitoring of protein oxidation using genetically encoded ratiometric fluorescent biosensors of methionine sulfoxide. Methods 109:149–157. doi:10.1016/j.ymeth.2016.06.022
16. Schindelin J, Rueden CT, Hiner MC, Eliceiri KW (2015) The ImageJ ecosystem: an open platform for biomedical image analysis. Mol Reprod Dev 82:518–529. doi:10.1002/mrd.22489
17. Schindelin J, Arganda-Carreras I, Frise E et al (2012) Fiji: an open-source platform for biological-image analysis. Nat Methods 9:676–682. doi:10.1038/nmeth.2019
18. Gennaris A, Ezraty B, Henry C et al (2015) Repairing oxidized proteins in the bacterial envelope using respiratory chain electrons. Nature 528:409–412. doi:10.1038/nature15764
19. Lavine TF (1947) The formation, resolution, and optical properties of the diastereoisomeric sulfoxides derived from L-methionine. J Biol Chem 169:477–491
20. Christensen BW, Kjær A (1965) The absolute configuration of methionine sulphoxide. Chem Commun (London):225–226. doi:10.1039/C19650000225
21. Holland HL, Andreana PR, Brown FM (1999) Biocatalytic and chemical routes to all the stereoisomers of methionine and ethionine sulfoxides. Tetrahedron Asymmetry 10:2833–2843. doi:10.1016/S0957-4166(99)00271-2
22. Vieira Dos Santos C, Cuiné S, Rouhier N, Rey P (2005) The Arabidopsis plastidic methionine sulfoxide reductase B proteins. Sequence and activity characteristics, comparison of the expression with plastidic methionine sulfoxide reductase A, and induction by photooxidative stress. Plant Physiol 138:909–922. doi:10.1104/pp.105.062430

Chapter 21

Selective Evaluation of Thioredoxin Reductase Enzymatic Activities

Elias S.J. Arnér

Abstract

Thioredoxin reductases are important oxidoreductases that keep the active site disulfide/dithiol motif of thioredoxins reduced using NADPH, thereby supporting many thioredoxin-dependent reductive pathways in cells. Mammalian thioredoxin reductases are selenoproteins that have several additional substrates beyond thioredoxins. This chapter first lists several different assays for measurement of thioredoxin reductase activities, before giving a protocol for a selective evaluation of these activities that can be used in either crude cell lysates as well as with purified enzymes. The same assay can also be easily adopted for the determination of thioredoxin activities.

Key words Thioredoxin reductase, Enzyme, Activity, Assay, Thioredoxin, Insulin, Spectrophotometry, NADPH

1 Introduction

Many different assays have been described for the measurement of thioredoxin reductase (TrxR) activities. Some of these are listed in Table 1 and are here only shortly commented upon. It must be noted that mammalian TrxRs are highly different from TrxRs found in bacteria, plants, or yeast. The mammalian enzymes are selenoproteins, i.e., depend upon a catalytic selenocysteine (Sec) residue for their activities, which is located within an additional redox active motif at the C-terminal part of a larger enzyme scaffold than that found in lower organisms [1–8]. Most important in the context of assays for activities of TrxRs is the fact that mammalian TrxRs can typically reduce a wide range of substrates in addition to thioredoxins (Trxs), while the bacterial, plant, and yeast species are much more restricted and normally only reduce Trx originating from related species [7, 9–11]. Thus, mammalian Trxs can typically not be used as substrates in assays to measure the activities of bacterial, plant, or yeast TrxRs. The many low molecular weight substrates of mammalian TrxRs, such as DTNB or some quinone

Laurent Chavatte (ed.), *Selenoproteins: Methods and Protocols*, Methods in Molecular Biology, vol. 1661,
DOI 10.1007/978-1-4939-7258-6_21, © Springer Science+Business Media LLC 2018

Table 1
Different assays for measurement of TrxR activities

Assay	Principle	Pros	Cons	References
Trx-linked insulin reduction (end-point assay) [Assay protocol given in this chapter.]	TrxR utilizes NADPH to reduce Trx, which in turn reduces disulfides in insulin, with subsequent determination of liberated thiol groups using DTNB	Highly specific assay for measurement of TrxR activities, compatible with crude lysates	Requires both Trx and insulin, species-restricted TrxR/Trx-combinations, not kinetic unless several time points are used	[9, 10]
Trx-linked insulin reduction (measuring NADPH consumption)	TrxR utilizes NADPH to reduce Trx, which in turn reduces disulfides in insulin, following NADPH consumption	Highly specific assay for measurement of TrxR activities	Requires both Trx and insulin, species-restricted TrxR/Trx-combinations, can be uncertain in use with crude lysates, reduced insulin precipitates	[9, 10, 12–14]
DTNB reduction	TrxR utilizes NADPH for direct reduction of DTNB	Simple and straightforward assay for pure mammalian TrxR, defines units of specific activity	Difficult to use with crude lysates, only works for mammalian TrxR and not the bacterial, plant, or yeast enzymes	[9, 10]
DTNB reduction in combination with gold compound inhibitors	TrxR utilizes NADPH for direct reduction of DTNB but is easily inhibited by gold, which is used to subtract background absorbance	Simple and straightforward assay for mammalian TrxR activities in crude lysates	Probably less specific assay than Trx-linked insulin reduction (end-point assay)	[15–17]
GHOST assay	TrxR utilizes NADPH to reduce Trx, which in turn is allowed to reduce glutathione disulfide	Simple assay as an alternative to Trx-linked insulin reduction	Interference will occur from any glutathione reductase present in the sample being analyzed	[18]

TRP14-linked cystine reduction	TrxR utilizes NADPH to reduce TRP14, which in turn is used to reduce cystine	Alternative to Trx-linked assay by using TRP14 instead, efficient in cystine reduction; can also be used as a fluorescent assay by exchanging cystine for bodypi-cystine	Not specific for TrxR because other enzyme systems, such glutathione reductase with glutathione, can also reduce cystine	[19]
Fluorescent substrates for TrxR	TrxR utilizes NADPH to reduce artificial fluorescent substrates, either directly or coupled to Trx	Can be utilized with either pure proteins or crude lysates, in some cases also with living cells	Requires high-sensitivity fluorescence determinations and quantifications of signals may be difficult	[20–24]
Juglone reduction	TrxR utilizes NADPH in reduction and redox cycling with juglone	Can be utilized as an assay for "SecTRAP" activities of mammalian TrxR that is compromised at the Sec residue	Complex kinetics, and unspecific assay if it would be performed in crude lysates	[25–29]
Phenanthrene quinone reduction	TrxR utilizes NADPH in reduction of phenanthrene quinone	Alternative direct assay for strict Sec-dependent activities of mammalian TrxR	Likely to be unspecific if used with crude lysates	[29]

compounds (*see* Table 1), can also not be used in assays for the determination of activities of TrxRs from non-mammalian species.

With regards to the mammalian TrxR enzymes it should also be noted that low-molecular weight species as well as different Trxs can vary in efficiency as substrates between the mitochondrial and cytosolic forms of TrxRs [25, 30]. It is therefore important to be aware of the specific kinetic characteristics of the exact TrxR isoform that is to be determined, when assays for its activity are to be performed. The assay method that is described in this chapter is the classic end-point Trx-linked insulin reduction assay, which is a specific assay for TrxR activity that can be performed with any type of biological sample. The assay can also be easily adopted for analyses of Trx activities, which may be determined at the same time in the same sample. The major aspects of this assay have previously been described elsewhere [10].

2 Materials

2.1 Reagents

1. TE buffer: 50 mM Tris–HCl, 2 mM EDTA, pH 7.5.
2. HEPES buffer, 1 M, pH 7.6.
3. EDTA, 0.2 M in water.
4. Insulin, 10 mg/ml (1.6 mM). Make a stock solution by suspending 50 mg insulin in 2.5 ml 50 mM Tris–HCl, pH 7.5; adjust pH to 2–3 with 1 M HCl to dissolve the protein completely, then titrate back to pH 7.5 with 1 M NaOH. Finally, adjust the volume with water to 5.0 ml. This clear stock solution of insulin (10 mg/ml) can be kept at −20 °C.
5. NADPH, 40 mg/ml (50 mM). Make stock vials by keeping 10 mg β-NADPH (dry) in small eppendorf tubes in −20 °C, then dissolve in 250 μl H_2O at the time of use. Store the stock solution at −20 °C less than 2 months and reuse only up to two times, since the reduced β-NADPH will oxidize (and become yellow) by repeated freezing-thawing or by time.
6. Pure control TrxR (of the homologous organism or, at least, mammalian TrxR for use with mammalian samples; *see* **Note 1**) at a dilution of about 1 U/ml (≈250 nM for the mammalian enzyme) in TE buffer, freshly made at the time of assay. Possible vendors include Cayman Chemicals or IMCO Corporation Ltd. AB.
7. Pure control Trx (preferentially from the homologous organism, alternatively *E. coli* Trx or the mutant human C62S/C73S Trx; *see* **Note 2**), at a working dilution of about 100 μM in TE buffer. Possible vendors include Cayman Chemicals or IMCO Corporation Ltd. AB.

8. 1 mM DTNB (5,5′-dithio-bis(2-nitrobenzoic acid) or Ellman's reagent) in 6 M guanidine hydrochloride in 0.2 M Tris–HCl, pH 8.0. First make 6 M guanidine hydrochloride in 0.2 M Tris–HCl, pH 8.0 (28.66 g guanidine hydrochloride to 50 ml Tris–HCl, pH 8.0—can be stored at room temperature) and 10 mM (4 mg/ml) DTNB in spectroscopic grade 99% ethanol (can be stored at −20 °C in well-sealed glass vials). At the time of assay, add 1 ml 10 mM DTNB to 9 ml 6 M guanidine hydrochloride in 0.2 M Tris–HCl, pH 8.0, which will be enough for 20 samples; keep on ice. CAUTION: Guanidine hydrochloride is highly corrosive.
9. Protein sample to be analyzed for TrxR activity, e.g., protein extracts from cells or tissues. Be sure to include EDTA in the sample, since TrxR is easily inhibited by metal ions, keep the pH at physiological range (7.0–8.0), and possibly include protease inhibitors. A useful protein concentration in the sample is usually 1–5 μg/μl.

2.2 Instrumentations

1. Spectrophotometer [either for cuvettes or microtiter plates] able to read absorbance at 412 nm and possibly 340 nm.
2. Transparent or UV compatible microplates (or cuvettes) for the reading at 412 or 340 nm wavelength, respectively.
3. Incubator set at 37 °C.

3 Methods

1. Make a master mixture as described in Table 2.
2. Two reaction mixes per sample are made in either semi-micro cuvettes or different wells of a microtiter plate. Disposable

Table 2
Composition of master mixture

	Volume	Final concentration in assay
HEPES buffer, 1 M	$0.130 \times V$	85 mM
Insulin, 1.6 mM	$0.285 \times V$	300 μM
NADPH, 50 mM	$0.02 \times V$	660 μM
EDTA, 0.2 M	$0.023 \times V$	3 mM
Water	$0.542 \times V$	
Total volume	V	

Preparation of a master mixture with the volume of V μl. The total volume required will depend upon the number of samples to be analyzed and the choice of final reaction volume for the assay. This can be estimated as follows: $n \times Z \times 2 + 4n$, where n = number of samples and Z = volume (in μl) of the reaction for each sample in the assay. If both Trx- and TrxR-activities are to be measured, then double the calculated volume once more. Note that each reaction vessel where absorbance will be measured has to hold a total volume of $Z \times 10$ μl plastic cuvettes can be used if many samples are to be analyzed at one time or, alternatively, the assay can be scaled down for use of 96-well microplates if a spectrophotometer for reading of such is available [31].

(a) *Optional*: For measurement of Trx activity (Table 3).

(b) For measurement of TrxR activity (Table 4).

3. Incubate all samples 20 min at 37 °C (*see* **Notes 3** and **4**).
4. Add $Z \times 10$ µl 0.4 mg DTNB/ml 6 M GuHCl in 0.2 M Tris–HCl, pH 8.0 to each sample (*see* Table 2 for information on volumes). This step will denature all the proteins in the samples and allow for the determination of thiol groups by reaction with DTNB, which releases TNB^- that displays high absorbance at 412 nm, with an extinction coefficient

Table 3
Description of Trx activity assay

	Reaction mix A Volume	Reaction mix B Volume
Master mixture	$0.66 \times Z$ µl	$0.66 \times Z$ µl
HEPES buffer, 85 mM pH 7.6	$0.14 \times Z - X\,\mu l$	$0.34 \times Z - X\,\mu l$
Pure control TrxR (to give ca. 50 nM)	$0.2 \times Z$ µl	–
Sample to be analyzed for the presence of Trx activity (usually 10–20 µg total protein in) a volume of X µl, where $X \leq 0.14 \times Z$)	$X\,\mu l$	$X\,\mu l$
Total volume	Z µl	Z µl

See Table 2 for information on volumes

Table 4
Description of TrxR activity assay

	Reaction mix A Volume	Reaction mix B Volume
Master mixture	$0.66 \times Z$ µl	$0.66 \times Z$ µl
HEPES buffer, 85 mM pH 7.6	$0.18 \times Z - X\,\mu l$	$0.34 \times Z - X\,\mu l$
Pure control Trx (to give 15 µM)	$0.16 \times Z$ µl	–
Sample to be analyzed for presence of TrxR activity (usually 10–20 µg total protein in) a volume of X µl, where $X \leq 0.18 \times Z$)	X	X
Total volume	Z µl	Z µl

See Table 2 for information on volumes

13,600 M^{-1} cm^{-1}. Every thiol group in the sample will give rise to one molecule of TNB^{-}.

5. Determine the absorbance at 412 nm of reaction mix A and reaction mix B for each sample. The difference (A – B) will correlate to the presence of Trx or TrxR in the protein sample, as determined in Subheading 3, **step 2a** or **2b**, respectively. The actual amount can be determined by comparison with a standard curve run in parallel using pure control Trx or TrxR. This standard curve should preferably be performed in the presence of sample protein, to make certain that no inhibitory or stimulatory factor in the sample affects the determination. As given here, the assay is optimized for 0–20 pmol (0–200 ng) Trx (Subheading 3, **step 2a**) or 0–2 pmol TrxR (0–250 ng of dimeric mammalian enzyme) (Subheading 3, **step 2a**).

4 Notes

1. It is important to observe the fact that mammalian TrxRs typically reduce Trxs of different species, albeit with different efficiency, while TrxRs of bacteria, such as the enzyme from *E. coli*, are more specific and typically do not reduce mammalian Trx. Thus, if mammalian Trxs are used as substrates in assays measuring bacterial TrxR activities, the assay will not work.
2. Wild-type human Trx has a pronounced tendency to lose activity due to aggregation as a result of oxidation of structural cysteine residues, which is a process that is exaggerated upon storage and freezing-thawing. This affects the assay and can be overcome by reducing the Trx sample with DTT (but take care that remaining traces of DTT can subsequently react with DTNB). The *E. coli* and mutant human C62S/C73S Trx are however preferred as substrates for TrxR in assays since these thioredoxins do not have the tendency to aggregate due to oxidation [32] and are thus more stable upon storage.
3. This incubation step is included to allow for TrxR-catalyzed support of Trx-mediated reduction of insulin. If activities are low, the 20 min incubation time can be prolonged. Note that the present protocol is made to include a reference sample lacking addition of Trx, while another background control can be to include a reference sample with addition of Trx but skipping the incubation step, thereby obtaining a zero time background value. Finally, in order to validate a time-dependent enzyme activity it is also possible to include several time points per sample, thereby enabling assessment of time-dependent insulin reduction in this assay.

4. If the instrumental setup allows it is possible to follow decrease of absorbance at 340 nm during this incubation step, which will reflect consumption of NADPH by time during the enzymatic reaction. However, as more insulin is being reduced the insulin A and B chains will precipitate and result in turbidity, which will subsequently appear as increasing absorbance values by time toward the latter parts of the assay.

Acknowledgments

The author acknowledges funding from The Swedish Cancer Society, The Swedish Research Council, Swedish Foundation for Strategic Research, Knut and Alice Wallenberg Foundation, and Karolinska Institutet.

References

1. Kuriyan J, Krishna TSR, Wong L, Guenther B, Pahler A, Williams CHJ, Model P (1991) Convergent evolution of similar function in two structurally divergent enzymes. Nature 352:172–174
2. Williams CH Jr (1992) Lipoamide dehydrogenase, glutathione reductase, thioredoxin reductase, and mercuric ion reductase - a family of flavoenzyme transhydrogenases. In: Müller F (ed) Chemistry and biochemistry of flavoenzymes, vol 3. CRC Press, Boca Raton, FL, pp 121–211
3. Arscott LD, Gromer S, Schirmer RH, Becker K, Williams CH Jr (1997) The mechanism of thioredoxin reductase from human placenta is similar to the mechanisms of lipoamide dehydrogenase and glutathione reductase and is distinct from the mechanism of thioredoxin reductase from Escherichia coli. Proc Natl Acad Sci U S A 94:3621–3626
4. Williams CH, Arscott LD, Muller S, Lennon BW, Ludwig ML, Wang PF, Veine DM, Becker K, Schirmer RH (2000) Thioredoxin reductase two modes of catalysis have evolved. Eur J Biochem 267:6110–6117
5. Gromer S, Johansson L, Bauer H, Arscott LD, Rauch S, Ballou DP, Williams CH Jr, Schirmer RH, Arnér ESJ (2003) Active sites of thioredoxin reductases—Why selenoproteins? Proc Natl Acad Sci U S A 100:12618–12623
6. Zhong L, Arnér ESJ, Ljung J, Åslund F, Holmgren A (1998) Rat and calf thioredoxin reductase are homologous to glutathione reductase with a carboxyl-terminal elongation containing a conserved catalytically active penultimate selenocysteine residue. J Biol Chem 273:8581–8591
7. Arner ES, Holmgren A (2000) Physiological functions of thioredoxin and thioredoxin reductase. Eur J Biochem 267:6102–6109
8. Zhong L, Arnér ESJ, Holmgren A (2000) Structure and mechanism of mammalian thioredoxin reductase: the active site is a redox-active selenolthiol/selenenylsulfide formed from the conserved cysteine-selenocysteine sequence. Proc Natl Acad Sci U S A 97:5854–5859
9. Arnér ESJ, Zhong L, Holmgren A (1999) Preparation and assay of mammalian thioredoxin and thioredoxin reductase. Methods Enzymol 300:226–239
10. Arnér ESJ, Holmgren A (2000) Measurement of thioredoxin and thioredoxin reductase. In: Maines M, Costa L, Reed D, Sassa S (eds) Current protocols in toxicology. Wiley, New York, pp 7.4.1–7.4.14
11. Arnér ESJ (2009) Focus on mammalian thioredoxin reductases – important selenoproteins with versatile functions. Biochim Biophys Acta 1790:495–526
12. Holmgren A (1977) Bovine thioredoxin system. Purification of thioredoxin reductase from calf liver and thymus and studies of its function in disulfide reduction. J Biol Chem 252:4600–4606
13. Holmgren A (1979) Reduction of disulfides by thioredoxin. Exceptional reactivity of insulin and suggested functions of thioredoxin in mechanism of hormone action. J Biol Chem 254:9113–9119

14. Holmgren A (1979) Thioredoxin catalyzes the reduction of insulin disulfides by dithiothreitol and dihydrolipoamide. J Biol Chem 254:9627–9632
15. Gromer S, Merkle H, Schirmer RH, Becker K (2002) Human placenta thioredoxin reductase: preparation and inhibitor studies. Methods Enzymol 347:382–394
16. Gromer S, Arscott LD, Williams CH, Schirmer RH, Becker K (1998) Human placenta thioredoxin reductase: isolation of the selenoenzyme, steady state kinetics, and inhibition by therapeutic gold compounds. J Biol Chem 273:20096–20101
17. Smith AD, Levander OA (2002) High-throughput 96-well microplate assays for determining specific activities of glutathione peroxidase and thioredoxin reductase. Methods Enzymol 347:113–121
18. Gromer S, Johansson L, Bauer H, Arscott LD, Rauch S, Ballou DP, Williams CH Jr, Schirmer RH, Arner ES (2003) Active sites of thioredoxin reductases: why selenoproteins? Proc Natl Acad Sci U S A 100:12618–12623
19. Pader I, Sengupta R, Cebula M, Xu J, Lundberg JO, Holmgren A, Johansson K, Arner ES (2014) Thioredoxin-related protein of 14 kDa is an efficient L-cystine reductase and S-denitrosylase. Proc Natl Acad Sci U S A 111:6964–6969
20. Montano SJ, Lu J, Gustafsson TN, Holmgren A (2014) Activity assays of mammalian thioredoxin and thioredoxin reductase: fluorescent disulfide substrates, mechanisms, and use with tissue samples. Anal Biochem 449:139–146
21. Ma H, Zhang J, Zhang Z, Liu Y, Fang J (2016) A fast response and red emission probe for mammalian thioredoxin reductase. Chem Commun (Camb) 52:12060–12063
22. Liu Y, Ma H, Zhang L, Cui Y, Liu X, Fang J (2016) A small molecule probe reveals declined mitochondrial thioredoxin reductase activity in a Parkinson's disease model. Chem Commun (Camb) 52:2296–2299
23. Zhang B, Ge C, Yao J, Liu Y, Xie H, Fang J (2015) Selective selenol fluorescent probes: design, synthesis, structural determinants, and biological applications. J Am Chem Soc 137:757–769
24. Zhang L, Duan D, Liu Y, Ge C, Cui X, Sun J, Fang J (2014) Highly selective off-on fluorescent probe for imaging thioredoxin reductase in living cells. J Am Chem Soc 136:226–233
25. Rackham O, Shearwood AM, Thyer R, McNamara E, Davies SM, Callus BA, Miranda-Vizuete A, Berners-Price SJ, Cheng Q, Arnér ES, Filipovska A (2011) Substrate and inhibitor specificities differ between human cytosolic and mitochondrial thioredoxin reductases: implications for development of specific inhibitors. Free Radic Biol Med 50:689–699
26. Xu J, Cheng Q, Arner ES (2016) Details in the catalytic mechanism of mammalian thioredoxin reductase 1 revealed using point mutations and juglone-coupled enzyme activities. Free Radic Biol Med 94:110–120
27. Salmon-Chemin L, Buisine E, Yardley V, Kohler S, Debreu MA, Landry V, Sergheraert C, Croft SL, Krauth-Siegel RL, Davioud-Charvet E (2001) 2- and 3-substituted 1,4-naphthoquinone derivatives as subversive substrates of trypanothione reductase and lipoamide dehydrogenase from Trypanosoma cruzi: synthesis and correlation between redox cycling activities and in vitro cytotoxicity. J Med Chem 44:548–565
28. Anestal K, Prast-Nielsen S, Cenas N, Arner ES (2008) Cell death by SecTRAPs: thioredoxin reductase as a prooxidant killer of cells. PLoS One 3:e1846
29. Cenas N, Nivinskas H, Anusevicius Z, Sarlauskas J, Lederer F, Arner ES (2004) Interactions of quinones with thioredoxin reductase: a challenge to the antioxidant role of the mammalian selenoprotein. J Biol Chem 279:2583–2592
30. Lothrop AP, Snider GW, Ruggles EL, Hondal RJ (2014) Why is mammalian thioredoxin reductase 1 so dependent upon the use of selenium? Biochemistry 53:554–565
31. Peng X, Gimenez-Cassina A, Petrus P, Conrad M, Ryden M, Arner ES (2016) Thioredoxin reductase 1 suppresses adipocyte differentiation and insulin responsiveness. Sci Rep 6:28080
32. Ren X, Bjornstedt M, Shen B, Ericson ML, Holmgren A (1993) Mutagenesis of structural half-cystine residues in human thioredoxin and effects on the regulation of activity by selenodiglutathione. Biochemistry 32:9701–9708

Part V

In Vivo Studies

Chapter 22

Association of Single Nucleotide Polymorphisms in Selenoprotein Genes with Cancer Risk

Catherine Méplan

Abstract

Genetic association studies have linked genetic variants in the Selenium (Se) metabolism with the development of complex diseases such as cancer and helped unravel novel mechanisms underlying cancer development. The chapter describes the specificity of genetic variants in the Se metabolism, the approaches used in association studies, and the limitations of such approaches.

Key words Biomarkers, Genomics, GWAs, Interaction, Nutrigenomics, Selenium, SNP

1 Introduction

Studying the association of genetic variants in the Selenium (Se) metabolism with complex diseases such as cancer has been a great focus of interest over the past decade, transforming our vision of the impact of Se nutrition on human health. The main objectives of these studies are to identify single nucleotide polymorphisms (SNPs) in genes involved in the Se metabolic pathway that correlate with a disease trait, and to determine whether these SNPs can serve as predictive biomarkers of disease risk and outcome and can inform decisions about medical treatment or nutritional intervention. These studies also evaluate the impact of interactions between clinically relevant SNPs and biomarkers of Se status on disease development and progression. The presence of such associations can reveal molecular pathways involved in carcinogenesis that are influenced by Se [1, 2].

Advances in this field have largely benefited from the development of genomics approaches; high-throughput, relatively low cost, genotyping technologies; and large public databases, providing unprecedented access to the vast amount of SNP data. The application of these technologies has revolutionized the selection of genetic variants and their investigation in disease-gene

Laurent Chavatte (ed.), *Selenoproteins: Methods and Protocols*, Methods in Molecular Biology, vol. 1661,
DOI 10.1007/978-1-4939-7258-6_22,

association studies, but not facilitated the data analysis or interpretation. In addition, the design of genetic association studies investigating the relationship between polymorphisms in the Se metabolism and cancer is determined by the unique characteristics of these SNPs and the challenges associated with SNP X nutrient interactions.

2 Material

2.1 Study Population

2.1.1 Ethical Considerations

Data collected in genetic association studies are extremely sensitive, requiring both subjects' consent and protection of participants' confidentiality. In most studies, ethical approval requests that genetic information is not shared with participants because genetic risk factors are often perceived as immutable and because of the predictive nature of genetic testing, the potential implications for other family members, and the risk of discrimination associated with such knowledge.

2.1.2 Population Study Size

The population study size is a limiting factor determining the number of SNPs that can be tested, the number of statistical tests that can be carried out, and the significant cut-off *p*-values for weak genetic effect (i.e., allele presents at a very low frequency) (*see* **Note 1**). A small population size can be underpowered to reliably detect a weak effect (association) and can lead to either an underestimation or, more likely, overestimation of the true effect of a variant, as a result of an imprecise estimation of the strength of the effect in the study, due to sampling variation. Thus, the study design requires a power calculation to determine the appropriate sample size for detecting a significant difference between genotypes (Table 1).

2.1.3 Case-Control Design and Population Stratification

Case-control studies are the most frequently used design in association studies. They rely on the critical assumption that differences between cases and controls in genotype frequency for a given SNP are solely due to difference in disease status, and not to distinct genetic backgrounds between cases and controls. However, it is difficult to ensure during the study design that the control and case populations do not correspond to a combination of ancestrally distinct populations, with different disease prevalence and SNP allele frequency. The structure of a population is made even more complicated by recent admixed populations of diverse ancestries. Yet population stratification may contribute only to a small proportion of the false positive or lack of replication in candidate SNP association studies. On the contrary, in the case of genome-wide associations studies (GWAs), the lack of homogeneity of the large population study can represent a huge challenge and may cause false negative or positive results (*see* **Notes 1** and **2**).

Table 1
Selection of SNP in the Se metabolism for disease-gene association

Study design	Considerations in study design	Design and interpretation
Number of SNPs	Population size determines statistical power and number of tests	• Power calculation • Correction for multiple testing
SNP selection	Presence in study population: – Present in different public databases (dbSNP, HGBASE, HAPMAP) – Known allele frequencies in target population	• Compare allele frequency with known frequencies from databases • Test for Hardy–Weinberg equilibrium in control population
Other SNPs in the neighborhood	Neighboring SNPs may interfere with primer design.	• Multiplex genotyping: need to assess whether primers' sets can be used simultaneously
Functional SNPs	Mechanistic studies providing evidence for functionality of a SNP in the same tissue or other tissue. SNP may affect: – Regulatory region – Protein function, regulation – Metabolic pathway – Other selenoproteins	• Evidence of causality: requires further functional studies as the SNP may *tag* for another functional SNP. The SNP may affect: – Corresponding selenoprotein's synthesis or activity – Selenoproteins' synthesis machinery, distribution of Sec between selenoproteins – Selenoproteome – Selenoproteins in the same metabolic pathway or sharing similar function – Metabolic pathway or stress level
Tagging SNP	Tagging SNP can help to capture effects of neighbouring SNPs in LD	No known functionality, may *tag* a neighbouring functional SNP, functional studies required
Interactions of SNPs with other genetic factors in other genes	Known interactions with other SNPs in other genes can help understand biological processes involved in diseases	• Determination of interaction for SNPs with a main effect – Correct for multiple testing – Validation or refutation of known interaction in current study population • Interactions could reflect: – Biological interactions between corresponding (seleno)proteins in a metabolic pathway or in disease development – A change in the selenoprotein hierarchy
Interactions of SNPs with Se biomarkers	– Plasma/serum Se – Plasma/serum Sepp – Selenoprotein activity or expression	• SNP could alter the selenoprotein hierarchy, Sec transport, Sec bioavailability, or selenoprotein synthesis • Interaction may explain discrepancy between distinct studies

The table presents key features that require consideration in the study design of genetic associations of SNPs in Se metabolism with cancer and the potential implication on data interpretation

2.1.4 Confounding Factors, Clinical Stratification, Inclusion and Exclusion Criteria, and Biomarkers of Selenium Status

Not only should the study design ensure that the case and control populations share a similar genetic background but also that both populations are controlled for a number of confounding factors such as gender, body mass index, age, alcohol and tobacco consumption. These factors are known to impact on cancer risk and therefore should be adjusted for in the statistical analysis. In addition, selenoprotein expression was shown to be affected by sex-hormones and alcohol intake [3, 4] and the impact of some SNPs in selenoprotein genes was found to be modulated by smoking and gender [5, 6]. Thus, these factors should be incorporated in the study design, statistical analysis, and accounted for data interpretation.

As SNPs may affect not only cancer risk but also cancer progression, the case population can be stratified according to clinical criteria, including the severity of the disease, survival and mortality data, age of onset, clinical and pathological characterization of the samples (i.e., grades and stages), and response to treatment. Differences in the definition of these clinical and pathological characteristics between studies may affect the replicability of results between different study populations. Similarly, difference in inclusion and exclusion criteria between studies can lead to misinterpretation and lack of replicability of the results (*see* **Note 1**).

Many studies have also demonstrated that the effect of genetic variants in the Se metabolism can be modulated by Se intake. Food questionnaires only provide a poor estimation of Se intake because the Se content of a given food varies greatly depending of its provenance, thus it is preferable to use biomarkers of Se status (Table 1), including serum or plasma Se concentration, plasma selenoprotein P (SePP) levels and glutathione peroxidase GPx3 activity, or erythrocyte GPx1 activity, to assess Se status [7, 8]. Theoretically, any selenoproteins can be used as a marker of Se status as their expression depends upon Se bioavailability; however, each biomarker provides a different information. To date, plasma SePP has been considered the best biomarker of active Se as its levels strongly correlate with plasma Se status in physiological conditions [9]. But Se exerts anti-carcinogenic properties other than the one carried out by selenoproteins and some selenoproteins play a particularly important role in certain tissues, thus choice of relevant biomarkers may depend upon the types of cancer or tissue (*see* **Notes 3** and **4**).

2.2 Source and Preparation of DNA Samples

Traditionally, blood sample and buffy coat have been used to prepare large amounts of DNA samples of good quality. However, if this approach is still regularly used, the increasing interest of SNP genotyping has led to the development of several, less invasive and cheaper technologies including a collection of kits to prepare DNA from buccal cells or the use of dried blood spots on filter paper (3 mm FTA card). Some studies but not all have suggested that the genotyping success rate can vary depending on the source of DNA (blood/ buffy coat versus buccal cells) [10–12].

2.3 Selection of SNPs

2.3.1 Specificity of SNP in Se Metabolism

The selection of polymorphisms in the Se metabolism for association studies requires an understanding of the specificity of these SNPs to fully comprehend their relationship with disease and interpret the results. Like for any other SNP, the association of a genetic variant in the Se pathway with cancer informs us on a potential role of a gene or protein in metabolic pathways involved in the disease development or progression (*see* **Notes 2** and **5**). In addition, SNPs in the Se pathway present unique characteristics due to functional interactions within metabolic pathways of several selenoproteins and to the existence of a shared biosynthesis complex machinery that supports the expression of all selenoproteins (Fig. 1). In particular, selenoproteins share the amino acid Selenocysteine (Sec) and specific translation factors such as SBP2 (SECIS binding Protein 2), a Sec-specific $tRNA^{[SerSec]}$ and the elongation factor EFSec [13]. The distribution of Sec between selenoproteins is governed by a prioritization mechanism, called the selenoprotein hierarchy that favors the synthesis of certain selenoproteins over others in condition of low Se supply [1, 2]. Thus, genetic and dietary factors affecting the bioavailability of Sec and the order of the selenoprotein hierarchy have the potential to not only affect individual selenoprotein expression but also the whole selenoproteome [1]. The combination of genetic associations and mechanistic work has highlighted the specific impact of polymorphisms affecting the 3′ UTR and SECIS element on Sec incorporation in the corresponding selenoprotein and on the overall selenoproteome, and the selenoprotein hierarchy [14–17]. The presence or absence of genetic associations should be cautiously interpreted in the light of these specific characteristics (*see* **Note 3**).

2.3.2 Criteria for SNP Selection

To date, three types of approaches (candidate SNP approach, unbiased GWAS, and pathway analysis) have been used to search for genetic associations between cancer and SNPs in the Se metabolic pathways. The selection of SNPs differs greatly between the three approaches (Table 1) (*see* **Notes 2**, **3** and **5**).

In candidate SNP approach, SNPs are selected based on knowledge derived from mechanistic studies demonstrating a functionality of the SNP. These functional variants can affect the protein function, activity, synthesis or stability, or regulatory gene region. Their association with a disease trait suggests that the corresponding selenoprotein may be implicated in the pathogenesis or in the normal tissue function. This approach has been widely used in combination with functional analysis of SNP in various in vitro, cell or animal model. Overall, the number of SNPs analyzed in candidate SNP approaches is small, allowing for interactions between these SNPs and with Se status to be tested (Table 1) (*see* **Notes 5–6**).

In GWAS, the selection of SNPs is not based on prior knowledge, and therefore considered unbiased. Theoretically, the

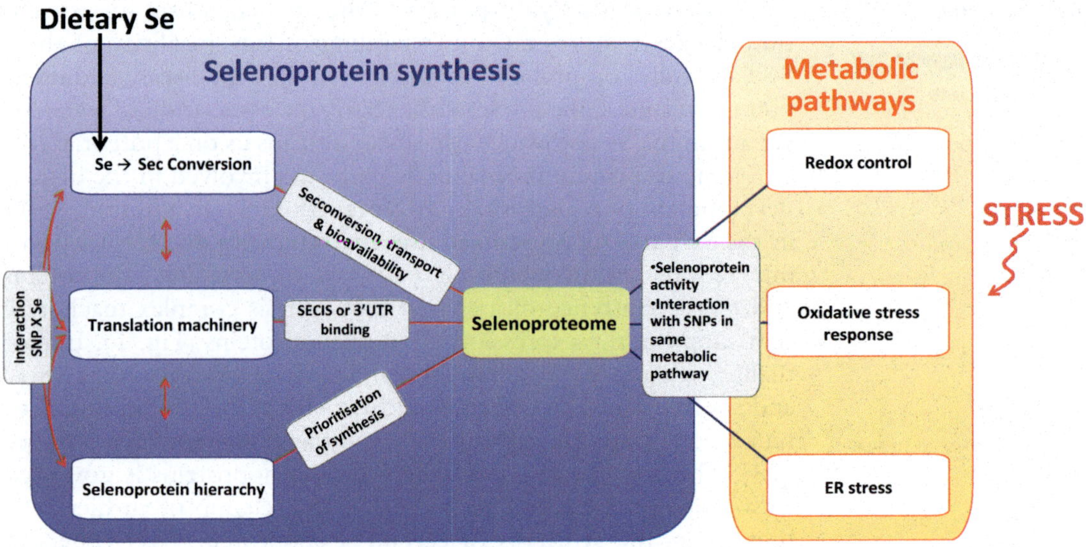

Fig. 1 Specificity of SNPs in the Se metabolism. The diagram illustrates how genetic variants in the Se metabolism can affect the whole selenoproteome, metabolic pathways in which selenoproteins play an essential role and the response to various stressors

selection covers all or most SNPs in the genome and depends on the use of tagging SNPs. Tagging SNPs exploit the fact that neighboring variants, present in the same haplotype and in high linkage disequilibrium (LD), are transmitted as a block. As a result, tagging SNPs can be used to capture the genotype of all variants in LD, reducing dramatically the number of SNPs to be genotyped. The number and selection of tagging SNPs depends here on the commercial platform used and on the determination of LD between SNPs within a study population (Table 1). However, as LD varies greatly between populations, the design of the platform should be suitable for the study population. Additionally, GWAs present some limitations. They do not provide evidence of functionality for the associated SNPs and they lack statistical power to test for interactions with Se biomarkers or with other genetic markers (*see* **Notes 4**, **6** and **7**).

In the pathways analysis, the selection of SNPs combines known functional SNPs and tagging SNPs in genes involved in the Se metabolic pathway, and SNPs in the non-selenoprotein genes that have previously been found to interact with functional variants in selenoprotein genes. This approach aims at identifying novel variants within the Se pathway and assessing interactions between linked variants and other genetic or environmental factors, exposing novel connections and molecular pathways involved in the carcinogenic or homeostatic processes [18]. Using this approach, the identification of several, genetically unlinked polymorphisms within the same gene associated with a disease trait strengthens the evidence of a role of a particular selenoprotein in the normal or diseased tissue but does not provide a functional mechanism. This approach allows for SNP x SNP and SNP X nutrients interactions to be investigated (*see* **Notes 4**, **5** and 7).

3 Methods

3.1 Genotyping

3.1.1 Genotyping Methods

In the past, in-house genotyping assays (such as RFLP, DNA-HPLC, TaqMan, or LightCycler assay) were used but these methods were expensive and time consuming, limiting the number of samples and SNPs that could be analyzed. The genotyping assay is then validated by sequencing a random 10% of the samples as duplicate quality controls. However with the decrease in sequencing costs and the continuous development of novel SNP genotyping methods, chemistries, and platforms, it is now relatively cheaper and faster to ask a company to genotype hundreds to thousands of SNPs in a large number of DNA samples. As reviewed in [19], the sensitivity, reproducibility, accuracy, capability of multiplexing, and level of throughput differ between platforms, and should be taken into consideration in the study design.

3.1.2 Genotype Frequency and Linkage Disequilibrium

Prior to the statistical analysis, a Pearson Chi^2 test is performed to compare observed genotype frequencies for the selected polymorphisms with expected frequency under conditions of Hardy-Weinberg equilibrium (HWE). Several HWE calculators are available online. Deviation from the HWE can sometimes indicate potential sequencing errors, but also can detect population structure deviating from the Hardy-Weinberg assumptions or can suggest an effect of natural selection on allele frequency. It is also important to compare genotype frequencies with reported frequencies in large databases such as Hapmap.org for a given population as this may reveal population substructure (Table 1).

Similarly, determination of LD (nonrandom transmission of alleles at different loci) should be made to inform the statistical analysis and the interpretation of the associations, as the tested SNP may not itself be the causal variant but may be tagging for a causal variant in LD.

3.2 Statistical Analysis

3.2.1 Odds Ratios and Relative Risks

For unmatched case-control studies, the association between individual SNPs and disease is determined using an unconditional logistic regression to calculate odds ratios (ORs, odds of having a disease or condition) and 95% confidence intervals (95% CI) for unmatched case-control studies. **For matched case-control studies**, the genetic association is determined by conditional logistic regression stratified by the matched case. In both cases, two models can be tested: a recessive model in which each genotype is compared with the homozygote for the frequent allele, taken as a reference; or a dominant model in which frequencies for heterozygotes and homozygotes for the rare allele are pooled together and compared with the homozygote for the frequent allele to assess the effect of the presence of at least one rare allele on the risk of developing the disease. **For prospective and retrospective cohort studies, and randomized controlled trials**, relative risks (RR, risk of developing a disease in the exposed population compared with another group of differing exposure) can be calculated. As a rule, for disease with low prevalence, the RR and the OR will be roughly the same. In addition, correction for multiple testing (e.g., Bonferroni correction) should be used to account for the potentially high number of statistical tests carried out.

3.2.2 Genetic Interactions: Two-Locus Interactions

Complex diseases such as cancers are polygenic with both disease risk and progression being influenced by the combination and interactions between multiple genes in complex biological networks. Thus, in addition to the characterization of the main effects of SNPs associated with a disease trait, the detection of genetic interactions between two markers has the potential to reveal biological pathways underpinning a disease and expose novel connections between proteins or metabolic pathways. The statistical analysis of such genetic interactions can be however challenging. In candidate gene and pathway approaches, traditional logistic regression models can be used to test for genetic interaction between a SNP having a main effect (i.e., a SNP significantly associated with a disease trait) with other SNPs (which are not necessarily statistically associated with the trait). This approach is tedious and requires a correction for multiple testing. When two single markers are genetically unlinked (not transmitted in the same haplotype), the tests are independent and a Bonferroni correction of the interaction p-values is conservative but pertinent. On the contrary, if the studied SNPs are in partial and complete LD (e.g., SNPs in the same gene or chromosomal region), the LD will induce a correlation and therefore the interaction might not be biologically relevant or statistically meaningful (*see* **Note 5**).

3.2.3 SNP X Se Status Interactions

The main effects of markers of Se status (e.g., serum/plasma Se, plasma SePP concentration, or GPx3 activity) can be calculated by unconditional logistic regression for unmatched case-control studies or conditional logistic regression for matched case-control studies, by using the biomarkers as continuous variables or as quantile based on the distribution among the controls.

To test for potential gene X environment interactions between a SNP with a main effect and markers of Se status, the ORs (and 95% CI) for the disease or disease trait are calculated using either the continuous Se variable stratified by genotype with unconditional logistic regression for unmatched case-control studies or unconditional logistic regression adjusting for the matching variables for matched case-control studies (*see* **Note 6**).

4 Notes

1. Several factors can contribute to the lack of replication between genetic association studies. Cancer is a complex, polygenic disease, caused by the interactions of multiple variants and environmental factors. The mechanisms by which a disease develops many differ between distinct study populations, with different genetic background, population structure, exposure to carcinogenic and stress factors, recruitment criteria, or confounding factors. The population size may affect the power to study the effect of genetic variants and may result in false positive or negative associations. A cancer for a given tissue usually comprises multiple types of cancers that differ in their etiology, and progression (grades, stages) and which may differ between the populations. In addition, suboptimal Se status is known to modulate the effect of many genetic variants in the Se metabolism on cancer risk, thus lack of replication between studies may reflect differences in Se status or biomarkers of Se status in distinct study populations.
2. The interpretation of the association between a genetic variant in the Se metabolism and a disease trait can be complex. It is determined by (a) the nature of the variant—whether the SNP is functional or tagging for a functional SNP in LD, (b) the impact of the polymorphism on the protein function or regulation and on the corresponding molecular pathway, (c) the environment, tissue specificity and disease context, and (d) the genetic background of the population. Genetic associations, supported by functional evidence of a causal link between the SNP and disease, provide insights into the disease mechanisms.
3. The presence of genetic associations between SNP in the Se metabolism and cancer can also reveal the specific impact of a variant in the Se metabolism on multiple selenoproteins, whether the SNP impacts on the synthesis of other selenoproteins, in particular when a SNP affects the selenoprotein hierarchy, or the SNP influences a biochemical pathway or function shared by other selenoproteins. For example, rs713041 in the *GPX4* gene affects GPx1 protein levels in erythrocyte in breast cancer patients, probably as a result of the stress associated with either the change in GPx4 expression or the redistribution of Sec toward other selenoproteins [3]. Similarly, rs3877899 and 7579 in *SEPP1* gene affect Se bioavailability and the expression of blood selenoproteins, in particular the ratio of lymphocyte GPx1 over GPx4 proteins [14]. Moreover, as most selenoproteins play a key role in stress response mechanisms, immune function, and cellular maintenance, the links between SNPs and diseases such as cancer may indicate that during evolution,

some of these variants have been selected as a result of exposure to various stresses and now affect processes involved in cancers.

4. The interaction of SNP with a main effect with either other genetic variants in other genes or with Se biomarkers is the key to the understanding of the etiology of complex, polygenic diseases such as cancer. These interactions modulate the initial main effect (association) of the SNP with the disease trait and provide clues on the mechanisms involved in the disease development. Interactions with other genetic variants may reflect biological interactions between the corresponding proteins within a metabolic pathway [3, 5, 6, 21].
5. The identification of several variants within the same gene, which are not in LD is providing strong evidence for a role of the corresponding protein in the normal tissue function or disease development.
6. A given variant may be associated with the same disease trait in several studies but the effect of the genotypes may differ between studies. As exemplified by the case of rs713041 in *GPX4* gene, several studies have identified this SNP to be associated with colorectal cancer risk. However, the TT genotype was found to confer a protection in a small Scottish population [20] but to increase colorectal cancer risk in a Czech cohort [6]. The apparent discrepancy between the two studies indicates that (a) GPx4 is likely to play a key role in colorectal cancer or colorectal function and (b) the effect of the SNP is likely to be modulated by another factor either genetic or environmental. Therefore, the identification in distinct association studies of the same SNP but different genotype reflects a more complex relationship between disease and polymorphisms than the one observed in monogenic disease.
7. Interactions of SNPs with biomarkers of Se status have been shown to modulate the disease risk, suggesting that some subgroup populations may benefit from additional intake of Se to reduce disease risk, offering preventive tools and accounting for some discrepancies between different study populations. In particular, a number of studies have shown an association of rs1050450 in the *GPX1* gene in studies carried out in Europeans populations with low Se status but the associations could not be replicated in several US populations with high Se status [3, 4, 22–24].

References

1. Meplan C (2015) Selenium and chronic diseases: a nutritional genomics perspective. Forum Nutr 7(5):3621–3651. doi:10.3390/nu7053621
2. Meplan C, Hesketh J (2014) Selenium and cancer: a story that should not be forgotten-insights from genomics. Cancer Treat Res 159:145–166. doi:10.1007/978-3-642-38007-5_9

3. Meplan C, Dragsted LO, Ravn-Haren G, Tjonneland A, Vogel U, Hesketh J (2013) Association between polymorphisms in glutathione peroxidase and selenoprotein P genes, glutathione peroxidase activity, HRT use and breast cancer risk. PLoS One 8(9):e73316. doi:10.1371/journal.pone.0073316
4. Ravn-Haren G, Olsen A, Tjonneland A, Dragsted LO, Nexo BA, Wallin H, Overvad K, Raaschou-Nielsen O, Vogel U (2006) Associations between GPX1 Pro198Leu polymorphism, erythrocyte GPX activity, alcohol consumption and breast cancer risk in a prospective cohort study. Carcinogenesis 27 (4):820–825. doi: bgi267 [pii] 10.1093/carcin/bgi267
5. Cooper ML, Adami HO, Gronberg H, Wiklund F, Green FR, Rayman MP (2008) Interaction between single nucleotide polymorphisms in selenoprotein P and mitochondrial superoxide dismutase determines prostate cancer risk. Cancer Res 68 (24):10171–10177
6. Meplan C, Hughes DJ, Pardini B, Naccarati A, Soucek P, Vodickova L, Hlavata I, Vrana D, Vodicka P, Hesketh JE (2010) Genetic variants in selenoprotein genes increase risk of colorectal cancer. Carcinogenesis 31(6):1074–1079. doi: bgq076 [pii] 10.1093/carcin/bgq076
7. Peters U, Takata Y (2008) Selenium and the prevention of prostate and colorectal cancer. Mol Nutr Food Res 52(11):1261–1272. doi:10.1002/mnfr.200800103
8. Burk RF, Norsworthy BK, Hill KE, Motley AK, Byrne DW (2006) Effects of chemical form of selenium on plasma biomarkers in a high-dose human supplementation trial. Cancer Epidemiol Biomark Prev 15(4):804–810
9. Combs GF Jr (2015) Biomarkers of selenium status. Forum Nutr 7(4):2209–2236. doi:10.3390/nu7042209
10. Ghantous A, Saffery R, Cros MP, Ponsonby AL, Hirschfeld S, Kasten C, Dwyer T, Herceg Z, Hernandez-Vargas H (2014) Optimized DNA extraction from neonatal dried blood spots: application in methylome profiling. BMC Biotechnol 14:60. doi:10.1186/1472-6750-14-60
11. Feigelson HS, Rodriguez C, Welch R, Hutchinson A, Shao W, Jacobs K, Diver WR, Calle EE, Thun MJ, Hunter DJ, Thomas G, Chanock SJ (2007) Successful genome-wide scan in paired blood and buccal samples. Cancer Epidemiol Biomark Prev 16(5):1023–1025. doi:10.1158/1055-9965.EPI-06-0859
12. Swanson JM, Moyzis RK, McGough JJ, McCracken JT, Riddle MA, Kollins SH, Greenhill LL, Abikoff HB, Wigal T, Wigal SB, Posner K, Skrobala AM, Davies M, Ghuman JK, Cunningham C, Vitiello B, Stehli A, Smalley SL, Grady D (2007) Effects of source of DNA on genotyping success rates and allele percentages in the preschoolers with attention-deficit/hyperactivity disorder treatment study (PATS). J Child Adolesc Psychopharmacol 17 (5):635–646. doi:10.1089/cap.2007.0076
13. Berry MJ (2005) Insights into the hierarchy of selenium incorporation. Nat Genet 37 (11):1162–1163
14. Meplan C, Crosley LK, Nicol F, Beckett GJ, Howie AF, Hill KE, Horgan G, Mathers JC, Arthur JR, Hesketh JE (2007) Genetic polymorphisms in the human selenoprotein P gene determine the response of selenoprotein markers to selenium supplementation in a gender-specific manner (the SELGEN study). FASEB J 21(12):3063–3074
15. Meplan C, Crosley LK, Nicol F, Horgan GW, Mathers JC, Arthur JR, Hesketh JE (2008) Functional effects of a common single-nucleotide polymorphism (GPX4c718t) in the glutathione peroxidase 4 gene: interaction with sex. Am J Clin Nutr 87(4):1019–1027
16. Penney KL, Schumacher FR, Li H, Kraft P, Morris JS, Kurth T, Mucci LA, Hunter DJ, Kantoff PW, Stampfer MJ, Ma J (2010) A large prospective study of SEP15 genetic variation, interaction with plasma selenium levels, and prostate cancer risk and survival. Cancer Prev Res (Phila) 3(5):604–610. doi:10.1158/1940-6207.capr-09-0216
17. YJ H, Korotkov KV, Mehta R, Hatfield DL, Rotimi CN, Luke A, Prewitt TE, Cooper RS, Stock W, Vokes EE, Dolan ME, Gladyshev VN, Diamond AM (2001) Distribution and functional consequences of nucleotide polymorphisms in the 3′-untranslated region of the human Sep15 gene. Cancer Res 61(5):2307–2310
18. Meplan C, Rohrmann S, Steinbrecher A, Schomburg L, Jansen E, Linseisen J, Hesketh J (2012) Polymorphisms in thioredoxin reductase and selenoprotein K genes and selenium status modulate risk of prostate cancer. PLoS One 7(11):e48709. doi:10.1371/journal.pone.0048709
19. Sobrino B, Brion M, Carracedo A (2005) SNPs in forensic genetics: a review on SNP typing methodologies. Forensic Sci Int 154 (2–3):181–194. doi:10.1016/j.forsciint.2004.10.020
20. Bermano G, Pagmantidis V, Holloway N, Kadri S, Mowat NA, Shiel RS, Arthur JR, Mathers JC, Daly AK, Broom J, Hesketh JE (2007) Evidence that a polymorphism within the 3′UTR of glutathione peroxidase 4 is

functional and is associated with susceptibility to colorectal cancer. Genes Nutr 2(2):225–232

21. Steinbrecher A, Meplan C, Hesketh J, Schomburg L, Endermann T, Jansen E, Akesson B, Rohrmann S, Linseisen J (2010) Effects of selenium status and polymorphisms in selenoprotein genes on prostate cancer risk in a prospective study of European men. Cancer Epidemiol Biomark Prev 19(11):2958–2968. doi:10.1158/1055-9965.EPI-10-0364
22. Jablonska E, Gromadzinska J, Reszka E, Wasowicz W, Sobala W, Szeszenia-Dabrowska N, Boffetta P (2009) Association between GPx1 Pro198Leu polymorphism, GPx1 activity and plasma selenium concentration in humans. Eur J Nutr 48(6):383–386
23. Hu J, Zhou GW, Wang N, Wang YJ (2010) GPX1 Pro198Leu polymorphism and breast cancer risk: a meta-analysis. Breast Cancer Res Treat 124(2):425–431. doi:10.1007/s10549-010-0841-z
24. Cox DG, Hankinson SE, Kraft P, Hunter DJ (2004) No association between GPX1 Pro198-Leu and breast cancer risk. Cancer Epidemiol Biomark Prev 13(11 Pt 1):1821–1822

Chapter 23

Identification of Genetic Disorders Causing Disruption of Selenoprotein Biosynthesis

Erik Schoenmakers and Krishna Chatterjee

Abstract

Disorders of selenoprotein biosynthesis in humans, due to mutations in three genes (*SECISBP2*, *TRU-TCA1-1*, and *SEPSECS*) involved in the selenocysteine insertion pathway, have been described. Patients with *SECISBP2* and *TRU-TCA1-1* defects manifest a multisystem disorder with a biochemical signature of abnormal thyroid function tests due to the impaired activity of deiodinase selenoenzymes, myopathic features linked to SEPN1 deficiency and phenotypes resulting from increased levels of reactive oxygen species attributable to lack of antioxidant selenoenzymes. In patients harboring *SEPSECS* mutations, severe, progressive, cerebello-cerebral atrophy (pontocerebellar hypoplasia type 2D) dominates the phenotype and it is not known whether the disorder is associated with thyroid dysfunction.

Key words SECISBP2, TRU-TCA1-1, SEPSECS, Selenocysteine, Selenoprotein, ROS, Thyroid, Western blot, Glutathione peroxidase activity

1 Introduction

Clinical presentation of patients is variable but includes growth retardation in childhood with thyroid function tests showing a distinctive abnormal pattern comprising raised thyroxine (T4), low or normal triiodothyronine (T3), elevated reverse T3 (rT3), and normal or high plasma thyroid-stimulating hormone (TSH) levels [1]. These thyroid hormone abnormalities are likely due to deficiencies of three, Sec-containing, deiodinase enzymes and low plasma selenium levels reflect deficiencies of circulating selenoproteins (SELENOP (previously referred to as SEPP1) and GPX3). Together, these abnormalities constitute a biochemical signature which is pathognomonic of the disorder. Once selenoprotein deficiency has been confirmed, further investigation, to delineate other features such as skeletal myopathy, cutaneous photosensitivity, hearing loss, neurodegeneration, or aortic root dilatation (recently recognized), which will vary in severity and time of onset, can be undertaken [2].

Laurent Chavatte (ed.), *Selenoproteins: Methods and Protocols*, Methods in Molecular Biology, vol. 1661,
DOI 10.1007/978-1-4939-7258-6_23,

The aim of this chapter is to describe initial assays that can be undertaken on blood samples from a patient, enabling detailed evaluation of selenoprotein levels (e.g., by western blotting) or functional activity. If it is possible to obtain a fresh blood sample, peripheral blood mononuclear cells (PBMCs) can be isolated and subsequently radiolabeled with ^{75}Se to profile the selenoproteome. If access to fresh blood samples from patients is limited, methodology used here to quantitate selenoprotein expression or activity or cellular oxidative stress in patient-derived PBMCs can also be applied to cell lines (e.g., primary dermal fibroblasts, EBV transformed B-cells) propagated from patients, but will not be discussed in detail in this chapter. Likewise, detailed methodology to identify genetic mutations will not be described as this involves use of standard protocols to sequence candidate genes (e.g., *SECISBP2*, *TRU-TCA1-1*, and *SEPSECS*) in genomic DNA isolated from patient's blood. As all the assays described below involve the use of human patient-derived material, suitable precautions should be taken to avoid exposure to blood-borne pathogens.

2 Materials

2.1 Collection of Blood Samples and Biochemical Measurements Using Serum or Plasma

1. Blood sample tubes: Sodium-Heparin Vacutainers and serum gel tubes, suitable for preparation of PBMCs and separation of plasma or serum.
2. Tourniquet.
3. Butterfly needles.
4. Disposable plastic hubs.
5. 70% isopropyl alcohol skin-cleansing swabs.
6. Cotton swabs and sterile plasters.

2.2 Isolation of Peripheral Blood Mononuclear Cells (PBMCs)

1. Gloves, Class 2 Safety Cabinet.
2. 1× PBS: 1 mM KH_2PO_4, 155 mM NaCl, 3 mM Na_2HPO_4, pH 7.4. Store at room temperature.
3. 50 mL/15 mL tubes.
4. Ficoll-paque™ plus (GE Healthcare).
5. Centrifuge able to spin without use of braking.
6. Disposable sterile Pasteur pipettes.
7. Cell counter (e.g., hemocytometer).
8. Erythrocyte lysis solution: 154 mmol/L NH_4Cl, 10 mmol/L $KHCO_3$, 0.1 mmol/L EDTA, pH 8.0.

2.3 ^{75}Se Labeling of Peripheral Blood Mononuclear Cells (PBMC)

1. ^{75}Se (University of Missouri research reactor facility).
2. ^{35}S-Methionine.

3. PBMC culture medium: RPMI-1640 + 10% fetal bovine serum (FBS) + penicillin/streptomycin/fungizone (PSF).
4. Methionine-free culture medium: Methionine-free DMEM +10% FBS + PSF.
5. Trypsin.
6. Cell lysis buffer: 20 mM Tris–HCl, 1% NP40, 200 mM NaCl, protease inhibitor, pH 7.4.
7. SDS-PAGE electrophoresis running system: 10% acryl amide BIS-Tris SDS-PAGE gel.
8. 2× Loading dye: 100 mM Tris–HCl, 2 mM DTT (add fresh), 4% SDS, 0.2% bromophenol blue, 20% glycerol, pH 6.8.
9. Ladder of protein markers of known molecular weight.
10. MES SDS-PAGE running buffer: 50 mM MES, 50 mM Tris Base, 0.1% SDS, 1 mM EDTA, pH 7.3.
11. Coomassie stain: 0.25 g Coomassie brilliant blue R250 in 90 mL methanol/H_2O (v/v: 1/1).
12. 3 mm chromatography paper.
13. Flat-bed gel dryer or cellophane frame system for gel drying.
14. Gel imaging system or X-ray film developing system.

2.4 Western Blotting of Plasma

1. Assay (e.g., Bradford) to measure protein concentration.
2. 2× Loading dye: 100 mM Tris–HCl, 2 mM DTT (add fresh), 4% SDS, 0.2% bromophenol blue, 20% glycerol, pH 6.8.
3. SDS-PAGE electrophoresis running system: 10% acryl amide BIS-Tris SDS-PAGE gel.
4. MES SDS-PAGE running buffer: 50 mM MES, 50 mM Tris Base, 0.1% SDS, 1 mM EDTA, pH 7.3.
5. Ladder of protein markers of known molecular weight.
6. Western blotting system.
7. Antibodies: SELENOP, AbFrontier; GPX3, Santa Cruz and ALB, Abcam.
8. Blot wash buffer: 1× PBS (1 mM KH_2PO_4, 155 mM NaCl, 3 mM Na_2HPO_4, pH 7.4) with 0.1% Tween20.
9. Blocking solution: blot wash buffer +5% skimmed milk.
10. Chemiluminescence detection reagent.
11. Gel imaging system or X-ray film system.

2.5 Measurement of Glutathione Peroxidase Activity in Serum

1. Glutathione peroxidase assay buffer: 50 mM Tris–HCl, 0.5 mM EDTA, pH 8.0.
2. NADPH assay reagent: 5 mM NADPH, 42 mM reduced glutathione, 10 U/mL Glutathione reductase in Glutathione peroxidase assay buffer.

3. Luperox TBH70X (SIGMA) (*see* **Note 1**).
4. Glutathione peroxidase (control).
5. 96-Well plate for colorimetric assay.
6. UV/VIS spectrophotometer with a kinetic program to measure absorbance at 340 nm.
7. A source of pure water. Glass-distilled water or HPLC-grade water is acceptable.

2.6 Detection of Lipid Peroxidation Using Bodipy581/591 C11

1. Isolated PBMCs (*see* Subheading 2.2).
2. 1× PBS: 1 mM KH_2PO_4, 155 mM NaCl, 3 mM Na_2HPO_4, pH 7.4.
3. Bodipy581/591 C11: 1 mM solution in methanol (ThermoFisher Scientific).
4. Methanol.
5. 37 °C incubator.
6. Fluorescence-activated cell sorter (FACS) with fluorescence detection at 530 and 585 nm, tubes, running buffer and software to analyze (*see* **Note 2**).

3 Methods

3.1 Collection of Blood and Biochemical Measurements in Serum

1. All investigations should be either part of an ethically approved protocol and/or indicated for clinical reasons, being undertaken with prior informed patient consent.
2. Blood collection is performed by a suitably trained or qualified individual. We recommend tubes containing sodium heparin as anticoagulant when collecting blood for the preparation of PBMCs and plasma. Serum is prepared by centrifugation (3000 × *g*, 10 min) of whole blood collected into serum gel tubes, after it has been allowed to clot (minimum 10 min).
3. Serum Selenium, free T4, free T3, reverse T3 and TSH are measured on using standard clinical biochemistry laboratory assays.

3.2 Isolation of Peripheral Blood Mononuclear Cells (PBMCs)

1. Work in a CL-2 Safety Cabinet. Carefully open sodium-heparin Vacutainer tubes (*see* **Note 3**).
2. Transfer the heparinized blood to a new tube and dilute it 1:1 with sterile 1× PBS pH 7.4. Mix well!
3. Aliquot a maximum of 15 mL Ficoll-Paque™ PLUS into a 50 mL tube.
4. Carefully layer 30 mL diluted blood over 15 mL of Ficoll-Paque™ PLUS in a 50 mL tube without disturbing the Ficoll-Paque™ PLUS (*see* **Note 4**). Handle the tube carefully

not to disturb the set phases. Adjust the volumes proportionately according to the volume of blood obtained.

5. Centrifuge at 2000 × *g* for 30 min at room temperature without braking in deceleration phase (*see* **Note 5**).
6. Handle the tube carefully to not disturb the set gradient. The layer of PBMCs is the white opaque ring located at the interface between the Ficoll-Paque™ PLUS (bottom liquid fraction) and the plasma (top liquid fraction, yellow). Red blood cells collect at the bottom of the tube. Transfer the PBMCs to a fresh 50 mL tube with a disposable sterile Pasteur pipette to dilute off residual Ficoll-Paque™ PLUS as soon as possible. The volume collected from each tube should be about 10–20 mL.
7. Fill the tube containing the collected cells up to ~40 mL with 1× PBS and centrifuge at 350 × *g* for 10 min at room temperature.
8. Decant the supernatant, loosen the pellet, and resuspend it in 40 mL 1× PBS pH 7.4.
9. Centrifuge at 350 × *g* for 10 min at RT (*see* **Note 6**).
10. Repeat **steps** 7 and **8** twice, and before your last spin remove an aliquot of cell suspension of sufficient volume to count the total number of cells.
11. Resuspend the cells in appropriate medium at the concentration required for planned experiments (*see* **Note 7**).

3.3 ^{75}Se Labeling of PBMC

1. Primary cells (PBMC: 2 × 10^6 cells) are cultured in 1 mL of medium in one well of a 12-well plate. Prepare one well for ^{75}Se-labeling and one for ^{35}S-Met labeling.
2. The appropriate number of cells is spun down after purification (*see* Subheading 3.2) and resuspended in fresh medium supplemented with ^{75}Se (0.037 MBq/well) or Methionine-free medium with ^{35}S-Met (0.037 MBq/well). Cells are incubated overnight at 37 °C with 5% CO_2 and 5% O_2 (*see* **Note 8**).
3. Cells are transferred to Eppendorf tubes, spun down (350 × *g*, 5 min) and washed once with PBS.
4. Cell pellets can either be frozen at −20 °C or lysed immediately in 50–100 μL lysis buffer keeping the suspension on ice. Leave cells on ice for 5 min, vortexing occasionally. Spin down insoluble fragments and then determine the protein concentration (*see* **Note 9**).
5. Proteins in cell lysates are separated by 10% acrylamide Bis-Tris SDS-PAGE using MES-buffer with appropriate protein standards. Run different gels with either ^{75}Se or ^{35}S-Met labeled cell lysates, since they require X-ray exposure for different periods.

6. The proteins are fixed by incubating for 10 min in 10% acetic acid solution.
7. Protein loading is analyzed by Coommassie staining, 1 h in Coommassie solution, destain until protein bands are clearly visible.
8. The gel is carefully positioned on Whatman filter paper and dried on a flatbed gel dryer for up to 2 h at 80 °C. Alternatively, the gel is dried using a cellophane frame overnight.
9. Dried gels are analyzed by autoradiography (X-ray film or Phospho-imager).

3.4 Western Blotting of Serum

1. Dilute the serum 1/10 with H_2O, mix well, spin down for 1 min at 15,000 × *g* at 4 °C, and measure the protein concentration.
2. Samples are analyzed by 10% acrylamide Bis-Tris SDS-PAGE for protein separation using MES-buffer with protein standards appropriate for MW of selenoproteins of interest: SELENOP (50 kDa); GPX3 (25 kDa); ALB (65 kDa). For the detection of SELENOP load 1–5 μg and for the detection of GPX3 load 25–50 μg of each sample per lane. To verify equal protein loading, ALB can be quantitated as an internal control, e.g., strip this blot and reprobe or run a separate gel.
3. Following electrophoresis, the proteins are transferred to a nitrocellulose membrane using your preferred Western blot method.
4. After blotting, immerse the membranes in blocking solution for 10 min, with gentle mixing.
5. Add appropriate anti-sera (in blocking solution) to the membranes (anti-SELENOP 1/500, anti-GPX3 1/500, anti-ALB 1/1000) and incubate at 4 °C overnight.
6. Discard the antibody solution and wash three times with blot wash buffer for 10 min each.
7. Add the appropriate secondary, alkaline phosphatase-conjugated, antibody (1:5000 dilution in blocking solution) and incubate for 1 h at room temperature or overnight at 4 °C.
8. Wash as in **step 6**.
9. Incubate with chemiluminescence detection reagent and visualize either by a Gel Imaging System or conventional autoradiography.

3.5 Glutathione Peroxidase (GPx) Activity in Serum

1. Dilute serum 1 in 2 with GPx-assay buffer and spin for 1 min at 15,000 × *g* 4 °C to remove insoluble fragments. Use 10 μL diluted serum sample per measurement in a 96-well plate and assay each sample in triplicate (*see* **Note 10**).

2. Prepare fresh NADPH assay buffer (*see* **Note 11**), 10 μL buffer per 96-well is required.
3. Prepare fresh 30 mM tert-Butyl hydroperoxide solution by diluting 4.3 μL Luperox to a total volume of 1 mL with water.
4. Distribute 10 μL diluted serum sample in each well, include a negative control (blank) and a positive control (diluted glutathione peroxidase (5–20 μL of a 0.25 U/mL solution).
5. For each well, dilute 10 μL NADPH assay buffer with GPx-assay buffer to a total volume of 200 μL and add 2 μL t-Bu-OOH, making one stock for the total number of samples (i.e., a total of 20 mL stock solution for 100 wells). Mix well and add immediately to your samples in the 96-well plate (*see* **Note 12**).
6. Measure the decrease in absorption at 340 nm using a kinetic program with an initial delay of 15 s and measure up to 5 min taking measurements every 10 s (at least 10 time points) (*see* **Note 13**).
7. Calculation of GPx-activity (*see* **Note 14**):

*Determine the change in absorbance ΔA_{340} per minute by plotting absorbance values as a function of time to obtain the slope of the linear portion of the curve with subtraction of the blanc or calculate $\Delta A_{340}/\text{min} = A_{340}/\text{min}_{(\text{blanc})} - A_{340}/\text{min}_{(\text{sample})}$.

$$*\text{GPx-activity (nmol/min/mL)} = \frac{\Delta A340/\text{min}}{0.00373\ \mu M^{-1}} \times \frac{0.2\ \text{mL}}{0.01\ \text{mL}} \times \text{sample dilution}$$

3.6 Measurement of Lipid Peroxidation Using Bodipy581/591 C11

1. Isolate PBMCs as described in Subheading 3.2, spin at 200 × *g* for 5 min at room temperature to pellet the cells, resuspend in 1× PBS, and transfer minimum 5×10^5 cells to each FACS tube (two tubes per patient sample).
2. From patient or control subject, one tube of cells with 1 μM bodipy581/591 C11 and a duplicate tube with methanol (same volume as bodipy solution, for background settings on FACS) in 1× PBS for 60 min at 37 °C, keeping the samples in the dark.
3. Spin at 200 × *g* for 5 min at room temperature to pellet the cells.
4. Resuspend the pellet in 1× PBS (200–500 μL).
5. Measure bodipy581/591 C11 fluorescence with FACS at 530 nm (green) and 585 nm (red) (*see* **Note 15**).
6. Disinfect and discard the waste.

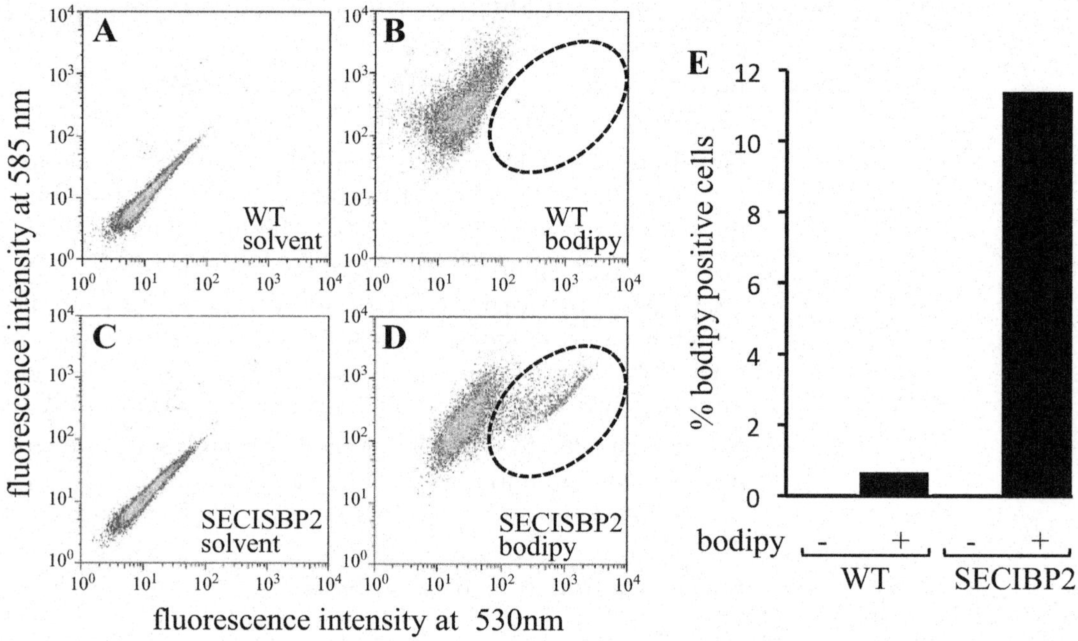

Fig. 1 Typical results of FACS analysis of PBMCs labeled with bodipy 581/591 11C. Panels (**a**) and (**c**) show traces from normal (WT) and patient (SECISBP2) cells without solvent, with fluorescence at the geometric mean of 530 and 585 nm being set at intensity 10. Panels (**b**) and (**d**) show traces from bodipy 581/591-labeled WT and SECISBP2 defect cells, showing increased fluorescence intensity at the geometric mean between 530 and 585 nM. A subset of cells (*circled*) from a SECISBP2 defect patient exhibit increased lipid peroxidation. Panel (**e**) quantifies the percentage of cells with increased lipid peroxidation from WT and SECISBP2 defect cells

7. Analyze the data with your preferred FACS software. An increase in red staining suggests good uptake of dye into the cells, shifting to green when both reactive oxygen and/or reactive nitrogen species, resulting in lipid peroxidation, are present (Fig. 1 and *see* **Note 16**). Some lipid peroxidation is present even in cells from healthy controls, illustrated by a general shift of all cells at 530 nm. With cells from patients, a subset clearly exhibits enhanced lipid peroxidation. The percentage of cell subset with increased lipid peroxidation compared to the total number of cells is shown in a graph (Fig. 1d).

4 Notes

1. Cumene hydroperoxide (0.25–1 mM) can be used as the substrate, and will measure total glutathione peroxidase activity mediated by both Se- and non-Se containing enzymes.
2. Most FACS machines will suffice. We use a BD FACS Calibur, all buffers are from BD and the tubes are from Falcon 5 mL, round-bottom tubes (REF352054).

3. Isolation of PBMCs from fresh blood samples works best (<2 h old). However, isolation from samples a few days old is possible, albeit with some loss of cell viability (store samples at room temperature). It is important to always include samples from healthy subjects (e.g., unaffected family members) as controls with these being treated exactly the same as patient-derived samples.
4. Start layering the diluted blood on the top of the ficoll, holding the pipette at a 45° angle against the tube wall, releasing the solution dropwise. Once a layer of blood is on the top of the ficoll you can speed up the release, but be careful not to disturb the interface.
5. Centrifugation with deceleration without braking adds about 30 min to this procedure before the samples are at standstill. However, this is absolutely essential, otherwise the separated phases will become mixed up leading to loss of cells.
6. Optional: Sometimes, the PBMC pellet contains a significant quantity of red blood cells (RBCs). To remove RBCs resuspend the pellet in 5 mL erythrocyte lysis buffer, incubate for 5 min at room temperature, then centrifuge 5 min at 350 × *g*.
7. For some experiments, e.g., western blotting, GPx activity, PMBCs can be frozen alive by resuspending PBMCs (5×10^6–10^7) in FBS + 10% DMSO, slowly cooling at −80 °C overnight in, e.g., a mister frosty and then storing in liquid nitrogen for future use. After thawing them (at 37 °C water bath, culture in PBMC medium) you will have sufficient cells to do a cell labeling experiment, or to increase the number of cells dramatically by using a specific mitogenic stimulus (incubation with anti-CD3 and anti-CD28 antibodies (with or without IL2) for 3–5 days) to promote T cell proliferation. However, T cells from patients with selenoprotein deficiency have limited proliferative capacity [2].
8. ^{75}Se is a very strong gamma emitter, make sure you take the proper safety precautions! The assay should be performed in a controlled radioactive area, using appropriate lead shielding/storage vials, with regular testing for spills using a scintillation counter and with personal monitoring using film badges and finger Thermo Luminescent Dosimeter. When labeling cells from selenoprotein deficient patients it is advisable to use an incubator at 37 °C, 5% CO_2 and 5% O_2. Lowering O_2 concentration and addition of the vitamin E homologue (e.g., Trolox 25 μM (Calbiochem CAT 648471) increases cell viability. Longer incubation (2–3 days) with radioisotope can improve selenoprotein radiolabeling, but ensure that the cells do not get to confluence early in the incubation process. Prolonged incubation of primary cells from selenoprotein-deficient patients reduces their viability, necessitating simultaneous ^{35}S-methionine labeling of cells to check this.

9. Determining the protein concentration of radioactive samples is not always possible. Counting cells carefully will make the pellets similar, although selenoprotein-deficient cells proliferate significantly more slowly. Analyzing the samples with ^{35}S-Met first will only require about 1/10 of the total sample per lane, and give you a good idea of the protein concentration/viability after overnight exposure. Then the concentration of cell suspension can be adjusted and the ^{75}Se labeled samples can be analyzed. ^{75}Se incorporation can be monitored easily with a scintillation counter, recognizing that significantly (<50%) lower signal will be obtained from selenoprotein deficient versus normal cells. Scintillation counting can be misleading when there is significant free ^{75}Se in cell medium, but an extra wash of cell pellets with PBS can overcome this. SDS-PAGE using samples containing 50–100 CPM per lane usually yield a good signal.
10. Perform several dilutions to obtain results within the linear range.
11. NAPDH assay buffer can be stored at 4 °C up to 24 h. Do not freeze the solution for reuse.
12. The reaction will start immediately, make sure the plate reader is set up to start measurements.
13. The initial absorbance at 340 nm should not be above 1.2 or below 0.5. To obtain reproducible results the absorbance decrease must be between 0.02 and 0.135 per min.
14. Unit definition: 1 unit of glutathione peroxidase will form 1 μmol NADP+ from NADPH per min at pH 8.0 at 25 °C. In the formula, 0.2 mL is total volume reaction, 0.01 mL is volume of the sample, 0.00373 μM^{-1} is the adjusted extinction coefficient for the path length of the solution in the 96-well (0.6 cm). The actual extinction coefficient for NADPH at 340 nM is 0.00622 μM. Note that assay kits from different manufacturers work equally well.
15. For each sample incubated with methanol set basal fluorescence of the geometric mean between 530 and 585 nm at intensity 10. The sample incubated with bodipy typically exhibits enhanced geometric mean fluorescence of 2- to 10-fold at 530 nm and 5- to 50-fold at 585 nm. When assaying samples ensure that you have twice the required amount of healthy control cell sample, to calibrate the machine correctly within the desired operating range. Once calibrated and settings saved the settings, the number of cells can be reduced for future experiments.
16. The bodipy 581/591 C11-labeled fatty acid is reportedly more sensitive to oxidation than endogenous lipids and can therefore overestimate lipid peroxidation, but is considered a very sensitive marker of free radical mediated membrane damage.

Acknowledgments

This work is supported by the Wellcome Trust (KC [095564/Z/11/Z]) and the National Institute for Health Research Biomedical Research Centre Cambridge (KC).

References

1. Dumitrescu AM, Liao XH, Abdullah MS et al (2005) Mutations in SECISBP2 result in abnormal thyroid hormone metabolism. Nat Genet 37:1247–1252
2. Schoenmakers E, Agostini M, Mitchell C et al (2010) Mutations in the selenocysteine insertion sequence-binding protein 2 gene lead to a multisystem selenoprotein deficiency disorder in humans. J Clin Invest 120:4220–4235

INDEX

Laurent Chavatte (ed.), *Selenoproteins: Methods and Protocols*, Methods in Molecular Biology, vol. 1661, DOI 10.1007/978-1-4939-7258-6, © Springer Science+Business Media LLC 2018

S

T

U

Printed by Printforce, the Netherlands